THEORY OF FUNCTIONS
OF A REAL VARIABLE

I. P. NATANSON

THEORY OF FUNCTIONS OF A REAL VARIABLE

(Teoria functsiy veshchestvennoy peremennoy, Chapters I to IX)

Translated from the Russian by
LEO F. BORON
Department of Mathematics, University of Michigan

With the editorial collaboration of, and with annotations by
EDWIN HEWITT
Professor of Mathematics, University of Washington

FREDERICK UNGAR PUBLISHING CO.
NEW YORK

Copyright 1955 by Frederick Ungar Publishing Co.

PRINTED IN THE UNITED STATES OF AMERICA

Library of Congress Catalog Card No. 54-7420

FOREWORD TO THE AMERICAN EDITION

This book treats the theory of functions of a real variable from a largely classical point of view. It is designed for use in the current Soviet university program, by students in their third year of university studies. The author is a distinguished Soviet scientist and teacher, who has evidently lectured for a number of years on the subject matter of the present book. The book has been through two earlier editions, one in 1941 under the title *Foundations of the Theory of Functions of a Real Variable*, and a second in the Ukrainian language, with various extensions by S. I. Zuhovickiĭ.

Prerequisites for understanding this book are, in general, the foundations of elementary analysis. For example, a student who has mastered Courant's *Differential and Integral Calculus* or de la Vallée Poussin's *Cours d'analyse infinitésimale* is well equipped to study the present book. The author assumes specifically that the reader is familiar with the theory of irrational numbers, the theory of limits (including *limes superior* and *limes inferior*), the principal properties of continuous functions, derivatives, Riemann integrals, and infinite series. It is evident that the author has taken great pains to make his definitions precise and his proofs lucid. His wide familiarity with the subject is illustrated by the unusual and interesting proofs presented for a number of well-known theorems (for example, Vitali's theorem (§8, Chapter III), and the equivalence of the Riemann and Lebesgue integrals when the former exists (Theorem 3, §4, Chapter V)).

It is hoped, accordingly, that this book will prove useful as a textbook in graduate courses in the theory of functions of a real variable in American universities, and that it may also serve as a reference work for persons studying analysis independently. Most of the chapters are provided with exercises, placed at the end of the chapter. These exercises are for the most part difficult. Like the exercises in, for example, Whittaker and Watson's *Modern Analysis*, they are designed to extend the subject matter of the text and to develop the student's abilities through hard thinking. It is earnestly recommended that the reader work through at least part of the exercises for each chapter.

Perhaps a few words may be permitted concerning the translation. A translator's task is an uneasy one at best. Even to give a full and faithful literal rendering of a foreign text may be troublesome. In the present case, it has been necessary to bear in mind and make allowances for national differences in terminology, in notation, and in readers' background. The current American and British terminology has been followed wherever possible, even at the cost of some violence to the original phraseology. It was impossible because of technical requirements to use the symbols " $\cup$ " and " $\cap$ " for union and intersection of sets; the Russian original from which the formulas were reproduced uses the signs $+$ and $\cdot$ instead. Accordingly, the latter signs were adopted, and the term *sum* was employed instead of *union*. The translation of proper names is a continual annoyance. The practice has been followed here of writing names of non-Russian origin, even when they are names of Soviet mathematicians, in the original spelling, French, German, Polish, etc. Russian names have been transliterated in accordance with the system adopted by the American Mathematical Society.

The present volume comprises the first nine chapters of the Russian text of seventeen chapters. Various liberties have been taken with the text. Most important of these consist in the addition of appendices to Chapters III, IV, VI, VII, and VIII, covering essential topics which are omitted, for no obvious reason, in the Russian edition. In addition, comments in the Russian text dealing exclusively with personalities have been deleted.

With full recognition, then, of the liberties taken with Natanson's text, for which the editor assumes full responsibility, and with the hope that these changes will have produced a text of real usefulness to a wide circle of readers, this translation is offered to the English-reading public.

EDWIN HEWITT
THE UNIVERSITY OF WASHINGTON

LEO F. BORON
THE UNIVERSITY OF NOTRE DAME

CONTENTS

CHAPTER I. INFINITE SETS   **11**

    1. Operations on Sets   11

    2. One-to-One Correspondences   15

    3. Denumerable Sets   17

    4. The Power of the Continuum   21

    5. Comparison of Powers   27

    Exercises   33

CHAPTER II. POINT SETS   **34**

    1. Limit Points   34

    2. Closed Sets   36

    3. Interior Points and Open Sets   41

    4. Distance and Separation   44

    5. The Structure of Bounded Open Sets and Bounded Closed Sets   47

    6. Points of Condensation; the Power of a Closed Set   50

    Exercises   54

CHAPTER III. MEASURABLE SETS   **55**

    1. The Measure of a Bounded Open Set   55

    2. The Measure of a Bounded Closed Set   59

    3. The Outer and Inner Measure of a Bounded Set   63

    4. Measurable Sets   66

CHAPTER III (continued)

    5. Measurability and Measure as Invariants under Isometries 71
    6. The Class of Measurable Sets 75
    7. General Remarks on the Problem of Measure 79
    8. Vitali's Theorem 81
    9. Editor's Appendix to Chapter III 84
    Exercises 88

CHAPTER IV. MEASURABLE FUNCTIONS   89

    1. The Definition and the Simplest Properties of Measurable Functions 89
    2. Further Properties of Measurable Functions 93
    3. Sequences of Measurable Functions. Convergence in Measure 95
    4. The Structure of Measurable Functions 101
    5. Two Theorems of Weierstrass 107
    6. Editor's Appendix to Chapter IV 112
    Exercises 114

CHAPTER V. THE LEBESGUE INTEGRAL OF A BOUNDED FUNCTION   116

    1. Definition of the Lebesgue Integral 116
    2. Fundamental Properties of the Integral 121
    3. Passage to the Limit under the Integral Sign 127
    4. Comparison of Riemann and Lebesgue Integrals 129
    5. Reconstruction of the Primitive Function 133

CHAPTER VI. SUMMABLE FUNCTIONS   136

    1. The Integral of a Non-negative Measurable Function 136

CONTENTS

CHAPTER VI (continued)

    2. Summable Functions of Arbitrary Sign   143

    3. Passage to the Limit under the Integral Sign   149

    4. Editor's Appendix to Chapter VI   159

    Exercises   162

CHAPTER VII. SQUARE-SUMMABLE FUNCTIONS   **185**

    1. Fundamental Definitions. Inequalities. Norm   165

    2. Mean Convergence   167

    3. Orthogonal Systems   175

    4. The Space l_2 184

    5. Linearly Independent Systems   192

    6. The Spaces L_p and l_p 196

    7. Editor's Appendix to Chapter VII   200

    Exercises   202

CHAPTER VIII. FUNCTIONS OF FINITE VARIATION. THE STIELTJES INTEGRAL   **204**

    1. Monotonic Functions   204

    2. Mapping of Sets. Differentiation of Monotonic Functions   207

    3. Functions of Finite Variation   215

    4. Helly's Principle of Choice   220

    5. Continuous Functions of Finite Variation   223

    6. The Stieltjes Integral   227

    7. Passage to the Limit under the Stieltjes Integral Sign   232

    8. Linear Functionals   236

    9. Editor's Appendix to Chapter VIII   238

    Exercises   241

Contents

CHAPTER IX. ABSOLUTELY CONTINUOUS FUNCTIONS. THE INDEFINITE LEBESGUE INTEGRAL   243

    1. Absolutely Continuous Functions   243

    2. Differential Properties of Absolutely Continuous Functions   246

    3. Continuous Mappings   248

    4. The Indefinite Lebesgue Integral   252

    5. Points of Density. Approximate Continuity   260

    6. Supplement to the Theory of Functions of Finite Variation and Stieltjes Integrals   263

    7. Reconstruction of the Primitive Function   266

    Exercises   270

INDEX   273

CHAPTER I

INFINITE SETS

§ 1. OPERATIONS ON SETS

The so-called *theory of sets* forms the basis of the theory of functions of a real variable. This discipline has a comparatively short history. The first serious works in this subject, due to G. Cantor, appeared at the end of the last century. Nonetheless, the theory of sets represents a very extensive area of mathematics at the present time. In this text, for which the theory of sets has only auxiliary significance, we shall limit ourselves to the elements of this discipline; the reader wishing to go more deeply into the theory of sets is referred to the books of A. Fraenkel and F. Hausdorff.[1]

The concept of a set is one of the basic concepts of mathematics and does not lend itself readily to a precise definition. Consequently, we shall limit ourselves to an informal description of this concept. A set is an assemblage, aggregate, or collection of objects combined according to any rule whatsoever. For example, one can speak of the set of all natural numbers, the set of all points on a line, the set of all polynomials with real coefficients, and so forth.

In speaking of a set, we assume with respect to every object that one and only one of the following is true: either this object belongs to our set as one of its elements, or it does not.

If A is a set and x is an object, then the fact that the object x belongs to the set A is denoted by the expression:

$$x \in A.$$

If, on the other hand, x does not belong to A, then this fact is written as follows:

$$x \notin A.$$

For example, if R is the set of all rational numbers, then

$$\tfrac{3}{4} \in R, \quad \sqrt{2} \notin R.$$

A set is never an element of itself:

$$A \notin A.^*$$

---

* This arbitrary convention is evidently introduced to avoid the famous paradox of Russell. Russell's paradox runs as follows. A set A is either a member of itself, $A \in A$, or it is not, $A \notin A$. Let R denote the set of all sets which are not members of themselves. Then if $R \in R$, it follows that $R \notin R$. If $R \notin R$, it follows that $R \in R$. Hence it cannot be that $R \in R$ or that $R \notin R$. The reader may decide for himself whether or not the convention that $A \notin A$ for all A avoids this paradox.—E. H.

[1] A. FRAENKEL, *Einleitung in die Mengenlehre*, Dover, New York, 1946; F. HAUSDORFF, *Mengenlehre*, 3rd edition, Dover, New York, 1944.

For the sake of generality and simplicity of statement, it is useful to introduce the so-called *void set*, which contains no elements. For example, the set of real roots of the equation:
$$x^2 + 1 = 0$$
is void. The void set is designated by the symbol 0; the danger of confusion with the number zero will not arise in the sequel, since it will be clear from the context which meaning is intended.

Besides the void set, we shall need one-element sets, *i.e.*, sets consisting of only one element. For example, the set of roots of the equation
$$2x - 6 = 0$$
consists of one element, the number 3. One must avoid confusing a one-element set with its single element.

If a general element of the set A is written as x, then we write
$$A = \{x\}.$$
If it is possible to write out all elements of a set, then we write them in order and put them into curly brackets; for example,
$$A = \{a, b, c, d\}.$$

DEFINITION 1. Let A and B be two sets. If every element of the set A is an element of the set B, then A is said to be a *subset* of the set B; we write
$$A \subset B, \quad B \supset A.$$
This relation is called *inclusion*. If $A \subset B$ and there exists a $b \in B$ which is not in A, then A is said to be a proper subset of B.

For example, let N be the set of all natural numbers and let R be the set of all rational numbers; then
$$N \subset R.$$
It is clear that every set is a subset of itself:
$$A \subset A.$$
The void set is a subset of every set A. In order that this statement be entirely clear, it suffices to state Definition 1 in the following form: an element not in B is not in A.

DEFINITION 2. Let A and B be two sets. If $A \subset B$ and $B \supset A$, then the sets A and B are said to be *equal*; we write
$$A = B.$$
For example, if $A = \{2, 3\}$ and B is the set of roots of the equation
$$x^2 - 5x + 6 = 0,$$
then $A = B$.

DEFINITION 3. Let A and B be two sets. The set S consisting of all elements which lie in A or in B or in both A and B is called the *sum* (or *union*) of the sets A and B and is denoted by
$$S = A + B.$$
We define in a similar way the sum of n sets $A_1, A_2, \ldots, A_n$, the sum of a sequence

of sets $A_1, A_2, A_3, \ldots$, and, in general, the sum of a family* of sets A_ξ, distinguished from each other by the index ξ, which takes on arbitrary distinct values. The corresponding symbols are as follows:

$$S = A_1 + A_2 + \cdots + A_n, \quad \text{or} \quad S = \sum_{k=1}^{n} A_k,$$

$$S = A_1 + A_2 + A_3 + \cdots, \quad \text{or} \quad S = \sum_{k=1}^{\infty} A_k,$$

$$S = \sum_{\xi} A_\xi.$$

For example, if S is the set of all positive numbers, then

$$S = \sum_{k=1}^{\infty} (k-1, k].^{**}$$

If $A \subset B$, then it is obvious that

$$A + B = B;$$

in particular,

$$A + A = A.$$

DEFINITION 4. Let A and B be two sets. The set P consisting of all elements which belong to both of the sets A and B is called the *intersection* of the sets A and B and is denoted by

$$P = AB.$$

For example, if $A = \{1, 2, 3, 4\}$ and $B = \{3, 4, 5, 6\}$, then

$$AB = \{3, 4\}.$$

We define in the same way the intersection of n sets $A_1, A_2, \ldots, A_n$, of a sequence of sets $A_1, A_2, A_3, \ldots$, and, generally, of a set of sets A_ξ which are distinguished from each other by index ξ. The corresponding symbols are:

$$P = A_1 A_2 \cdots A_n, \quad \text{or} \quad P = \prod_{k=1}^{n} A_k,$$

$$P = A_1 A_2 A_3 \cdots, \quad \text{or} \quad P = \prod_{k=1}^{\infty} A_k,$$

$$P = \prod_{\xi} A_\xi.^{***}$$

For example,

---

* We shall use the words "family" and occasionally "system" in referring to sets whose elements are sets. Thus one can speak of the family of all subsets of the positive integers, or of a system of sets whose sum contains a given set.—E. H.

** The symbols $[a, b]$, $(a, b]$, $[a, b)$, and (a, b) denote the sets of real numbers x such that $a \leqslant x \leqslant b$, $a < x \leqslant b$, $a \leqslant x < b$, and $a < x < b$, respectively. The set $[a, b]$ is non-void if and only if $a \leqslant b$. It is called a closed interval. The set (a, b), which is non-void if and only if $a < b$, is called an open interval. The other two sets mentioned are called half-open intervals.—E. H.

*** The sum of A and B is also often written as $A \cup B$, and the intersection of A and B as $A \cap B$. The symbols $\bigcup_\xi A_\xi$ and $\bigcap_\xi A_\xi$ have similar interpretations as sums and intersections.—E. H.

$$\prod_{k=1}^{\infty}\left(-\frac{1}{k},\frac{1}{k}\right)=\{0\} \qquad \text{(a set consisting of one element);}$$

$$\prod_{k=1}^{\infty}\left(0,\frac{1}{k}\right)=0 \qquad \text{(the void set).}$$

If $A \subset B$, then it is obvious that
$$AB = A$$
and, in particular, $AA = A$.

The condition that the sets A and B have no common elements can be written as
$$AB = 0.$$
In this case, we say that A and B have *void intersection* or are *disjoint*. It will often occur in the sequel that we must deal with families $\{A_\xi\}$ of sets with the property that $A_\xi A_{\xi'} = 0$ if $\xi \neq \xi'$. In this case, we say that the sets of this family are *pairwise disjoint*.

THEOREM 1. *Let A be an arbitrary set and let $\{E_\xi\}$ be any family of sets. Then*
$$A \sum_\xi E_\xi = \sum_\xi A E_\xi. \qquad (1)$$

Proof. We write
$$S = A \sum_\xi E_\xi, \quad T = \sum_\xi A E_\xi.$$

Let $x \in S$. This means that $x \in A$ and $x \in \sum_\xi E_\xi$. The latter relation means $x \in E_{\xi_0}$ for some index ξ_0. But then $x \in A E_{\xi_0}$ and, necessarily, $x \in T$. Thus
$$S \subset T.$$

Conversely, suppose that $x \in T$. This implies that $x \in A E_{\xi_0}$ for some index ξ_0, and thus $x \in A$ and $x \in E_{\xi_0}$. But if $x \in E_{\xi_0}$, then, $x \in \sum_\xi E_\xi$ and then (since $x \in A$), $x \in S$. Therefore
$$T \subset S$$
which, together with what was proved above, gives us
$$S = T.$$

It follows in particular from the theorem just proved that
$$A(B \dotplus C) = AB \dotplus AC.$$

DEFINITION 5. Let A and B be two sets. The set R consisting of all those elements of the set A which are *not* elements of the set B is called the *difference* of the sets A and B and is denoted by the expression
$$R = A - B.$$

For example, if $A = \{1, 2, 3, 4\}$, $B = \{3, 4, 5, 6\}$, then
$$A - B = \{1, 2\}.$$

THEOREM 2. *If A, B, C are sets, then*
$$A(B - C) = AB - AC.$$
The proof is left to the reader.

One is struck by the analogy between properties of operations on sets and the properties of arithmetic operations. However, this analogy is not complete. We have already seen that $A + A = A$, $AA = A$; in general, these relations are not valid in arithmetic. We give one more example which violates the asserted analogy.

THEOREM 3. *The relation*

$$(A - B) \dotplus B = A \tag{2}$$

holds if and only if $B \subset A$.

Proof. Suppose that (2) holds. Since a summand is always contained in the sum, we have $B \subset A$. Now suppose that $B \subset A$. Then obviously $(A-B) + B \subset A$. But the reverse inclusion $(A-B) + B \supset A$, as can easily be seen, holds without restriction; consequently, (2) follows.

§ 2. ONE-TO-ONE CORRESPONDENCES

Let A and B be two finite sets. It is natural to ask whether or not the number of elements in these sets is the same. In order to answer this question, we can *count* the elements of each set and then observe whether or not the numbers obtained as a result of the counting are the same. However, our question can also be answered without actually counting the elements of the sets. For instance, let A be a set of Latin letters,

$$A = \{a, b, c, d, e\},$$

and B be a set of Greek letters,

$$B = \{\alpha, \beta, \gamma, \delta, \varepsilon\}.$$

If we arrange these sets as follows :

A:	a	b	c	d	e
B:	α	β	γ	δ	ε

then we see without any counting that A and B have the same number of elements. What is characteristic of this method of comparing sets ? For each element of one set, there appears one and only one element corresponding to it in the other set, and conversely.

The power of this second method of comparison lies in the fact that it can be applied when the sets to be compared are *infinite*. For instance, if N is the set of all natural numbers and M is the set of all numbers of the form $\frac{1}{n}$, then the second method of comparison shows at once that the *number* of elements (in some sense) is the same in both the sets N and M; to convince ourselves of this, it is sufficient to arrange our sets as follows :

N:	1	2	3	4	...
M:	1	$\frac{1}{2}$	$\frac{1}{3}$	$\frac{1}{4}$	...

and pair off the numbers n and $\frac{1}{n}$. We turn now to precise definitions.

DEFINITION 1. Let A and B be two sets. A rule φ, which associates with each element a of the set A exactly one element b of the set B, and under which each element $b \in B$ corresponds to exactly one $a \in A$, is called a *one-to-one correspondence* between the sets A and B.

DEFINITION 2. If it is possible to establish a one-to-one correspondence between two sets A and B, these sets are said to be *equivalent* or to have the same *power*, and we write

$$A \sim B.$$

It is easily seen that two finite sets are equivalent if and only if they consist of the same number of elements ; so the concept of equivalence is a direct generalization of the concept of having the same number of elements, for finite sets.

Fig. 1.

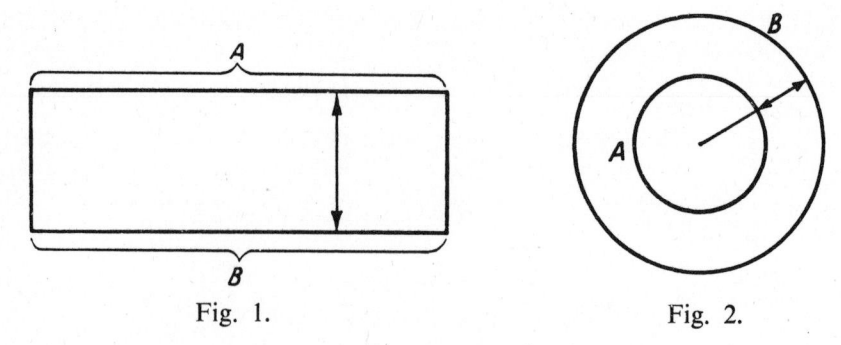

Fig. 2.

We present a few examples of pairs of equivalent sets. Let A and B be the sets of points on two parallel sides of a rectangle (Fig. 1). It is easy to see that $A \sim B$. Next, let A and B be the sets of points of two concentric circles (Fig. 2). Here also it is clear that $A \sim B$. The second example is less trivial than the first. If we straighten out our circles, one of them is transformed into a shorter line segment than the other. It would seem that there ought to be *more* points on the longer segment. We see that this is not so.

Here is an example where the above paradox is still more startling. Let A be the set of points of the hypotenuse and B the set of points of a leg of a right triangle. As is clear from Fig. 3, $A \sim B$, despite the fact that the leg is shorter than the hypotenuse.

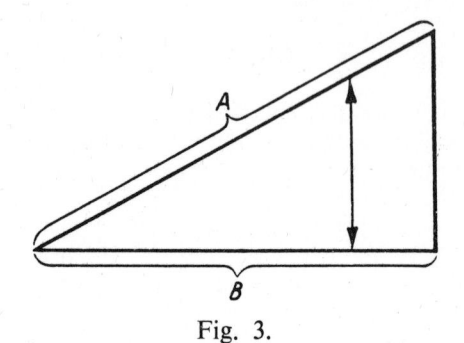

Fig. 3.

If we lay off the leg on the hypotenuse, the set B appears to be a proper *subset* of the set A and hence different from A itself. In the last example, we encounter a set containing a proper subset equivalent to itself. It is clear that a finite set cannot contain proper subsets equivalent to itself. It is thus the infiniteness of the set A which produces this curious phenomenon. Further on we shall see that *every* infinite set contains proper subsets equivalent to itself. In the meantime, we illustrate this fact by one more example. Let N be the set of all natural numbers and let M be the set of all even numbers :

$$N = \{n\}, \ M = \{2n\}.$$

Arranging these sets in the following way,

$$\begin{array}{c|c|c|c|c|c|c} N: & 1 & 2 & 3 & 4 & 5 & \ldots \\ \hline M: & 2 & 4 & 6 & 8 & 10 & \ldots \end{array},$$

we are immediately convinced of their equivalence, even though M is a proper subset of N. We may therefore say that *there are as many even numbers as there are natural numbers*.

We set down certain simple properties of equivalence, the proofs of which should cause the reader no difficulty.

THEOREM 1. a) $A \sim A$ for all sets A.
 b) If $A \sim B$, then $B \sim A$.
 c) If $A \sim B$ and $B \sim C$, then $A \sim C$.

THEOREM 2. Let $A_1, A_2, A_3, \ldots$ and $B_1, B_2, B_3, \ldots$ be two sequences of sets. If the A_n are pairwise disjoint and the sets B_n are pairwise disjoint, that is, if

$$A_n A_{n'} = 0, \quad B_n B_{n'} = 0 \qquad (n \neq n')$$

and if for every n

$$A_n \sim B_n, \qquad (n = 1, 2, 3, \ldots)$$

then

$$\sum_{k=1}^{\infty} A_k \sim \sum_{k=1}^{\infty} B_k.$$

§ 3. DENUMERABLE SETS

DEFINITION 1. Let N be the set of all natural numbers,

$$N = \{1, 2, 3, 4 \ldots\}.$$

Every set A equivalent to the set N is said to be *denumerable*.* Sometimes one says that the set A has the power a. An infinite set which is not equivalent to the set N is said to be *non-denumerable*. We shall encounter a number of examples of non-denumerable sets in the sequel. A set which is finite or equivalent to N will often be referred to as *at most denumerable*.

It is clear that all denumerable sets are equivalent among themselves. Here are some examples of denumerable sets:

$$A = \{1, 4, 9, 16, \ldots, n^2, \ldots\},$$
$$B = \{1, 8, 27, 64, \ldots, n^3, \ldots\},$$
$$C = \{2, 4, 6, 8, \ldots, 2n, \ldots\},$$
$$D = \left\{1, \frac{1}{2}, \frac{1}{3}, \frac{1}{4}, \ldots, \frac{1}{n}, \ldots\right\}.$$

THEOREM 1. *A set A is denumerable if and only if it is possible to enumerate it, i.e., to put it into the form of an ordinary infinite sequence:*

$$A = \{a_1, a_2, a_3, \ldots, a_n, \ldots\}. \tag{1}$$

Proof. If the set A is represented in the form (1), then it is sufficient to associate with each of its elements the index, n, of this element in order to obtain a one-to-one

---

* There is no general agreement in English usage governing the meaning attached to the words *denumerable* and *countable*. Both of these terms are used by various authors to mean finite or equivalent to N, while others use the narrower interpretation of the text for both terms. We follow in this translation the convention that *denumerable* means equivalent to N and that *countable* means finite or equivalent to N. However, the term *countable* will be used very little.—E. H.

correspondence between A and N. Hence A is denumerable. Conversely, if A is denumerable, then there exists a one-to-one correspondence φ between A and N. It is sufficient to designate by a_n that element of the set A which in the correspondence φ corresponds to the number n in order to represent A in the form (1).

THEOREM 2. *Every infinite set A contains a denumerable subset D.*

Proof. Let A be an infinite set. We select from A an arbitrary element a_1. Since A is infinite, it is not exhausted by the extraction of the element a_1, and we can extract an element a_2 from the remaining set $A - \{a_1\}$. For the same reason the set $A - \{a_1, a_2\}$ is non-void and we can extract an element a_3 from it. Since the set A is infinite, we can continue this process indefinitely, obtaining, as a result, a sequence of elements

$$a_1, a_2, \ldots, a_n, \ldots,$$

which forms the required set D.

THEOREM 3. *Every infinite subset of a denumerable set is denumerable.*

Proof. Let A be a denumerable set and let B be an infinite subset of A. Arrange the elements of A in a sequence

$$a_1, a_2, a_3, \ldots, a_n, \ldots.$$

and consider the elements of A, in the order of their occurrence. In doing this, we shall meet from time to time elements of the set B, and sooner or later we shall encounter each element of B. We enumerate the set B by making the number of meeting of the element correspond to this element of B, and, consequently, because B is infinite, we must use all the natural numbers in this enumeration.

COROLLARY. *If a finite subset M is removed from a denumerable set A, then the remaining set $A - M$ is denumerable.*

THEOREM 4. *The sum of a finite set and a denumerable set without common elements is a denumerable set.*

Proof. Let

$$A = \{a_1, a_2, \ldots, a_n\},$$

and

$$B = \{b_1, b_2, b_3, \ldots\},$$

where $AB = 0$. If $A + B = S$, then S can be written in the form

$$S = \{a_1, a_2, \ldots, a_n, b_1, b_2, b_3, \ldots\};$$

now it is obvious that S can be enumerated.

The condition that the sets be disjoint can be omitted in this theorem as well as in the following theorems.

THEOREM 5. *The sum of a finite number of pairwise disjoint denumerable sets is a denumerable set.*

Proof. We carry out the proof for the case of the sum of three sets ; the complete generality of the reasoning will be clear. Let A, B, C be three denumerable sets :

$$A = \{a_1, a_2, a_3, \ldots\},$$
$$B = \{b_1, b_2, b_3, \ldots\},$$
$$C = \{c_1, c_2, c_3, \ldots\}.$$

Then the sum $S = A + B + C$ can be written in the form of a sequence

$$S = \{a_1, b_1, c_1, a_2, b_2, c_2, a_3, \ldots\},$$

and its denumerability is obvious.

3. Denumerable Sets

Theorem 6. *The sum of a denumerable family of pairwise disjoint finite sets is a denumerable set.*

Proof. Let A_k ($k = 1, 2, 3, \ldots$) be pairwise disjoint finite sets :

$$A_1 = \{a_1^{(1)}, a_2^{(1)}, \ldots, a_{n_1}^{(1)}\},$$
$$A_2 = \{a_1^{(2)}, a_2^{(2)}, \ldots, a_{n_2}^{(2)}\},$$
$$A_3 = \{a_1^{(3)}, a_2^{(3)}, \ldots, a_{n_3}^{(3)}\},$$
$$\cdots\cdots\cdots\cdots\cdots$$
$$\cdots\cdots\cdots\cdots\cdots$$

In order to arrange their sum in the form of a sequence, it is sufficient to write out in order all elements of the set A_1, then the elements of the set A_2, and so on.

Theorem 7. *The sum of a denumerable family of pairwise disjoint denumerable sets is a denumerable set.*

Proof. Let the sets A_k ($k = 1, 2, 3, \ldots$) be pairwise disjoint and denumerable. We write these sets as follows :

$$A_1 = \{a_1^{(1)}, a_2^{(1)}, a_3^{(1)}, \ldots\},$$
$$A_2 = \{a_1^{(2)}, a_2^{(2)}, a_3^{(2)}, \ldots\},$$
$$A_3 = \{a_1^{(3)}, a_2^{(3)}, a_3^{(3)}, \ldots\},$$
$$\cdots\cdots\cdots\cdots\cdots$$
$$\cdots\cdots\cdots\cdots\cdots$$

If we first write the element $a_1^{(1)}$, then the two elements $a_2^{(1)}$ and $a_1^{(2)}$ in which the sum of the upper and lower indices equals 3, then the elements for which this sum of indices equals 4, etc., the sum $S = \sum_{k=1}^{\infty} A_k$ is arranged in the form of a sequence

$$S = \{a_1^{(1)}, a_2^{(1)}, a_1^{(2)}, a_3^{(1)}, a_2^{(2)}, a_1^{(3)}, a_4^{(1)}, \ldots\},$$

and from this the denumerability of S is obvious.

Using the symbol a to denote the power of a denumerable set,[*] as noted in Definition 1 above, we can write the theorems just proved with the aid of the following mnemonic schemes :

$$a - n = a, \quad a + n = a, \quad a + a + \cdots + a = na = a,$$
$$n_1 + n_2 + n_3 + \cdots = a, \quad a + a + a + \cdots = aa = a.$$

Theorem 8. *The set R of all rational numbers is denumerable.*

Proof. The set of fractions of the form $\frac{p}{q}$ with given denominator q, *i. e.*, the set

$$\frac{1}{q}, \frac{2}{q}, \frac{3}{q}, \ldots$$

is obviously denumerable. But the denominator can also assume a denumerable set of natural values 1, 2, 3, ... By Theorem 7, this implies that the set of fractions $\frac{p}{q}$ is denumerable ; removing from it all reducible fractions and applying Theorem 3, we are convinced of the denumerability of the set R_+ of all *positive* rational numbers.

---

[*] The symbol $\aleph_0$ (aleph-nought) is often used instead of a.—E. H.

Since the set R_- of negative rational numbers is obviously equivalent to the set R_+, it is also denumerable and so the set R is denumerable since

$$R = R_- + \{0\} + R_+.$$

COROLLARY. *The set of rational numbers lying in an arbitrary interval $[a, b]$ is denumerable.*

THEOREM 9. *If we add a finite or denumerable set A of elements to an infinite set M, the power of M is not changed, i. e.,*

$$M + A \sim M.$$

Proof. Using Theorem 2, extract from M a denumerable subset D and let $M - D = P$; then $M = P + D$, $M + A = P + (D + A)$.

Since $P \sim P$ and $D + A \sim D$ (Theorems 4 and 5), it follows that $M + A \sim M$.

THEOREM 10. *If the infinite set S is non-denumerable and A is a finite or denumerable subset of A, then*

$$S - A \sim S.$$

Proof. The set $M = S - A$ cannot be finite, since otherwise the original set S would be finite or denumerable. But then, by virtue of Theorem 9, $M + A \sim M$ and this means that $S \sim S - A$.

COROLLARY. *Every infinite set contains a proper subset equivalent to itself.*

In fact, by removing an arbitrary finite subset from an infinite set, we do not change its power, in view of Theorems 3 and 10.

As we have already noted, a finite set does not possess the last-mentioned property. This circumstance allows us to give an exact definition of an infinite set.

DEFINITION 2. *A set is called* infinite *if it contains a proper subset equivalent to itself.*

Finally, we prove the following very general theorem:

THEOREM 11. *If the elements of a set A are defined by n symbols each of which, independently of the others, runs through a denumerable set of values*

$$A = \{a_{x_1, x_2, \ldots, x_n}\} \qquad (x_k = x_k^{(1)}, x_k^{(2)}, \ldots;\ k = 1, 2, 3, \ldots, n),$$

then the set A is denumerable.

Proof. We prove the theorem by the method of mathematical induction.

The theorem is obvious for $n = 1$, i.e., if there is only one symbol. Let us suppose that the theorem is true for $n = m$. We prove that it is true for $n = m + 1$. Thus, let

$$A = \{a_{x_1, x_2, \ldots, x_m, x_{m+1}}\}.$$

Denote by A_i the set of all elements of A such that

$$x_{m+1} = x_{m+1}^{(i)}.$$

where $x_{m+1}^{(i)}$ is one of the possible values of the $(m+1)$-st symbol, i.e., put

$$A_i = \{a_{x_1, x_2, \ldots, x_m, x_{m+1}^{(i)}}\}$$

By virtue of our hypothesis, the set A_i is denumerable, and since

$$A = \sum_{i=1}^{\infty} A_i,$$

A is also denumerable. Thus the theorem is proved.

Here are some consequences of this theorem.

1) The set of points (x, y) of the plane, such that both x and y are rational, is denumerable.

2) The set of complexes $(n_1, n_2, \ldots, n_k)$ consisting of k natural numbers, is denumerable.

The following fact is more interesting:

3) The set of polynomials

$$a_0 x^n + a_1 x^{n-1} + \ldots + a_{n-1} x + a_n$$

with integral coefficients is denumerable.

In fact, this follows immediately from Theorem 11, if we consider only polynomials of fixed degree n. To complete the proof, we must apply Theorem 7.

Since every polynomial has only a finite number of roots, the following theorem is a consequence of the preceding observation.

THEOREM 12. *The set of algebraic numbers is denumerable.*

(We recall that an algebraic number is a number which is the root of a polynomial with integral coefficients.)

§ 4. THE POWER OF THE CONTINUUM

One should not assume that all infinite sets are denumerable. We prove this with the following important example.

THEOREM 1. *The closed interval $U = [0, 1]$ is non-denumerable.*

Proof. Let us assume, on the contrary, that the interval U is a denumerable set. Then all of its points can be arranged in a sequence

$$x_1, x_2, x_3, \ldots \qquad (*)$$

We assume, then, that every point $x \in U$ occurs in the sequence (*). Divide U into three equal parts by means of the points $\frac{1}{3}$ and $\frac{2}{3}$. It is clear that the point x_1 cannot belong to all three subintervals

$$\left[0, \frac{1}{3}\right], \quad \left[\frac{1}{3}, \frac{2}{3}\right], \quad \left[\frac{2}{3}, 1\right] \qquad (1)$$

and that at least one of these intervals fails to contain x_1 (Figure 4). We denote this interval by U_1 (if the point x_1 is exterior to two of the intervals (1), then U_1 may be either one of them, for example, the one lying to the left of the other). We now divide the interval U_1 into three sub-intervals of equal length and denote by U_2 one of the new segments which does not contain the point x_2. We then divide the segment U_2 into three equal segments and designate by U_3 that one which does not contain x_3, and so on.

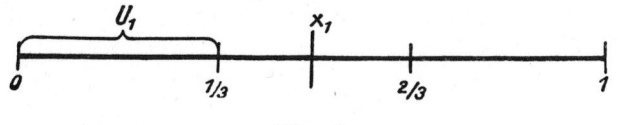

Fig. 4.

As a result, we obtain an infinite sequence of nested intervals

$$U \supset U_1 \supset U_2 \supset U_3 \supset \ldots$$

which possess the property that

$$x_n \overline{\in} U_n.$$

Since the length of the interval U_n is $\frac{1}{3^n}$, it is clear that this length tends to zero as

$n\to\infty$. Then, in accordance with a well-known limit therorem, there exists exactly one point ξ belonging to all of the intervals U_n:

$$\xi \in U_n \qquad (n = 1, 2, 3, \ldots).$$

Being a point of the interval U, the point ξ must appear in the sequence (*), but this is clearly impossible, because for every n, we have

$$x_n \overline{\in} U_n, \quad \xi \in U_n;$$

it follows that

$$\xi \neq x_n,$$

i.e., ξ is not a point of the sequence (*). This contradiction proves the theorem.

The theorem just proved suggests the following definition.

DEFINITION. If the set A is equivalent to the segment $U = [0, 1]$,

$$A \sim U,$$

then A is said to have the *power of the continuum*, or, more briefly, *the power of A is c*.

THEOREM 2. *Every closed interval $[a, b]$, every open interval (a, b) and every half-open interval $(a, b]$ or $[a, b)$ such that $a < b$ has the power c.*

Proof. Let

$$A = [a, b], \quad U = [0, 1].$$

The function

$$y = a + (b-a)x$$

establishes a one-to-one correspondence between the sets $A = \{y\}$ and $U = \{x\}$, from which it follows that A has the power of the continuum. Since the removal of one or two elements from an infinite set leaves a set equivalent to the original one, the intervals

$$(a, b), \quad (a, b], \quad [a, b)$$

have the same power as the closed interval $[a, b]$, *i.e.*, they have the power c.

THEOREM 3. *The sum of a nite finumber of pairwise disjoint sets of power c has power c.*

Proof. Let

$$S = \sum_{k=1}^{n} E_k \qquad (E_k E_{k'} = 0, \ k \neq k')$$

where each of the sets E_k has power c. Divide the half-open interval $[0, 1)$ into n half-open intervals

$$[c_{k-1}, c_k) \qquad (k = 1, 2, \ldots, n)$$

by means of the points

$$c_0 = 0 < c_1 < c_2 < \ldots < c_{n-1} < c_n = 1.$$

Each of these subintervals has the power c, so we can establish a one-to-one correspondence between the set E_k and the interval $[c_{k-1}, c_k)$. It is easy to see that this also establishes a one-to-one correspondence between the sum S and the interval

$$[0, 1) = \sum_{k=1}^{n} [c_{k-1}, c_k).$$

This proves the theorem.

THEOREM 4. *The sum of a denumerable family of pairwise disjoint sets each of power c has power c.*

4. The Power of the Continuum

Proof. Let
$$S = \sum_{k=1}^{\infty} E_k \qquad (E_k E_{k'} = 0, \ k \neq k')$$
where each of the sets E_k has power c. Take a monotonic increasing sequence
$$c_0 = 0 < c_1 < c_2 < \ldots,$$
in the interval [0, 1) such that
$$\lim_{k \to \infty} c_k = 1.$$

Establishing a one-to-one correspondence between the sets E_k and $[c_{k-1}, c_k)$ for all k, we establish a one-to-one correspondence between S and [0, 1).

COROLLARY 1. *The set Z of all real numbers has the power c.*
In fact,
$$Z = \sum_{k=1}^{\infty} \{[k-1, k) + [-k, -k+1)\}.$$

COROLLARY 2. *The set of all irrational numbers has power c.*
COROLLARY 3. *Transcendental numbers* [2] *exist.*
THEOREM 5. *The set Q of all sequences of natural numbers*
$$Q = \{(n_1, n_2, n_3, \ldots)\}$$
has power c.

Proof. We establish a one-to-one correspondence between Q and the set of all irrational numbers of the interval (0, 1) (the latter set obviously has the power of the continuum) by setting the sequence
$$(n_1, n_2, n_3, \ldots) \in Q$$
into correspondence with the irrational number x whose continued fraction expansion has the form
$$x = \cfrac{1}{n_1 + \cfrac{1}{n_2 + \cfrac{1}{n_3 + \cdots}}}.$$

The theorem is proved by observing that this correspondence carries Q in a one-to-one way onto the given set of irrational numbers. This proof requires that the reader be familiar with the theory of continued fractions.[3]

One can give another proof based on the theory of *binary* expansions. For this purpose, we recall some facts from this theory, which will also be useful to us for other purposes. The properties of binary expansions which we shall need are the following:

1) A binary expansion is the sum of the series
$$\sum_{k=1}^{\infty} \frac{a_k}{2^k}, \qquad a_k = \begin{cases} 0 \\ 1 \end{cases}.$$

---

[2] *I.e.*, non-algebraic.
[3] See, *e.g.*, HARDY AND WRIGHT, *An Introduction to the Theory of Numbers*, Oxford University Press, 1938, Chapt. X.

This sum is designated by the symbol
$$0.a_1a_2a_3\ldots \qquad (1)$$

2) Every number $x \in [0, 1]$ can be represented in the form
$$x = 0.a_1a_2a_3\ldots$$

This representation is *unique* if x is not a number of the form $\frac{m}{2^n}$ ($m = 1, 3, \ldots, 2^n - 1$). The numbers 0 and 1 are written (uniquely) as the fractions
$$0 = 0.000\ldots, \quad 1 = 0.111\ldots.$$

If $x = \frac{m}{2^n}$ ($m = 1, 3, \ldots, 2^n - 1$), then x has *two* binary expansions of the form (1). In these expansions, the numbers $a_1, a_2, \ldots, a_{n-1}$ coincide and the symbol a_k equals 1 in one of them and 0 in the other. All remaining symbols in the first development are zeros, and in the other they are ones. For example,

$$\frac{3}{8} = \begin{cases} 0.011000\ldots \\ 0.010111\ldots \end{cases}$$

3) Every binary expansion (1) is equal to some number x in the closed interval $[0, 1]$. If the expansion (1) is equal to 0 from some point on, or is equal to 1 from some point on, then it is a number of the form $\frac{m}{2^n}$ ($m = 1, 3, \ldots, 2^n - 1$) (the fractions $0.000\ldots$ and $0.111\ldots$ are exceptions). In this case, there exists a second binary development for its sum x in addition to the original development. If the expansion (1) is not ultimately equal to 0 or 1, then $x \neq \frac{m}{2^n}$, and x has only one binary expansion.

Having noted these facts, we return to Theorem 5. We agree not to use expansions (1) in which all a_k are 1 from a certain point on. Then every number of the interval $[0, 1)$ is uniquely representable in the form
$$0.a_1a_2a_3\ldots \qquad (1)$$

where for every natural number N an a_k can be found such that
$$a_k = 0, \quad k > N.$$

Conversely, there is a point in $[0, 1)$ corresponding to every expansion (1) having this property. However, an expansion (1) can be specified by indicating those k for which
$$a_k = 0.$$

These k's form an increasing sequence of natural numbers
$$k_1 < k_2 < k_3 < \ldots \qquad (2)$$

and to every such sequence, there corresponds an expansion (1). This implies that the set H of sequences (2) has power c. A one-to-one correspondence can easily be established between the sets H and Q. To do this, it is sufficient to pair off the sequence
$$(n_1, n_2, n_3, \ldots)$$

from Q such that
$$n_1 = k_1, \quad n_2 = k_2 - k_1, \quad n_3 = k_3 - k_2, \quad \ldots$$

with the sequence (2). This proves the theorem.

4. The Power of the Continuum

THEOREM 6. *If the elements of the set A are defined by n symbols, each of which, independently of the other symbols, assumes c values,*[4]

$$A = \{a_{x_1, x_2, \ldots, x_n}\}$$

then the set A has power c.

Proof. It suffices to consider the case of three symbols, since the reasoning can easily be generalized. Let

$$A = \{a_{x, y, z}\}.$$

Denote by X, Y, and Z the sets of values of the symbols x, y, and z respectively, where each of the symbols varies independently of the others and each of the sets X, Y, Z has power c. We establish a one-to-one correspondence between each of the sets X, Y, Z and the set Q of all sequences of natural numbers. This will allow us to establish a similar correspondence between A and Q. Indeed, let ξ be any element of A. Then

where
$$\xi = a_{x_0, y_0, z_0},$$

$$x_0 \in X, \quad y_0 \in Y, \quad z_0 \in Z.$$

In the correspondences between X, Y, Z and Q, certain elements of Q correspond to the elements x_0, y_0, z_0; let

the sequence $(n_1, n_2, n_3, \ldots)$ correspond to the element x_0,
the sequence $(p_1, p_2, p_3, \ldots)$ correspond to the element y_0,
the sequence $(q_1, q_2, q_3, \ldots)$ correspond to the element z_0.

We associate the sequence

$$(n_1, p_1, q_1, n_2, p_2, q_2, n_3, \ldots)$$

which obviously occurs in Q, with the element ξ. It is easy to see that we have indeed produced a one-to-one correspondence between A and Q.

A number of important consequences follow from the theorem just proved.

COROLLARY 1. *The set of all points of the plane has power c.*

COROLLARY 2. *The set of all points of three-dimensional space has power c.*

In other words, 1-, 2-, 3-, ... , and n- dimensional spaces all have the same power, namely, c.

COROLLARY 3. *The sum of c pairwise disjoint sets of power c has power c.*

One can establish a one-to-one correspondence between the set of component sets and the set of all straight lines of the xy-plane parallel to the x-axis. Then, if we establish a one-to-one correspondence between each of the component sets and the line corresponding to it, we obviously obtain a one-to-one correspondence between the sum and the xy-plane.

Theorems 3, 4 and Corollary 3 can be written schematically as

$$c + c + \ldots + c = cn = c, \quad c + c + c + \ldots = ca = c, \quad cc = c.$$

THEOREM 7. *If the elements of the set A are defined by a denumerable set of symbols*

$$A = \{a_{x_1, x_2, x_3, \ldots}\}$$

---

[4] Here and in the sequel, we use such expressions as "c values," "there are c elements in the set," and so on, instead of saying "the set of values has power c," etc.; we trust that this will cause the reader no difficulty.

each of which takes on c values, independently of the other symbols, then the set A has power c.

Proof. Let X_k be the set of values of the symbol x_k. We relate it by a one-to-one correspondence with the set Q of all sequences of natural numbers. Let this correspondence be denoted by φ_k ($k = 1, 2, 3, \ldots$). Having done this, we choose an arbitrary element $\xi \in A$. Thus

$$\xi = a_{x_1^{(0)},\; x_2^{(0)},\; x_3^{(0)},\; \ldots}$$

where

$$x_k^{(0)} \in X_k \qquad\qquad (k = 1, 2, 3, \ldots).$$

Let the sequence

$$(n_1^{(k)},\; n_2^{(k)},\; n_3^{(k)}, \ldots) \in Q$$

correspond to the value $x_k^{(0)}$ of the symbol x_k in the correspondence φ_k. Then the infinite matrix

$$\left|\begin{array}{llll} n_1^{(1)}, & n_2^{(1)}, & n_3^{(1)}, & \ldots \\ n_1^{(2)}, & n_2^{(2)}, & n_3^{(2)}, & \ldots \\ n_1^{(3)}, & n_2^{(3)}, & n_3^{(3)}, & \ldots \\ \cdot\;\cdot\;\cdot\;\cdot\;\cdot\;\cdot\;\cdot\;\cdot \\ \cdot\;\cdot\;\cdot\;\cdot\;\cdot\;\cdot\;\cdot\;\cdot \\ \cdot\;\cdot\;\cdot\;\cdot\;\cdot\;\cdot\;\cdot\;\cdot \end{array}\right. \qquad (*)$$

with positive integral elements is put into correspondence with the element $\xi \in A$. It is easily seen that the correspondence between A and the set L of matrices (*) is one-to-one. It remains to show that the set L has power c. This is almost obvious, since we can associate with the matrix (*) the sequence

$$(n_1^{(1)},\; n_2^{(1)},\; n_1^{(2)},\; n_3^{(1)},\; n_2^{(2)},\; n_1^{(3)},\; n_4^{(1)},\; \ldots)$$

(constructed exactly as we made the construction in proving Theorem 7, §3). We obtain at once a one-to-one correspondence between L and Q.

THEOREM 8. *The set T of all sequences of the form*

$$(a_1, a_2, a_3, \ldots),$$

where the a_k assume the values 0 and 1 independently, has power c.

Proof. Let S be the set of those sequences in T such that all of the a_k are equal to 1 from some place on. With each sequence $(a_1, a_2, a_3, \ldots)$ in S, one can associate a number having the binary expansion $0.a_1 a_2 a_3 \ldots$; this number will be either 1 or $\frac{m}{2^n}$ ($m = 1, 3, \ldots, 2^n - 1$). The correspondence obtained between S and the set of numbers of the indicated form is obviously one-to-one; from this it follows that S is a denumerable set.

On the other hand, if we associate the number having binary development $0.a_1 a_2 a_3 \ldots$ with the sequence $(a_1, a_2, a_3, \ldots)$, then we obtain a one-to-one correspondence between $T-S$ and the interval $[0, 1)$, from which it follows that $T-S$, and hence T, has power c.

COROLLARY. *If the elements of a set A are defined by means of a denumerable set of* *symbols, each of which, independently of the others, assumes two values, then the set A has power c.*

In fact, if $A = \{a_{x_1}, a_{x_2}, a_{x_3}, \ldots\}$ and $x_k = \begin{cases} l_k \\ m_k \end{cases}$,

then it is sufficient to associate with each element of A the sequence $(a_1, a_2, a_3, \ldots)$, where a_k equals 0 or 1 according as $x_k = l_k$ or $x_k = m_k$, in order to obtain a one-to-one correspondence between A and T.

§ 5. COMPARISON OF POWERS

We have defined above the expressions "two sets have the same power," "a set has power a," "a set has power c." So, when we meet the word "power" in one of these expressions, we know what it means, but we have not yet defined this concept by itself. What, then, is the power of a set?

G. Cantor tried to define this concept by the aid of such cloudy expressions as: "The power of a given set A is that general idea which remains with us when, thinking of this set, we abstract from all properties of its elements as well as from their order." In this connection, Cantor denoted the power of the set A by the symbol

$$\bar{\bar{A}}$$

(the two lines indicate "double" abstraction).

At the present time, Cantor's method of defining the concept of power is not considered satisfactory (although the notation $\bar{\bar{A}}$ has proved very felicitous).

Instead, the following formal definition has been set forth.

DEFINITION 1. *Let all sets be divided into families such that two sets fall into one family if and only if they are equivalent. To every such family of sets, we assign some arbitrary symbol and we call this symbol the* power* *of every set of the given family.***
If the power of a set A is α, we write

$$\bar{\bar{A}} = \alpha.$$

Once the definition has been stated in this manner, it is clear that equivalent sets do have the same power.

We use the letter a to denote the power of the set N of all natural numbers. Hence every denumerable set has power a. The letter "c" will be used to denote the power of the interval $U = [0, 1]$, and we therefore say that all sets equivalent to U have power c.

We give one more example in which Definition 1 is applied. Let the symbol "3" correspond to the family containing the set

---

* The term *cardinal number* is commonly used as synonymous with the term *power* as defined here.—E. H.

** The author here passes over the fact that this definition of the power of a set is quite unacceptable from a strictly rigorous point of view. The difficulty lies in making precise what is meant by the term "all sets." If sufficiently many objects are admitted as sets, then paradoxes (*e.g.*, Russell's paradox) are immediately introduced. For the purposes of the student of analysis, the definition given in the text is sufficient, but one should be aware that there is a very difficult and subtle point involved.—E. H.

$$A = \{a, b, c\} \ .$$

Then we can say that an arbitrary set equivalent to the set A (i.e., in simple terms, an arbitrary set containing three elements), has power 3. We thus see that the concept of number of elements of a finite set is a particular case of the more general concept of power. Finally, 0 is the power of the void set, and 1 is the power of an arbitrary set consisting of exactly one element.

Having the definition of the concept of power, it is natural to ask if powers can be compared.

DEFINITION 2. Let A and B be sets having powers α and β respectively:

$$\bar{\bar{A}} = \alpha, \quad \bar{\bar{B}} = \beta.$$

If 1) the sets A and B are not equivalent but 2) there is a subset B^* of the set B equivalent to the set A, then the set B is said to have the *greater* and the set A to have the *smaller* power. We write

$$\alpha < \beta, \quad \beta > \alpha.$$

For example, if
$$A = \{a_1, a_2, \ldots, a_{32}\}, \quad \bar{\bar{A}} = 32,$$
$$B = \{b_1, b_2, \ldots, b_{49}\}, \quad \bar{\bar{B}} = 49$$

then A is not $\sim B$, but $A \sim B^*$ where $B^* = \{b_1, b_2, \ldots, b_{32}\}$, and hence

$$32 < 49.$$

Similarly, every natural number n is less than both of the powers a and c.
Finally, if
$$N = \{1, 2, 3, \ldots\}, \quad \bar{\bar{N}} = a,$$
$$U = [0, 1], \quad \bar{\bar{U}} = c,$$

then N is not $\sim U$ (see Theorem 1, §4), but $N \sim U^*$, where $U^* = \{1, \tfrac{1}{2}, \tfrac{1}{3}, \ldots\} \subset U$. Hence

$$a < c.$$

The problem whether there exists a power μ intermediate between a and c, i.e., such that

$$a < \mu < c$$

is not yet solved,[5] in spite of the fact that much research has been devoted to it. On the other hand, it is easy to construct a set having power greater than c.

THEOREM 1. *The set F of all real-valued functions defined on the segment $[0, 1]$ has power greater than c.*

Proof. We first show that

$$F \text{ is not } \sim U$$

where $U = [0, 1]$. Let us assume, on the contrary, that $F \sim U$ and let φ be a one-to-one correspondence between F and U. We denote by $f_t(x)$ the function in F which corresponds to the number $t \in [0, 1]$ under the correspondence φ. We set

---

[5] The assertion that there are no such sets is called the continuum hypothesis.

5. Comparison of Powers

$$F(t, x) = f_t(x).$$

This is a completely defined function of the two variables x and t, defined in the region $0 \leq t \leq 1, 0 \leq x \leq 1$. We now set

$$\psi(x) = F(x, x) + 1.$$

This function is defined for $0 \leq x \leq 1$, i.e., $\psi(x) \in F$. But then the function $\psi(x)$ corresponds to some number $a \in U$ under the correspondence φ, i.e., $\psi(x) = f_a(x)$, or

$$\psi(x) = F(a, x).$$

In other words,

$$F(x, x) + 1 = F(a, x)$$

for all x in [0, 1]. But this is impossible, as we see on setting $x = a$. It follows that F is not $\sim U$. But if we consider the set of functions

$$F^* = \{\sin x + c\} \qquad (0 \leq c \leq 1),$$

which is a subset of F, then we see at once that $F^* \sim U$, and the theorem is proved.

DEFINITION 3. *The power of the set F of all real-valued functions defined on the segment [0, 1] is designated by the symbol f.*

With the aid of this notation, we can formulate Theorem 1 as follows:

$$c < f.$$

Do there exist powers greater than f? It happens that such powers do exist. More than that, we can show that, starting from a set of arbitrary power, it is possible to construct a set of greater power.

THEOREM 2. *Let M be an arbitrary set. If we let T denote the family of all subsets of the set M, then we have*

$$\overline{\overline{T}} > \overline{\overline{M}}.$$

Proof. We notice that *all* subsets of M, and, in particular, M itself, the void set and all one-element subsets of M, are elements of the family T. We first show that

$$T \text{ is not } \sim M.$$

Let us assume, on the contrary, that $T \sim M$, and let φ be a one-to-one correspondence carrying M onto T. To every $m \in M$, there corresponds a definite element of T under the correspondence φ which we designate by $\varphi(m)$, and every element of T is $\varphi(m)$ for one and only one $m \in M$. Let us call an element $m \in M$ *good* if

$$m \in \varphi(m);$$

and *bad* otherwise. The element under the correspondence φ which corresponds to the set M itself is clearly good, and the element corresponding to the void set is bad. Let S be the set of all bad elements of M. Since $S \in T$, there is an element $m_0 \in M$ corresponding to the set S under the correspondence φ:

$$S = \varphi(m_0).$$

Is this element m_0 good or bad? Let us assume that m_0 is a good element. This means that

$$m_0 \in \varphi(m_0) = S$$

but since S consists of bad elements only, the element m_0 is bad, which contradicts our hypothesis that m_0 is a good element. Hence, m_0 is a bad element. But then

$$m_0 \overline{\in} \varphi(m_0) = S;$$

and this means that m_0 is a good element. We thus see that the element m_0 can be neither good nor bad, and since each element is either good or bad, we have here an absurdity, which shows that

$$T \text{ is not} \sim M.$$

But, if T^* is the set of all one-element subsets of M, then, obviously, $T^* \sim M$, and since $T^* \subset T$, the theorem is proved.

Remark. Let M be a finite set consisting of n elements. Then the set T contains 2^n elements. In fact, T contains the void set, C_n^1 one-element sets, C_n^2 two-element sets, etc., and so there are

$$1 + C_n^1 + C_n^1 + \ldots + C_n^n = 2^n$$

elements in T. We notice that this result is true also for the cases when M is the void set ($n = 0$) or a one-element set ($n = 1$), for in the first case, T consists of M only and in the second case of M and the void set. In this connection, it is natural to make the following definition.

DEFINITION 4. *If the set M has power μ and the set T of all its subsets has power τ, we say that*

$$\tau = 2^\mu.$$

Theorem 2 shows that

$$2^\mu > \mu.$$

THEOREM 3. *The following formula is valid:*

$$c = 2^a.$$

Proof. Let T be the set of all subsets of the set N of natural numbers, and let L be the set of all sequences of the form

$$(a_1, a_2, a_3, \ldots) \qquad a_k = \begin{cases} 0 \\ 1 \end{cases}.$$

Then (Theorem 8, §4)

$$\overline{\overline{T}} = 2^a, \quad \overline{\overline{L}} = c.$$

Take an arbitrary element $N^* \in T$. N^* is a set of natural numbers. We assign the sequence $(a_1, a_2, a_3, \ldots)$ to N^*, according to the following rule. If $k \in N^*$, then $a_k = 1$, and if $k \overline{\in} N^*$, then $a_k = 0$. It is clear that we obtain in this manner a one-to-one correspondence between T and L, and this proves the theorem.

We obtain from Theorems 2 and 3 a second proof that

$$c > a.$$

The following two theorems are of great importance.

THEOREM 4. *Let $A \supset A_1 \supset A_2$. If $A_2 \sim A$, then also*

$$A_1 \sim A.$$

5. Comparison of Powers

Proof. Let φ be a one-to-one correspondence between A and A_2. To each element of A there corresponds some element of A_2 under φ. In particular, those elements of A_2 which correspond to elements of A_1 form a definite set $A_3 \subset A_2$.
Thus A_1 is related to A_3 by a one-to-one correspondence. But since $A_2 \subset A_1$, it follows that those elements of A_3 which correspond to the elements of A_2 under φ form a definite set $A_4 \subset A_3$. Now, since $A_3 \subset A_2$, and A_2 and A_4 are related by the one-to-one correspondence φ, we can form the set $A_5 \subset A_4$ consisting of those elements of A_4 which correspond to elements of A_3. Continuing this process, we obtain a sequence of sets

$$A \supset A_1 \supset A_2 \supset A_3 \supset A_4 \supset A_5 \supset \ldots$$

such that

$$A \sim A_2$$
$$A_1 \sim A_3$$
$$A_2 \sim A_4$$
$$A_3 \sim A_5$$
$$\ldots\ldots$$
$$\ldots\ldots$$

We notice that the following relations are also valid.

$$\left.\begin{aligned} A - A_1 &\sim A_2 - A_3 \\ A_1 - A_2 &\sim A_3 - A_4 \\ A_2 - A_3 &\sim A_4 - A_5 \\ &\ldots\ldots\ldots \\ &\ldots\ldots\ldots \end{aligned}\right\} \quad (*)$$

These follow from the very definition[6] of the sets A_n.
Let

$$D = AA_1A_2A_3\ldots$$

It is easy to see that

$$A = (A - A_1) + (\underline{A_1 - A_2}) + (\underline{\underline{A_2 - A_3}}) + (\underline{A_3 - A_4}) + (\underline{\underline{A_4 - A_5}}) + \ldots + D$$
$$A_1 = (A_1 - A_2) + (\underline{\underline{A_2 - A_3}}) + (A_3 - A_4) + (\underline{\underline{A_4 - A_5}}) + \ldots + D,$$

where the terms in parentheses on each line are pairwise disjoint. In view of (*), the underlined terms of both sums are equivalent. But the remaining terms of these sums are mutually identical, from which it follows that A and A_1 are equivalent.

THEOREM 5. *Let A and B be two sets. If each of them is equivalent to a subset of the other, then they are equivalent.*

Proof. Let

$$A \sim B^*, \quad B^* \subset B,$$
$$B \sim A^*, \quad A^* \subset A.$$

We establish a one-to-one correspondence between B and A^* under which those elements of A^* which correspond to elements of the set B^* form a set A^{**}. Ob-

---

[6] We call the reader's attention to the fact that the relations $A^* \subset A$, $B^* \subset B$, $A^* \sim B^*$, $A \sim B$ do *not* imply $A - A^* \sim B - B^*$.

viously, $A \supset A^* \supset A^{**}$ and $A \sim A^{**}$ (since $A \sim B^*$, $B^* \sim A^{**}$). Hence, by Theorem 4, $A \sim A^*$, and since $A^* \sim B$, it follows that $A \sim B$.

Theorems 4 and 5 have a number of important corollaries.

COROLLARY 1. *If α and β are two powers, then the relations*

$$\alpha = \beta, \quad \alpha < \beta, \quad \alpha > \beta$$

are mutually exclusive.

It is first obvious that the relation $\alpha = \beta$ excludes the other two. Let us now assume that both of the relations $\alpha < \beta$ and $\alpha > \beta$ occur simultaneously. Let A and B be two sets of power α and β respectively:

$$\bar{\bar{A}} = \alpha, \quad \bar{\bar{B}} = \beta.$$

Since $\alpha > \beta$,
1) A and B are not equivalent;
2) $A \sim B^*$, where $B^* \subset B$.

But $\alpha > \beta$ implies that
3) $B \sim A^*$, where $A^* \subset A$.

Now 2) and 3) imply that $A \sim B$, which contradicts 1).

COROLLARY 2. *If α, β, γ are three powers and*

$$\alpha < \beta, \quad \beta < \gamma,$$

then

$$\alpha < \gamma,$$

i.e., the relation $<$ is transitive.

In fact, if A, B, C are three sets of powers α, β, γ, then $A \sim B^* \subset B$, $B \sim C^* \subset C$; from this it follows that $A \sim C^{**} \subset C$, where C^{**} is the set of those elements of C^* which correspond to the elements of B^* in the correspondence between B and C^*. It remains to show that A is not $\sim C$.

If A were $\sim C$, then we would have $C^{**} \sim C$; then by Theorem 4, we would have $C^* \sim C$, whence $B \sim C$ and $\beta = \gamma$.

REMARK. *It follows from Definition 2 that if $A \sim B^* \subset B$, then either $\bar{\bar{A}} = \bar{\bar{B}}$ or $\bar{\bar{A}} < \bar{\bar{B}}$.*

In this connection, we frequently write the relation $A \sim B^* \subset B$ in the form

$$\bar{\bar{A}} \leqslant \bar{\bar{B}}.$$

With the aid of this notation, Theorem 5 can be formulated as follows.

If $\alpha \geqslant \beta$ and $\alpha \leqslant \beta$, then $\alpha = \beta$.

If m and n are two natural numbers, then one and only one of the three relations

$$m = n, \quad m < n, \quad m > n$$

holds. It can be shown that for arbitrary powers α and β, one of the three mutually exclusive relations

$$\alpha = \beta, \quad \alpha < \beta, \quad \alpha > \beta$$

necessarily holds.

This property of powers is called *trichotomic*.

We give an application of Theorem 5 in the following example.

THEOREM 6. *The set Φ of all continuous real-valued functions defined on the closed interval $[0, 1]$ has power c.*

Proof. Let $\Phi^* = \{\sin x + k\}$, where k assumes all real values. Obviously, $\Phi^* \subset \Phi$ and $\overline{\overline{\Phi^*}} = c$, whence it follows that

$$\overline{\overline{\Phi}} \geqslant c. \tag{1}$$

It remains to show that

$$\overline{\overline{\Phi}} \leqslant c. \tag{2}$$

For this purpose, let H be the set of all sequences of the form

$$[u_1, u_2, u_3, \ldots],$$

where the u_k assume all real values independently of each other. By Theorem 7, §4, $\overline{\overline{H}} = C$. Enumerate all rational numbers of the segment $[0, 1]$ as a sequence:

$$r_1, r_2, r_3, \ldots$$

and let the sequence

$$a_f = [f(r_1), f(r_2), f(r_3), \ldots]$$

correspond to the function $f(x) \in \Phi$. It is obvious that $a_f \in H$. If the continuous functions $f(x)$ and $g(x)$ are not identical, then

$$a_f \neq a_g.$$

In fact, if we have $a_f = a_g$, then the equality

$$f(x) = g(x)$$

holds for all rational x in $[0, 1]$; from this, owing to the continuity of both functions it follows that $f(x) = g(x)$ for every x in $[0,1]$, and the functions $f(x)$ and $g(x)$ are identical.

This implies that the set Φ is equivalent to the set $H^* = \{a_f\}$. Since $H^* \subset H$ and $\overline{\overline{H}} = c$, relation (2) is proved, and with it the present theorem.

Exercises for Chapter I

1. Prove that the set of points of discontinuity of a monotonic function is finite or denumerable.
2. Establish a one-to-one correspondence between $(0, 1)$ and $[0, 1]$.
3. Prove that $f = 2^c$.
4. If $A = B + C$, $\overline{\overline{A}} = c$, then at least one of the sets B or C has power c. Prove this statement.
5. Let $f(x)$ be a function possessing the property that to every x_0 there corresponds a $\delta > 0$ such that $f(x) > f(x_0)$ whenever $|x - x_0| < \delta$. Prove that the set of values of $f(x)$ is finite or denumerable.
6. Prove that in Theorems 4, 5, 6, 7, §3 and Theorems 3, 4, §4, it is possible to omit the condition that the component sets are pairwise disjoint.
7. Prove the formula $AB + C = (A + C)(B + C)$. Generalize this relation.
8. Let $A_1, A_2, A_3, \ldots$ be a sequence of sets. Denote by $\overline{A}$ the set of elements belonging to an infinite number of the sets A_n, and by $\underline{A}$ the set of elements belonging to all but a finite number of the sets A_n. Prove that

$$\overline{A} = \prod_{n=1}^{\infty} \sum_{k=n}^{\infty} A_k, \quad \underline{A} = \sum_{n=1}^{\infty} \prod_{k=n}^{\infty} A_k.$$

9. If $A = \sum_{n=1}^{\infty} A_n$ and $\overline{\overline{A}} = c$, prove that at least one of the sets A_n has power c.

CHAPTER II

POINT SETS

In this chapter, we shall study sets of points on the real line. We shall denote the set of all real numbers by the symbol Z. We note that all terms encountered in the sequel such as "point," "closed interval," "open interval," and so on, will be used in a purely arithmetic sense; when we state, for instance, that the point y lies to the right of the point x, we have in mind that $y > x$, and so on.

§1. LIMIT POINTS

DEFINITION 1. *The point x_0 is called a* limit point [1] *of a point set E if every open interval containing this point contains at least one point of E distinct from the point x_0.*

Remarks. 1) The point x_0 itself may or may not belong to the set E.

2) If the point x_0 belongs to the set E but is not a limit point of E, it is called an *isolated* point of the set E.

3) If x_0 is a limit point of the set E, then every open interval (α, β) containing this point contains an infinite number of points of E.

We prove the last remark. Assume the contrary, i.e., that the interval (α, β) containing x_0 contains only a finite number of points of E. Let the points of the set $E \cdot (\alpha, \beta)$ which are distinct from x_0 be $\xi_1, \xi_2, \ldots \xi_n$. Designate by δ the minimum of the positive numbers $|x_0 - \xi_1|, |x_0 - \xi_2|, \ldots, |x_0 - \xi_n|, x_0 - \alpha, \beta - x_0$. Consider the open interval $(x_0 - \delta, x_0 + \delta)$. None of the points ξ_k ($k = 1, 2, \ldots, n$) belongs to this interval, and since

$$(x_0 - \delta, x_0 + \delta) \subset (\alpha, \beta)$$

the interval $(x_0 - \delta, x_0 + \delta)$ contains no points of E distinct from x_0; this contradicts the assumption that x_0 is a limit point of the set E.

One can approach the concept of limit point from another point of view. For this purpose, we prove the following assertion.

THEOREM 1. *In order that the point x_0 be a limit point of the set E, it is necessary and sufficient that it be possible to select from the set E a sequence of distinct points $x_1, x_2, \ldots, x_n, \ldots$ such that*

$$x_0 = \lim_{n \to \infty} x_n.$$

Proof. The sufficiency of the condition is quite obvious. We shall prove its necessity. Let x_0 be a limit point of the set E. Choose a point $x_1 \in E$ in the open interval $(x_0 - 1, x_0 + 1)$ such that x_1 is different from x_0. Next, choose a point $x_2 \in E$ in the interval $(x_0 - \frac{1}{2}, x_0 + \frac{1}{2})$, different from both x_0 and x_1, and so on. At the n-th step of the process, we choose a point $x_n \in E$ in the interval $(x_0 - \frac{1}{n}, x_0 + \frac{1}{n})$, different from

---

[1] Or *point of accumulation*.

1. Limit Points

$x_0, x_1, \ldots, x_{n-1}$. We thus have a sequence $\{x_n\}$ of distinct points of the set E for which, clearly enough,

$$\lim x_n = x_0.$$

The theorem just proved allows us to state Definition 1 in a different form.

DEFINITION 2. *The point x_0 is said to be a* limit point *of the set E if it is possible to select a sequence of distinct points*

$$x_1, x_2, x_3, \ldots$$

from the set E such that

$$x_0 = \lim_{n \to \infty} x_n.$$

THEOREM 2 (BOLZANO-WEIERSTRASS). *Every bounded infinite set E has at least one limit point (which may or may not belong to E).*

Proof. Since the set E is bounded, we can find a closed interval $[a, b]$ containing it. Set

$$c = \frac{a+b}{2}$$

and consider the closed subintervals $[a, c]$ and $[c, b]$. It is impossible that both of these segments contain only a finite number of points of E, because, in that case, the entire set E would be finite. This implies that at least one of the subintervals contains an infinite number of points of E. Denote it by $[a_1, b_1]$ (if both of the intervals $[a, c]$ and $[c, b]$ contain an infinite set of points of E, then denote one of them by $[a_1, b_1]$; it makes no difference which one). Set

$$c_1 = \frac{a_1 + b_1}{2}$$

and denote by $[a_2, b_2]$ one of the intervals $[a_1, c_1]$ and $[c_1, b_1]$ which contains an infinite number of points of E (the existence of such a set is established as above). Continuing this process, we construct an infinite sequence of nested intervals

$$[a, b] \supset [a_1, b_1] \supset [a_2, b_2] \supset \ldots$$

each of which contains an infinite number of points of E. Since

$$b_n - a_n = \frac{b-a}{2^n}$$

the length of the interval $[a_n, b_n]$ tends to zero as $n \to \infty$, and by a known theorem from the theory of limits, there exists a point x_0 contained in all of the intervals $[a_n, b_n]$, such that

$$\lim a_n = \lim b_n = x_0.$$

We shall show that x_0 is a limit point of the set E. To do this, we take an arbitrary open interval (α, β) containing x_0. It is obvious that if n is sufficiently large, then

$$[a_n, b_n] \subset (\alpha, \beta),$$

It follows that (α, β) contains an infinite number of points of E, and this proves our assertion.

We note that boundedness of the set E is essential for the validity of Theorem 2.

The set N of all natural numbers serves as an example. Although N is infinite, it has no limit points at all.

In applications, another form of the Bolzano-Weierstrass theorem frequently turns out to be useful; in this form, one is concerned not with sets but with sequences of numbers.

We say that we have a sequence of numbers

$$x_1, x_2, x_3 \ldots ,\qquad (*)$$

if to every n, there corresponds a definite number x_n; here, distinct terms of the sequence may be equal to each other. Such, for example, is the sequence

$$0, 1, 0, 1, 0, 1, \ldots .$$

Considered as a point set, this set is finite, since it consists of only two points, 0 and 1; but, considered as a sequence, it is infinite.

The sequence (*) is said to be *bounded* if there exists a number K such that for all n,

$$|x_n| < K .$$

The second form of the Bolzano-Weierstrass theorem, mentioned above, is as follows.

THEOREM 2*. *From every bounded sequence*

$$x_1, x_2, x_3 \ldots ,\qquad (*)$$

it is possible to select a convergent subsequence

$$x_{n_1}, x_{n_2}, x_{n_3}, \ldots \qquad (n_1 < n_2 < n_3 < \ldots)$$

Proof. Consider the set E of all numbers appearing in the sequence (*). If this set is finite, then at least one of its points must occur an infinite number of times in the sequence (*); let this point be ξ, and let

$$x_{n_1} = x_{n_2} = x_{n_3} = \ldots = \xi,$$

then the required sequence is $\{x_{n_k}\}$. If the set E is infinite, the Bolzano-Weierstrass theorem can be applied to it. Let x_0 be a limit point of the set E; then we can select from E a sequence

$$x_{m_1}, x_{m_2}, x_{m_3}, \ldots \qquad (**)$$

converging to the point x_0, where all of its terms, and hence also the indices $m_1, m_2, m_3, \ldots$ are distinct. Set $n_1 = m_1$, and denote by n_2 the first of the numbers $m_1, m_2, m_3, \ldots$ which is larger than n_1; then denote by n_3 the first of these numbers which is greater than m_2, and so on. We obtain in this way a sequence

$$x_{n_1}, x_{n_2}, x_{n_3}, \ldots$$

with *increasing* indices. Inasmuch as this sequence is a subsequence of (**), it is clear that

$$\lim x_{n_k} = x_0 .$$

This proves the theorem.

§ 2. CLOSED SETS

We now define a number of concepts closely related to the concept of limit point.
DEFINITIONS. Let E be a point set.

2. CLOSED SETS

1. The set of all limit points of E is called the *derived set* of the set E and is denoted by E'.
2. If $E' \subset E$, the set E is said to be *closed*.
3. If $E \subset E'$, the set E is said to be *dense in itself*.
4. If $E = E'$, the set E is said to be *perfect*.
5. The set $E + E'$ is called the *closure* of the set E and is denoted by $\bar{E}$.

Thus a set is closed if it contains all its limit points. A set which is dense in itself has no isolated points. A perfect set is one which is closed and dense in itself.

We illustrate the above definitions by means of examples.

EXAMPLES.
1. $E = \{1, \frac{1}{2}, \frac{1}{3}, \ldots, \frac{1}{n}, \ldots\}$, $E' = \{0\}$. This set is neither closed nor dense in itself.
2. $E = (a, b)$, $E' = [a, b]$. This set is dense in itself but not closed.
3. $E = [a, b]$, $E' = [a, b]$. This set is perfect.
4. $E = Z$, $E' = Z$, i.e., the set of all real numbers is perfect.
5. $E = \{1, \frac{1}{2}, \frac{1}{3}, \ldots, \frac{1}{n}, \ldots, 0\}$, $E' = \{0\}$. This set is closed but is not dense in itself.
6. $E = R$ (the set of all rational numbers), $E' = Z$; this set is dense in itself but is not closed.
7. $E = 0$, $E' = 0$; i.e., the empty set is perfect.
8. E is a finite set, $E' = 0$; i.e., a finite set is closed but not dense in itself.

As we proceed, we shall encounter more complicated and interesting examples of closed sets and perfect sets.

THEOREM 1. *The derived set E' of an arbitrary point set E is closed.*

Proof. The theorem is trivial if E' is void. Let E' be non-void and let x_0 be a limit point of E'. Choose an arbitrary open interval (α, β) containing the point x_0. By the definition of limit point, there exists a point $z \in E'$ in this interval. This implies that the interval (α, β) contains a point of the initial set E (Fig. 5), and hence it contains an infinite number of points of E.

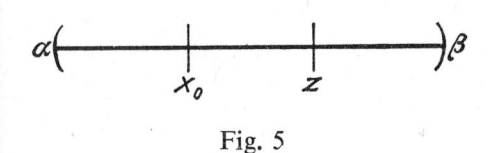

Therefore, every interval which contains the point x_0 contains also an infinite number of points of E. That is, the point x_0 is a limit point of E. In symbols, $x_0 \in E'$. It follows that the set E' contains all of its limit points and is in fact closed.

Fig. 5

THEOREM 2. *If $A \subset B$, then $A' \subset B'$.*
This is obvious.

THEOREM 3. *The following formula holds for all $A, B \subset Z$:*
$$(A + B)' = A' + B'.$$

Proof. The inclusion
$$A' + B' \subset (A + B)'$$
follows from Theorem 2. We establish the reverse inclusion
$$(A + B)' \subset A' + B'. \tag{*}$$
Let
$$x_0 \in (A + B)'.$$

Then one can find a sequence of distinct points $x_1, x_2, x_3, \ldots$ in $A + B$ such that
$$\lim x_n = x_0.$$
If an infinite number of the points x_n lie in A, then x_0 is a limit point of the set A, and $x_0 \in A' \subset A' + B'$. If only a finite number of the points x_n belong to A, then $x_0 \in B' \subset A' + B'$. We therefore have $x_0 \in A' + B'$ in every case, whence (*) follows; this completes the proof.

COROLLARY 1. *The closure $\bar{E}$ of every set E is closed.*
In fact,
$$(\bar{E})' = (E + E')' = E' + (E')' \subset E' + E' = E' \subset \bar{E}.$$

COROLLARY 2. *A set E is closed if and only if it coincides with its closure :*
$$E = \bar{E}.$$

The sufficiency of this condition follows from Corollary 1. Conversely, let the set E be closed; then
$$\bar{E} = E + E' \subset E \subset \bar{E},$$
it follows from these relations that
$$\bar{E} = E .$$

The following theorem is also a consequence of Theorem 3.

THEOREM 4. *The sum of a finite number of closed sets is closed.*

Proof. We consider first the case of a sum of two terms :
$$\Phi = F_1 + F_2.$$
By Theorem 3, we have
$$\Phi' = F_1' + F_2',$$
but since $F_1' \subset F_1$ and $F_2' \subset F_2$,
$$\Phi' \subset \Phi,$$
and from this the theorem follows.

The general case is handled by means of mathematical induction.

Remark. The sum of an infinite set of closed sets need not be closed.
For example, let
$$F_n = \left[\frac{1}{n}, 1\right] \qquad (n = 1, 2, 3, \ldots).$$
Then all of the F_n are closed, but their sum
$$\sum_{n=1}^{\infty} F_n = (0, 1]$$
is not closed.

For the *intersection* of closed sets, we have the following theorem.

THEOREM 5. *The intersection of an arbitrary family of closed sets is closed.*

Proof. Let the closed sets F_ξ be distinguished one from another by the indices ξ, which run through some set of values. Let
$$\Phi = \prod_\xi F_\xi$$

2. CLOSED SETS

be their intersection. Then $\Phi \subset F_\xi$ for all of the indices ξ, from which it follows that $\Phi' \subset F'_\xi$ and consequently, $\Phi' \subset \bar{F}_\xi$. Since this is true for all ξ, we infer that

$$\Phi' \subset \prod_\xi \bar{F}_\xi,$$

i.e., $\Phi' \subset \Phi$, which was to be proved.

LEMMA. *Let the set E be bounded above (below) and $\beta = \sup E$ ($\alpha = \inf E$). Then*

$$\beta \in \bar{E} \quad (\alpha \in \bar{E}).$$

Proof. If $\beta \in E$, then $\beta \in \bar{E}$, since $E \subset \bar{E}$. Suppose that $\beta \bar{\in} E$. For every $\epsilon > 0$, there exists a point $x \in E$ with $x > \beta - \epsilon$ by the definition of sup E, and therefore an arbitrary open interval containing the point β also contains points of the set E, which must be different from β, because $\beta \bar{\in} E$. Hence, β is a limit point of the set E, and we have $\beta \in E' \subset \bar{E}$. It follows that $\beta \in \bar{E}$.

THEOREM 6. *In a closed set F which is bounded above (below), there is a right (left) end point.*

In fact, let $\beta = \sup F$. Then

$$\beta \in \bar{F} = F.$$

DEFINITION 6. Let E be a point set and let $\mathfrak{M}$ be a system of open intervals. If for every $x \in E$, there exists an interval $\delta \in \mathfrak{M}$ such that

$$x \in \delta$$

then we say the set E is *covered* by the system of open intervals $\mathfrak{M}$.

THEOREM 7 (E. BOREL). *If the closed bounded set F is covered by an infinite system $\mathfrak{M}$ of open intervals, then it is possible to select a finite subsytem $\mathfrak{M}^*$ of $\mathfrak{M}$ which also covers the set F.*

Proof. We prove this theorem by contradiction. Assume that it is impossible to choose any finite system of intervals from $\mathfrak{M}$ which covers the set F (this implies, among other things, that the set F is infinite). Enclose F in a closed interval $[a, b]$ (which is certainly possible, since F is bounded) and set

$$c = \frac{a+b}{2}.$$

It is not the case that both of the sets $F \cdot [a, c]$ and $F \cdot [c, b]$ can be covered with a finite number of intervals of the system $\mathfrak{M}$, since then the entire set F would be covered with a finite number of these intervals. Therefore, at least one of the closed intervals $[a, c]$ and $[c, b]$ contains a subset of F which cannot be covered with any finite subsystem of $\mathfrak{M}$. Denote by $[a_1, b_1]$ one of these closed intervals which contains such a subset of F. In doing this, if both segments $[a, c]$ and $[c, b]$ contain subsets of F which cannot be covered by any finite subsystem of $\mathfrak{M}$, then let $[a_1, b_1]$ be either one of them. It is clear that the set $F \cdot [a_1, b_1]$ is infinite.

Now set

$$c_1 = \frac{a_1 + b_1}{2}$$

and denote by $[a_2, b_2]$ one of the closed intervals $[a_1, c_1]$ and $[c_1, b_1]$ which contains a subset of the set F which cannot be covered with a finite number of intervals of the system $\mathfrak{M}$; at least one of the segments $[a_1, c_1]$ and $[c_1, b_1]$ possesses this property just as above (if both of them possess this property, then we select either one of them as $[a_2, b_2]$).

Continuing this process, we construct a sequence of nested closed intervals
$$[a, b] \supset [a_1, b_1] \supset [a_2, b_2] \supset \ldots$$
having the property that none of the sets
$$F \cdot [a_n, b_n] \qquad (n = 1, 2, 3, \ldots)$$
can be covered with a finite number of intervals of the system $\mathfrak{M}$ (hence each of these sets is infinite). Since the length of the closed interval $[a_n, b_n]$, which is equal to $\frac{b-a}{2^n}$, tends to zero as $n \to \infty$, there exists exactly one point x_0 belonging to all of these closed intervals. We clearly have
$$\lim a_n = \lim b_n = x_0.$$

Let us show that the point x_0 belongs to the set F. For this purpose, choose a point x_1, from the set $F \cdot [a_1, b_1]$, then a point x_2 distinct from x_1 from the (infinite) set $F \cdot [a_2, b_2]$, then a point x_3 distinct from x_1 and x_2 from the set $F \cdot [a_3, b_3]$, and so on. As a result, we obtain the sequence
$$x_1, x_2, x_3, \ldots$$
of distinct points of the set F, where
$$a_n \leqslant x_n \leqslant b_n.$$
But then, obviously,
$$x_0 = \lim x_n,$$
so that x_0 is a limit point of the set F. Since the set F is closed and x_0 is a limit point of F, we have
$$x_0 \in F.$$

It is now easy to complete the proof. Since the set F is covered by the system $\mathfrak{M}$, there is an open interval δ_0 of the system $\mathfrak{M}$ such that
$$x_0 \in \delta_0.$$
If n is sufficiently large, then it is clear that
$$[a_n, b_n] \subset \delta_0$$
(see Fig. 6) and consequently
$$F \cap [a_n, b_n] \subset \delta_0,$$
i.e., the set $F \cdot [a_n, b_n]$ is covered by a *single* interval of $\mathfrak{M}$. This contradicts the very definition of the interval $[a_n, b_n]$ and completes the proof.

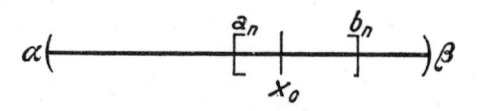

Fig. 6

REMARK. *The theorem ceases to be true if the condition that the set F be bounded and closed is omitted.*

Consider once more the set N of all natural numbers. It is closed (because $N' = 0$), but unbounded. Consider the system $\mathfrak{M}$ of all intervals of the form

$$\left(n - \frac{1}{3}, n + \frac{1}{3}\right) \qquad (n = 1, 2, 3, \ldots),$$

covering the set N. Since each of the intervals of the system $\mathfrak{M}$ contains *only one* point of the set N, it is clear that no finite system of these intervals can possibly cover the *infinite* set N. The condition of boundedness is therefore essential.

As a second example, we consider the set E of all numbers of the form $\frac{1}{n}$:

$$E = \left\{1, \frac{1}{2}, \frac{1}{3}, \ldots\right\}.$$

This set is bounded but not closed. Construct an interval δ_n about each point $\frac{1}{n}$, containing this point, but so small that it contains no other point of the set E, and denote the system of all intervals δ_n by $\mathfrak{M}$. Plainly, the system $\mathfrak{M}$ covers the set E, but the same reasoning as in the preceding example shows that E cannot be covered by any finite subsystem of $\mathfrak{M}$. Thus the condition of closure is also essential.

In concluding this paragraph, we point out a certain property of closed sets, the application of which would shorten somewhat the proof of Theorem 7.

THEOREM 8. *Let F be a closed set and let*

$$x_1, x_2, x_3, \ldots \tag{*}$$

be a sequence of points of F. If

$$\lim x_n = x_0,$$

then $x_0 \in F$.

In fact, if the sequence (*) contains an infinite set of distinct points, then x_0 is a limit point of F and $x_0 \in F$; if the sequence (*) contains only a finite number of distinct points, then, as is easy to see, all elements of the sequence, beginning at some stage, coincide with x_0, and $x_0 \in F$.

§ 3. INTERIOR POINTS AND OPEN SETS

DEFINITION 1. A point x_0 is called an *interior point* of the set E if there exists an open interval, contained entirely in the set E, which contains this point:

$$x_0 \in (\alpha, \beta) \subset E.$$

It is clear from the definition that every interior point of a set E *belongs* to E.

DEFINITION 2. A set E is said to be *open* if all of its points are interior points.

Examples.
1. Every interval (a, b) is an open set.
2. The set Z of all real numbers is open.
3. The void set 0 is open.
4. A closed interval $[a, b]$ is not an open set because its end points are not interior points.

THEOREM 1. *The sum of an arbitrary family of open sets is an open set.*

Proof. Let

$$S = \sum_\xi G_\xi,$$

where all of the sets G_ξ are open. If $x_0 \in S$, then $x_0 \in G_{\xi_0}$ for some ξ_0. Since G_{ξ_0} is

an open set, there exists an interval (α, β) such that
$$x_0 \in (\alpha, \beta) \subset G_{\xi_0}$$
but then
$$x_0 \in (\alpha, \beta) \subset S,$$
and x_0 is an interior point of S. Since x_0 is an arbitrary point of S, the theorem is proved.

COROLLARY. *A set which can be represented in the form of a sum of intervals is open.*

THEOREM 2. *The intersection of a finite number of open sets is open.*

Proof. Let
$$P = \prod_{k=1}^{n} G_k,$$
where all G_k are open. If P is void, the theorem is trivial. Suppose that P is non-void, and let $x_0 \in P$. Then
$$x_0 \in G_k \qquad (k = 1, 2, \ldots, n),$$
and for every k ($k = 1, 2, \ldots, n$) we can find an interval (α_k, β_k) such that
$$x_0 \in (\alpha_k, \beta_k) \subset G_k.$$
Set
$$\lambda = \max(\alpha_1, \alpha_2, \ldots, \alpha_n); \quad \mu = \min(\beta_1, \beta_2, \ldots, \beta_n);$$
obviously,
$$x_0 \in (\lambda, \mu) \subset P,$$
i.e., x_0 is an interior point of P. This proves the theorem.

Remark. The intersection of an infinite number of open sets need not be an open set. In fact, if
$$G_n = \left(-\frac{1}{n}, \frac{1}{n}\right) \qquad (n = 1, 2, 3, \ldots),$$
then all G_n are open, but their intersection
$$\prod_{n=1}^{\infty} G_n = \{0\}$$
is not an open set.

DEFINITION 3. Let E and S be two point sets. If $E \subset S$, then the set $S - E$ is called the complement of the set E with respect to the set S and is denoted by the symbol
$$C_S E.$$
In particular, the set $C_Z E$ [where $Z = (-\infty, +\infty)$] is called the *complement* of the set E and is designated by
$$CE.*$$

---

* In the same way, if one is dealing with subsets of the plane, 3-, ... or n-dimensional space, one uses the symbol CE to denote the complement of the set E with respect to the space under consideration. Given an abstract set X, and a subset E of X, one writes CE for $X - E$, provided that there is no possibility for misunderstanding.—E.H.

3. Interior Points and Open Sets

With the aid of the notion of complement, it is easy to set forth the relation between closed and open sets.

Theorem 3. *If the set G is open, then its complement CG is closed.*

Proof. Let $x_0 \in G$; then there exists an interval (α, β) such that

$$x_0 \in (\alpha, \beta) \subset G.$$

This interval contains no points at all of CG; therefore, x_0 is not a limit point of the set CG. Hence any point which is a limit point of the set CG cannot lie in G. It follows that CG contains all its limit points and is accordingly closed.

Theorem 4. *If the set F is closed, then its complement is open.*

Proof. Let $x_0 \in CF$. Then x_0 is not a limit point of the set F, and consequently there exists an open interval (α, β) containing the point x_0 and containing no points of F distinct from x_0. But since x_0 is not in F, there are no points at all of F in (α, β), so that $(\alpha, \beta) \subset CF$ and x_0 is an interior point of CF.

As an example, we note that both of the mutually complementary sets Z and 0 are closed *and* open.

It is easy to see that 1) if G is an open set and $[a, b]$ is a closed interval containing G, then the set $[a, b] - G$ is closed, and 2) if F is a closed set and (a, b) is an open interval containing F, then the set $(a, b) - F$ is open.

These assertions follow from the obvious identities

$$[a, b] - G = [a, b] \cdot CG;$$

$$(a, b) - F = (a, b) \cdot CF.$$

On the other hand, if F is closed and $[a, b] \supset F$, then the set $[a, b] - F$ is, in general, not open. For example, let $F = [0, 1]$ and let $[a, b] = [0, 2]$. Then

$$[a, b] - F = (1, 2].$$

In connection with this, the following definition is useful.

Definition 4. Let E be a non-void bounded set and let $a = \inf E$, $b = \sup E$. The closed interval $S = [a, b]$ is called the *smallest closed interval containing E*.

Theorem 5. *If S is the smallest closed interval containing the bounded closed set F, then the set*

$$C_S F = [a, b] - F$$

is open.

Proof. It is obviously sufficient to verify that the following identity holds:

$$C_S F = (a, b) \cdot CF.$$

Let $x_0 \in C_S F$; this means that

$$x_0 \in [a, b], \quad x_0 \overline{\in} F.$$

If $x_0 \overline{\in} F$, then $x_0 \neq a$ and $x_0 \neq b$ (because, by Theorem 6, §2, a and b are in F). It follows that $x_0 \in (a, b)$. Besides, it is obvious that x_0 is in CF, so that

$$C_S F \subset (a, b) \cdot CF.$$

The reverse inclusion is obvious.

§ 4. DISTANCE AND SEPARATION

DEFINITION 1. Let x and y be two points of the real line. The number
$$|x-y|$$
is called the *distance between the points x and y* and is denoted by the expression
$$\varrho(x, y).$$
It is obvious that $\rho(x, y) = \rho(y, x) \geqslant 0$, and that
$$\varrho(x, y) = 0$$
if and only if $x = y$.

DEFINITION 2. Let x_0 be a point and E a non-void point set. The greatest lower bound of distances between x_0 and points of the set E is called the *distance between the point x_0 and the set E*; it is denoted by $\rho(x_0, E)$ or $\rho(E, x_0)$:
$$\varrho(x_0, E) = \inf\{\varrho(x_0, x)\} \qquad (x \in E).$$
Obviously, $\rho(x_0, E)$ always exists and is non-negative. If $x_0 \in E$, then
$$\varrho(x_0, E) = 0$$
but the converse is not generally true. For example, if $x_0 = 0$ and $E = (0, 1)$, then $\rho(x_0, E) = 0$, but $x_0 \bar{\in} E$.

DEFINITION 3. Let A and B be two non-void point sets. The greatest lower bound of distances between points of the set A and points of the set B is called the *distance between the sets A and B* and is denoted by $\rho(A, B)$:
$$\varrho(A, B) = \inf\{\varrho(x, y)\}.$$
Obviously, $\rho(A, B)$ always exists and $\rho(A, B) = \rho(B, A) \geqslant 0$.
If the sets A and B intersect, then
$$\varrho(A, B) = 0$$
but the converse assertion is not generally true. For example, if $A = (-1, 0)$, $B = (0, 1)$, then $\rho(A, B) = 0$, but $AB = 0$.

We note that the distance between the point x_0 and the set E is merely the distance between the set E and the set $\{x_0\}$, whose only point is x_0. This observation will be very useful.

THEOREM 1. *Let A and B be two non-void closed sets, at least one of which is bounded. Then there exist points*
$$x^* \in A, \quad y^* \in B,$$
such that
$$\varrho(x^*, y^*) = \varrho(A, B).$$

Proof. From the definition of greatest lower bound, we see that for every natural number n there exist two points $x_n \in A$, $y_n \in B$, such that
$$\varrho(A, B) \leqslant |x_n - y_n| < \varrho(A, B) + \frac{1}{n}. \tag{1}$$
By hypothesis, one of the sets A and B is bounded. Suppose, for example, that A is

4. Distance and Separation

bounded. Then the sequence $\{x_n\}$ is bounded, and by the Bolzano-Weierstrass theorem, a convergent subsequence $x_{n_1}, x_{n_2}, x_{n_3}, \ldots$ with

$$\lim x_{n_k} = x^*$$

can be selected from it.

Since the set A is closed, the point x^* must belong to this set,

$$x^* \in A.$$

Consider the sequence $\{y_{n_k}\}$. If $|x_{n_k}| < C$, then

$$|y_{n_k}| \leqslant |x_{n_k}| + |y_{n_k} - x_{n_k}| < C + \varrho(A, B) + \frac{1}{n_k} \leqslant C + \varrho(A, B) + 1.$$

From this, it is clear that the sequence $\{y_{n_k}\}$ also is bounded; hence a convergent subsequence can also be selected from it, having the limit y^*:

$$y_{n_{k_1}}, y_{n_{k_2}}, y_{n_{k_3}}, \ldots \quad \lim y_{n_{k_i}} = y^*.$$

Since the set B is closed, we have $y^* \in B$. It is easy to see that

$$|y^* - x^*| = \lim |y_{n_{k_i}} - x_{n_{k_i}}| = \varrho(A, B)$$

whereby the theorem is proved.

We give an example to show that the theorem may fail if both of the sets A and B are unbounded. Let $N = \{n\}$ and $M = \{n + \frac{1}{2n}\}$. Both of these sets are closed ($N' = M' = 0$) and $\rho(N, M) = 0$, but since $N \cdot M = 0$, there do not exist two points $x^* \in N$, $y^* \in M$ for which $\rho(x^*, y^*) = 0$. It is also clear that if at least one of the sets A and B is not closed, the theorem also may fail; this is clear from the example $A = [1, 2)$, $B = [3, 5]$, where $\rho(A, B) = 1$.

We note some consequences of the above theorem.

COROLLARY 1. *If A and B are closed, at least one of them is bounded and $\rho(A, B) = 0$, then A and B intersect.*

COROLLARY 2. *Let x_0 be an arbitrary point and let F be a non-void closed set. Then there is a point x^* in F for which*

$$\varrho(x_0, x^*) = \varrho(x_0, F).$$

COROLLARY 3. *If the point x_0 and the closed set F are such that $\rho(x_0, F) = 0$, then $x_0 \in F$.*

We proceed to establish an important *separation theorem*. As a preliminary, we prove two simple lemmas.

LEMMA 1. *Let A be a non-void point set and let d be a positive number. Set*[2]

$$B = Z(\varrho(x, A) < d).$$

Then $A \subset B$ and B is an open set.

Proof. The inclusion $A \subset B$ is obvious. We have only to prove that the set B is open. Let $x_0 \in B$. Then $\rho(x_0, A) < d$, and we can find a point x^* in A such that

$$\varrho(x_0, x^*) < d.$$

We set $d - \rho(x_0, x^*) = h$ and show that the interval $(x_0 - h, x_0 + h)$ is contained in B. It will follow from this that x_0 is an interior point of B and hence that B is open.

Consider an arbitrary point

$$y \in (x_0 - h, x_0 + h).$$

---

[2] This notation has the following meaning: "B is the set of all those points of Z for which $\rho(x, A) < d$."

Then $|y - x_0| < h$, and since $|x_0 - x^*| = d - h$, we have
$$|y - x^*| \leq |y - x_0| + |x_0 - x^*| < h + (d - h) = d.$$
This implies that $\rho(y, x^*) < d$ and that accordingly
$$\varrho(y, A) < d,$$
so that $y \in B$. Thus
$$(x_0 - h, x_0 + h) \subset B,$$
and the lemma is proved.

LEMMA 2. *Let A_1 and A_2 be two non-void sets such that*
$$\varrho(A_1, A_2) = r > 0.$$
Set
$$B_1 = Z\left(\rho(x, A_1) < \frac{r}{2}\right) \text{ and } B_2 = Z\left(\rho(x, A_2) < \frac{r}{2}\right).$$
Then
$$B_1 B_2 = 0.$$

Proof. Assume that $B_1 B_2 \neq 0$, and let
$$z \in B_1 B_2.$$
Then
$$\varrho(z, A_1) < \frac{r}{2}, \quad \varrho(z, A_2) < \frac{r}{2},$$
and points $x_1 \in A$, and $x_2 \in A_2$ can be found such that
$$|z - x_1| < \frac{r}{2}, \quad |z - x_2| < \frac{r}{2}$$
therefore
$$|x_1 - x_2| < r$$
and in consequence $\rho(A_1, A_2) < r$, which is an absurdity.

THEOREM 2. (THE SEPARATION PROPERTY.) *Let F_1 and F_2 be two disjoint closed, bounded sets. There exist open sets G_1 and G_2 such that*
$$G_1 \supset F_1, \quad G_2 \supset F_2, \quad G_1 G_2 = 0.$$

Proof. By Corollary 1 to Theorem 1, we have
$$\varrho(F_1, F_2) = r > 0.$$
It remains to set
$$G_i = Z\left(\varrho(x, F_i) < \frac{r}{2}\right) \qquad (i = 1, 2)$$
and apply Lemmas 1 and 2.

We note that the condition that the sets F_1 and F_2 be bounded can be removed without destroying the validity of the theorem, but we shall not delay over this point. On the other hand, the condition that both sets be closed is essential, as is clear from the example
$$A = [0, 1), \quad B = [1, 2].$$

§ 5. THE STRUCTURE OF BOUNDED OPEN SETS AND BOUNDED CLOSED SETS

DEFINITION 1. Let G be an open set. If the open interval (a, b) is contained in G and its end points do not belong to G,
$$(a, b) \subset G, \quad a \overline{\in} G, \quad b \overline{\in} G,$$
then this interval is said to be a *component interval* of the set G.

THEOREM 1. *If G is a non-void bounded, and open set, then each of its points belongs to a component interval of G.*

Proof. Let
$$x_0 \in G.$$
Set
$$F = [x_0, +\infty) \cdot CG.$$

Both of the sets $[x_0, +\infty)$ and CG are closed, and hence the set F is closed. Since G is bounded, F is non-void. Finally, none of the points of the set F lies to the left of the point x_0, so that the set F is bounded below. F therefore contains a left end point μ; it is clear that $\mu \geq x_0$. But $x_0 \in G$, and hence $x_0 \overline{\in} F$, so that $x_0 \neq \mu$, i.e.,
$$x_0 < \mu.$$
We note further that $\mu \overline{\in} G$ (since $\mu \in F \subset CG$). We next establish that
$$[x_0, \mu) \subset G$$
We assume the contrary. Then there exists a point y such that
$$y \in [x_0, \mu), \quad y \overline{\in} G.$$
From these relations it follows that
$$y \in F, \quad y < \mu$$
and this contradicts the definition of the point μ.

We have thus established the existence of a point μ with the following three properties:

1) $\mu > x_0$, 2) $\mu \overline{\in} G$, 3) $[x_0, \mu) \subset G$.

Analogously, we can show the existence of a point λ such that

1) $\lambda < x_0$, 2) $\lambda \overline{\in} G$, 3) $(\lambda, x_0] \subset G$.

From this it follows that (λ, μ) is a component interval of the set G containing the point x_0, which was to be proved.

The existence of component intervals for every non-void bounded, open set follows from the theorem just proved.

THEOREM 2. *If (λ, μ) and (σ, τ) are two component intervals of an open set G, then they are either disjoint or identical.*

Proof. Let us suppose that there exists a point x lying in both of the intervals (λ, μ) and (σ, τ):
$$\lambda < x < \mu, \quad \sigma < x < \tau.$$
Assume that
$$\tau < \mu.$$

Then, obviously, $\tau \in (\lambda, \mu)$, but this is plainly impossible, since

$$(\lambda, \mu) \subset G, \quad \tau \overline{\in} G .$$

This implies that

$$\mu \leqslant \tau.$$

Since μ and τ are interchangeable, it follows from the same reasoning that

$$\tau \leqslant \mu$$

and hence $\tau = \mu$. We show similarly that $\sigma = \lambda$, and therefore the intervals (λ, μ) and (σ, τ) are identical.

COROLLARY. *The set of distinct component intervals of a non-empty bounded open set G is finite or denumerable.*

In fact, if we choose a rational point in each of these intervals, the set of component intervals is put into one-to-one correspondence with a subset of the set R of all rational numbers.

The foregoing can be summarized as follows.

THEOREM 3. *Every non-void bounded open set G can be represented in the form of a sum of a finite or denumerable number of pairwise disjoint open intervals, the end points of which do not belong to the set G.*

$$G = \sum_k (\lambda_k, \mu_k) \qquad (\lambda_k \overline{\in} G, \ \mu_k \overline{\in} G).$$

We have already noted that the converse holds : every set which is representable as a sum of intervals is open.

THEOREM 4. *Let G be a non-void bounded open set and let (a, b) be an open interval contained in G. Then there exists among the component intervals of G one which contains the open interval (a, b).*

Proof. Let $x_0 \in (a, b)$. Then $x_0 \in G$, and among the component intervals of the set G, there exists an interval (λ, μ) such that

$$x_0 \in (\lambda, \mu).$$

Assuming that

$$\mu < b,$$

we would find that $\mu \in (a, b)$, which is impossible, since $\mu \overline{\in} G$. This implies that

$$b \leqslant \mu.$$

In the same way, we find that

$$\lambda \leqslant a$$

and hence

$$(a, b) \subset (\lambda, \mu),$$

which was to be proved.

We turn now to the study of the structure of closed bounded sets. Let F be such a set and let S be the smallest closed interval containing F. As we know, the set $C_S F$ is open. If this set is non-void, we can apply Theorem 3 to it. For this reason, we can assert the following theorem.

THEOREM 5. *A non-void bounded closed set F either is a closed interval or is obtained from some closed interval by removing a finite or denumerable family of pairwise disjoint open intervals whose end points belong to the set F.*

5. The Structure of Bounded Sets

It is perfectly clear that, conversely, every set obtained from a closed interval by removing a family of open intervals is closed.

The component intervals the of set $C_S F$ are called *complementary* intervals of the set F.

Since a perfect set is closed, Theorem 5 holds for all bounded perfect sets. It remains to describe the requirements that the complementary intervals of a closed set must satisfy in order that it be perfect. The following theorem answers this question.

THEOREM 6. *Let F be a non-void bounded closed set and let $S = [a, b]$ be the smallest closed interval containing F. We then have:*

1. *A point x_0 which is an end point of two distinct complementary intervals of F is an isolated point of F;*
2. *if the point a (or b) is an end point of a complementary interval of F, then it is an isolated point of F;*
3. *no isolated points of F exist besides those mentioned in 1 and 2.*

Proof. Assertions 1. and 2. are obvious. We shall prove assertion 3. Let x_0 be an isolated point of F, and suppose first that $a < x_0 < b$. By the definition of an isolated point, there exists an open interval (α, β) containing x_0 and containing no points of F other than x_0; it is clear that $(\alpha, \beta) \subset [a, b]$.

The interval (x_0, β) thus contains no points of F at all, and we have $(x_0, \beta) \subset C_S F$. By Theorem 4, there exists a complementary interval (λ, μ) of the set F containing the interval (x_0, β). If λ were $< x_0$, the point x_0 would not belong to the set F; so it is necessary that $\lambda \geqslant x_0$. But the inequality $\lambda > x_0$ would contradict the fact that $(x_0, \beta) \subset (\lambda, \mu)$.

Therefore $\lambda = x_0$, i.e., x_0 is the left end point of some complementary interval of the set F.

In exactly the same way, we establish that x_0 is the right end point of some complementary interval of F, and this proves assertion 3. The cases $x_0 = a$ and $x_0 = b$ are handled in the same way.

From this theorem, we infer the following result.

THEOREM 7. *Every non-void bounded perfect set P is either a closed interval or is obtained from some closed interval by removing a finite or denumerable number of pairwise disjoint open intervals which have no common end points with each other or with the original closed interval. Conversely, every set, obtained in this manner is perfect.*

We now introduce an interesting and important example of a perfect set.

The Cantor Sets G_0 and P_0. Divide the closed interval $U = [0, 1]$ into three parts by the points $\frac{1}{3}$ and $\frac{2}{3}$, and remove the open interval $(\frac{1}{3}, \frac{2}{3})$ from U. Divide each of the remaining closed intervals $[0, \frac{1}{3}]$ and $[\frac{2}{3}, 1]$ into three parts (by the points $\frac{1}{9}$ and $\frac{2}{9}$ for the first segment and the points $\frac{7}{9}, \frac{8}{9}$ for the second) and remove the middle open intervals $(\frac{1}{9}, \frac{2}{9})$, $(\frac{7}{9}, \frac{8}{9})$. Next, divide each of the remaining four segments into three equal parts and remove the middle intervals from them (Fig. 7). Continue this process indefinitely. By this process, we remove an open set G_0 from $[0, 1]$; the set G_0 is the sum of a denumerable family of intervals

$$G_0 = \left(\frac{1}{3}, \frac{2}{3}\right) + \left[\left(\frac{1}{9}, \frac{2}{9}\right) + \left(\frac{7}{9}, \frac{8}{9}\right)\right] + \cdots$$

The remaining set $P_0 = C_{[0, 1]}(G_0)$ is perfect (as Theorem 7 shows). The sets G_0 and P_0 are called Cantor sets.

It is easy to give an arithmetic characterization of these sets. For this purpose, we introduce the apparatus of ternary fractions. What points fall into the first of the removed intervals, *i.e.*, into the interval $(\frac{1}{3}, \frac{2}{3})$? It is clear that in developing each of these points in a ternary "decimal"

$$x = 0.\, a_1 a_2 a_3 \ldots \qquad (a_k = 0, 1, 2)$$

we must have

$$a_1 = 1$$

Each of the end points of this interval allows of two representations

$$\frac{1}{3} = \begin{cases} 0.100000\ldots; \\ 0.022222\ldots; \end{cases} \qquad \frac{2}{3} = \begin{cases} 0.12222\ldots \\ 0.20000\ldots \end{cases}$$

No remaining point of the segment [0, 1], in its development as a ternary "decimal," can have 1 immediately following the decimal point.

Fig. 7

So, at the first step of the process of constructing the set P_0 from the closed interval U, those and only those points are removed whose ternary expansions must have 1 immediately following the decimal point.

We establish in a similar way that at the second step those and only those points x are removed for which $a_2 = 1$ necessarily, and so on.

For this reason, after completing the process, those and only those points remain which *can be* expressed as ternary "decimals"

$$0.\, a_1 a_2 a_3 \ldots$$

in which every a_k equals 0 or 2. Briefly, the set G_0 consists of points whose ternary development is impossible without the use of 1, and P_0 consists of points for which such a development is possible.

COROLLARY. *The Cantor perfect set P_0 has power c.*

In fact,

$$P_0 = \{0.\, a_1 a_2 a_3 \ldots\} \qquad a_k = \begin{cases} 0 \\ 2 \end{cases}$$

and the assertion follows from the corollary to Theorem 8, §4, Chapt. I.

The result obtained shows that the Cantor set contains points in addition to the end points of the removed intervals (which form a denumerable set). An example of a point which is not an end point is an arbitrary ternary expansion

$$0.\, a_1 a_2 a_3 \ldots, \qquad a_k = \begin{cases} 0 \\ 2 \end{cases}$$

containing an infinite number of 0's and an infinite number of 2's.

§ 6. POINTS OF CONDENSATION; THE POWER OF A CLOSED SET

At the end of §5, we established that the power of the Cantor set P_0 is c. It happens that this is an inherent property of perfect sets.

6. Points of Condensation; Power of Closed Set

THEOREM 1. *Every non-void perfect set P has power c.*

Proof. Let P be a non-void perfect set. Consider a point

$$x \in P$$

and an open interval δ containing x. Since x is not an isolated point of P, the set $P\delta$ is infinite. We choose two distinct points x_0 and x_1 in $P\delta$ and choose open intervals δ_0 and δ_1 such that for $i = 0, 1$,

1) $x_i \in \delta_i$, 2) $\delta_i \subset \delta$, 3) $\overline{\delta}_0 \, \overline{\delta}_1 = 0$, 4) $m\delta_i < 1$

($\overline{\delta}$ is the closure of the interval δ, and $m\delta$ is the length of δ). Since x_0 is a limit point of the set P, there is an infinite set of points of P in the interval δ_0. We choose two distinct points $x_{0,0}$ and $x_{0,1}$ from among them and construct intervals $\delta_{0,0}$ and $\delta_{0,1}$ such that for $k = 0, 1$ we have

1) $x_{0,k} \in \delta_{0,k}$, 2) $\delta_{0,k} \subset \delta_0$, 3) $\overline{\delta}_{0,0} \cdot \overline{\delta}_{0,1} = 0$, 4) $m\delta_{0,k} < \frac{1}{2}$.

We make a similar construction starting with the point x_1. That is, we select points $x_{i,k}$ ($i, k = 0, 1$) and open intervals $\delta_{i,k}$ such that

1) $x_{i,k} \in P \cdot \delta_{i,k}$, 2) $\delta_{i,k} \subset \delta_i$, 3) $\overline{\delta}_{i,k} \cdot \overline{\delta}_{i',k'} = 0$, if $(i, k) \neq (i', k')$,

4) $m\delta_{i,k} < \frac{1}{2}$.

We continue the process of selection further. After the n-th step, we have selected points

$$x_{i_1, \ldots, i_n} \qquad (i_k = 0, 1; \quad k = 1, 2, \ldots, n)$$

and open intervals

$$\delta_{i_1, i_2, \ldots, i_n}$$

such that 1) $x_{i_1, i_2, \ldots, i_n} \in P \cdot \delta_{i_1, i_2, \ldots, i_n}$, 2) $\delta_{i_1, i_2, \ldots, i_{n-1}, i_n} \subset \delta_{i_1, \ldots, i_{n-1}}$,

3) $\overline{\delta}_{i_1, i_2, \ldots, i_n} \, \overline{\delta}_{i'_1, i'_2, \ldots, i'_n} = 0$ (if $(i_1, i_2, \ldots, i_n) \neq (i'_1, i'_2, \ldots, i'_n)$),

4) $m\delta_{i_1, \ldots, i_n} < \frac{1}{n}$.

Since each of the points $x_{i_1, \ldots, i_n}$ is a limit point of the set P, the set $P\delta_{i_1, \ldots, i_n}$ contains two distinct points

$$x_{i_1, \ldots, i_n, 0} \text{ and } x_{i_1, \ldots, i_n, 1}.$$

We can also find open intervals

$$\delta_{i_1, \ldots, i_n, 0} \text{ and } \delta_{i_1, \ldots, i_n, 1}$$

such that for $i_{n+1} = 0, 1$,

1) $x_{i_1, \ldots, i_n, i_{n+1}} \subset \delta_{i_1, \ldots, i_n, i_{n+1}}$, 2) $\delta_{i_1, \ldots, i_n, i_{n+1}} \subset \delta_{i_1, \ldots, i_n}$,

3) $\overline{\delta}_{i_1, \ldots, i_n, 0} \cdot \overline{\delta}_{i_1, \ldots, i_n, 1} = 0$, 4) $m\delta_{i_1, \ldots, i_n, i_{n+1}} < \frac{1}{n+1}$.

Suppose that this process has been carried out for all natural numbers n. With each infinite sequence

$$(i_1, i_2, i_3, \ldots) \qquad (i_k = 0, 1)$$

we associate the point
$$z_{i_1, i_2, i_3, \ldots}$$
which is the unique point in the set
$$\overline{\delta}_{i_1} \cdot \overline{\delta}_{i_1, i_2} \cdot \overline{\delta}_{i_1, i_2, i_3} \ldots$$
It is easy to see that the points $z_{i_1, i_2, i_3, \ldots}$ and $z_{i'_1, i'_2, i'_3, \ldots}$, corresponding to two different sequences
$$i_1, i_2, i_3, \ldots \text{ and } i'_1, i'_2, i'_3, \ldots$$
are necessarily distinct. In fact, if n is the smallest of the m's for which $i_m \neq i'_m$, then
$$i_1 = i'_1, \quad i_2 = i'_2, \quad \ldots, \quad i_{n-1} = i'_{n-1}, \quad i_n \neq i'_n$$
and the closed intervals
$$\overline{\delta}_{i_1, \ldots, i_n} \text{ and } \overline{\delta}_{i'_1, \ldots, i'_n}$$
are disjoint. From this, it follows that
$$z_{i_1, i_2, i_3, \ldots} \neq z_{i'_1, i'_2, i'_3, \ldots}$$
Let
$$S = \{z_{i_1, i_2, i_3, \ldots}\}.$$
By Theorem 8, § 4, Chapt. I, we have
$$\overline{\overline{S}} = c.$$
It is easy to see that $S \subset P$, which relation implies that
$$\overline{\overline{P}} \geqslant c.$$
On the other hand, it is clear that
$$\overline{\overline{P}} \leqslant c$$
and hence $\overline{\overline{P}} = c$, as was to be proved.

Our next task is to generalize the result just obtained to arbitrary closed sets. For this purpose, it is useful to introduce the concept of *point of condensation*.

DEFINITION. *The point x_0 is called a point of condensation of the set E if every open interval (a, b) containing x_0 contains a non-denumerable set of points of E.*

It is obvious that every point of condensation of a set is a limit point of that set,

THEOREM. 2. *If no point of the set E is a point of condensation of E, then the set E is finite or denumerable.*

Proof. We call an open interval (r, R) proper if : 1) its endpoints r and R are rational ; 2) at most a denumerable set of points of the set E is contained in this interval. Obviously, there exists at most a denumerable set of proper intervals, because there are only a denumerable number of pairs (r, R) of rational numbers.

We will show that every point of the set E (we may naturally suppose that E is non-void) is contained in some proper interval. In fact, let $x \in E$. Since x is not a point of condensation of the set E, there exists an open interval (a, b) containing x and such that there is at most a denumerable set of points of E in (a, b). If we take rational numbers r and R such that
$$a < r < x < R < b$$

the interval (r, R) is a proper interval containing the point x. We enumerate all proper intervals

$$\delta_1, \delta_2, \delta_3, \ldots .$$

It follows from the statement just proved that

$$E = \sum_{k=1}^{\infty} E\delta_k.$$

Here we have a denumerable family of sets each of which is at most denumerable. From this, it follows that the set E is at most denumerable.

COROLLARY 1. *If the set E is non-denumerable, then E contains at least one point of condensation of itself.*

It is interesting to compare this corollary with the Bolzano-Weierstrass theorem. The Bolzano-Weierstrass theorem is valid for every infinite set whereas this corollary refers only to non-denumerable sets. For this reason (in contrast to the Bolzano-Weierstrass theorem) there is no need here to require that the set E be bounded and, besides the fact that condensation points exist, it is possible to guarantee the existence of points of condensation which are elements of the set E.

COROLLARY 2. *Let E be a point set and let P be the set of all points of condensation of E. Then the set $E - P$ is at most denumerable.*

In fact, no point of $E - P$ can be a point of condensation of E, and *a fortiori*, no point of $E - P$ can be a point of condensation of the set $E - P$.

COROLLARY 3. *Let E be a non-denumerable set and let P be the set of all its points of condensation. Then the set EP is non-denumerable.*

In fact, $EP = E - (E - P)$. The assertion follows from Theorem 10, §3, Chapt. I.

We note that Corollary 1 is a very special case of Corollary 3.

THEOREM 3. *Let the set E be non-denumerable. Then the set P of all points of condensation of the set E is a perfect set.*

Proof. We establish first that the set P is closed. Let x_0 be a limit point of P. We take an arbitrary open interval (a, b) containing the point x_0. There is at least one point z of the set P in it. But then the interval (a, b), as an interval containing a point of condensation of the set E, contains a non-denumerable set of points of E. Since (a, b) is an arbitrary interval containing x_0, x_0 is a point of condensation of E and hence belongs to P. The set P is therefore closed.

It remains to verify that P has no isolated points. Let $x_0 \in P$ and let (a, b) be an open interval containing the point x_0. Then the set

$$Q = E \cdot (a, b)$$

is non-denumerable and hence, by Corollary 3 of Theorem 2, a non-denumerable set of points of condensation of the set Q is contained in Q. But $Q \subset E$, and all points of condensation of the set Q are obviously points of condensation of E, so that a non-denumerable set of points of P is contained in Q (and hence also in (a, b)). Thus every open interval containing the point x_0 contains a non-denumerable set of points of P; and therefore $x_0 \in P'$. This completes the proof.

THEOREM 4. *Every non-denumerable closed set F can be represented in the form*

$$F = P + D,$$

where P is a perfect set and D is at most a denumerable set.

Proof. If P is the set of points of condensation of the set F, then $P \subset F$ and $D = F - P$ is at most denumerable.

COROLLARY. *A non-denumerable closed set has power c.*

II. Point Sets

Exercises for Chapter II

1. If $f(x)$ is a continuous function defined on $[a, b]$, then the set of points at which $f(x) \geq d$ is closed, for all numbers d.

2. Every closed set is the intersection of a denumerable set of open sets.

3. Generalize the separation theorem to unbounded closed sets.

4. Prove that the set of points of $[0, 1]$ having a decimal development not using the number 7 is perfect.

5. Represent $[0, 1]$ as the sum of c perfect sets which are pairwise disjoint.

6. Prove that it is impossible to represent the set of irrational numbers in the closed interval $[0, 1]$ as the sum of a denumerable family of closed sets.

7. Construct a function $\varphi(x)$ on $[0, 1]$ which is discontinuous at every r tional point and continuous at every irrational point.

8. Prove the impossibility of constructing a function on $[0, 1]$ which is continuous at every rational point, and is discontinuous at every irrational point.

9. If the function $f(x)$ defined on $[a, b]$ has the property that the sets $Z(f(x) \geq d)$ and $Z(f(x) \leq d)$, for arbitrary d, are closed, then $f(x)$ is continuous.

10. If the set E is covered with an arbitrary system of open intervals $\mathfrak{M}$, then it is possible to extract from the latter a denumerable subsystem $\mathfrak{M}^*$ which also covers E.

11. Prove that the set of interior points of an arbitrary set is an open set.

CHAPTER III

MEASURABLE SETS

§ 1. THE MEASURE OF A BOUNDED OPEN SET

In the theory of functions of a real variable, the concept of measure of a point set plays a vital rôle ; this notion generalizes the concept of length of interval, area of a rectangle, volume of a parallelepiped, and so on. In this chapter, we set forth the theory of measure of linear bounded point sets due to H. Lebesgue.

Since open sets possess a very simple structure, it is natural to begin the study of measure with these sets.

DEFINITION 1. *The length of an open interval (a, b), i.e., $b - a$, is called the measure of the interval (a, b).* This number is written as

$$m(a, b) = b - a.$$

Obviously, $m(a, b)$ is always > 0.

LEMMA 1. *If a finite number of pairwise disjoint open intervals $\delta_1, \delta_2, \ldots, \delta_n$ are contained in the open interval Δ, then*

$$\sum_{k=1}^{n} m\delta_k \leqslant m\Delta.$$

Proof. Let

$$\Delta = (A, B), \quad \delta_k = (a_k, b_k) \qquad (k = 1, 2, \ldots, n).$$

Without loss of generality, we may suppose that the open intervals δ_k are listed in the order of increasing left end points, *i.e.*, so that

$$a_1 < a_2 < \ldots < a_n.$$

Then, obviously,

$$b_k \leqslant a_{k+1} \qquad (k = 1, 2, \ldots, n-1),$$

because otherwise the intervals δ_k and δ_{k+1} would have common points. Therefore the sum

$$Q = (B - b_n) + (a_n - b_{n-1}) + \ldots + (a_2 - b_1) + (a_1 - A)$$

is non-negative. Since it is obvious that

$$m\Delta = \sum_{k=1}^{n} m\delta_k + Q,$$

the lemma follows.

COROLLARY. *If a denumerable family of pairwise disjoint open intervals δ_k ($k = 1, 2, 3, \ldots$) are contained in the open interval Δ, then*

$$\sum_{k=1}^{\infty} m\delta_k \leqslant m\Delta.$$

[Given a divergent series of positive terms, we assign to it the sum $+ \infty$; thus every series of positive terms has a sum. The inequality

$$\sum_{k=1}^{\infty} a_k < C$$

(for a series of positive terms) guarantees its convergence.]

DEFINITION 2. The measure mG of a non-void bounded open set G is the sum of the lengths of all its component intervals δ_k:

$$mG = \sum_k m\delta_k.$$

(Since the set $\{\delta_k\}$ may be finite or denumerable, we will use the notation $\sum_k m\delta_k$ meaning either $\sum_{k=1}^{n} m\delta_k$ or $\sum_{k=1}^{\infty} m\delta_k$, according to the situation at hand).

By the above corollary,

$$mG < +\infty.$$

If the set G is void, we set

$$mG = 0$$

by definition, so that mG is always $\geqslant 0$. If Δ is an open interval containing an open set G, then

$$mG \leqslant m\Delta,$$

as the above corollary shows.

Example (*the Cantor Set* G_0). The construction of the Cantor set G_0 proceeded in a series of steps. At the first step, we selected the interval $(\frac{1}{3}, \frac{2}{3})$, of length $\frac{1}{3}$. At the second step, two intervals, $(\frac{1}{9}, \frac{2}{9})$ and $(\frac{7}{9}, \frac{8}{9})$, each of length $\frac{1}{9}$ were added to it. At the third step, we added four more intervals, each of length $\frac{1}{27}$, and so on. It follows that

$$mG_0 = \frac{1}{3} + \frac{2}{9} + \frac{4}{27} + \ldots$$

Summing this series by the usual formula, we obtain

$$mG_0 = 1.$$

THEOREM 1. *Let G_1 and G_2 be two bounded open sets. If $G_1 \subset G_2$, then*

$$mG_1 \leqslant mG_2.$$

Proof. Let δ_i $(i = 1, 2, \ldots)$ and Δ_k $(k = 1, 2, \ldots)$ be the component intervals of the sets G_1 and G_2, respectively. By Theorem 4, §5, Chapter II, each of the intervals δ_i is contained in one and only one of the intervals Δ_k. Hence the family of sets $\{\delta_i\}$ can be divided into pairwise disjoint subfamilies $A_1, A_2, A_3, \ldots$, where we put δ_i in A_k if

$$\delta_i \subset \Delta_k.$$

Then, using elementary properties of double series, we can write

$$mG_1 = \sum_i m\delta_i = \sum_k \left(\sum_{\delta_i \in A_k} m\delta_i \right).$$

But by the corollary of Lemma 1,

$$\sum_{\delta_i \in A_k} m\delta_i \leqslant m\Delta_k,$$

whence

$$mG_1 \leqslant \sum_k m\Delta_k = mG_2$$

as we wished to prove.

COROLLARY. *The measure of a bounded open set G is the greatest lower bound of the measures of all bounded open sets containing G.*

1. The Measure of a Bounded Open Set

Theorem 2. *If the bounded open set G is the sum of a finite or denumerable family of pairwise disjoint open sets,*

$$G = \sum_k G_k \qquad (G_k G_{k'} = 0, \; k \neq k'),$$

then

$$mG = \sum_k mG_k.$$

This property of the measure m is called countable additivity.*

Proof. Let $\delta_i^{(k)}$ ($i = 1, 2, \ldots$) be component intervals of the set G_k. Let us show that each of them is a component interval of the sum G. It is first of all clear that

$$\delta_i^{(k)} \subset G.$$

We must show that the end points of the interval $\delta_i^{(k)}$ do not belong to G. Assume for example, that the right end point μ of the interval $\delta_i^{(k)}$ belongs to G. Then the point μ must belong to one of the component sets. Let

$$\mu \in G_{k'}.$$

(It is obvious that $k' \neq k$ since we know that the point μ does not belong to the set G_k.) The set $G_{k'}$ is open and accordingly the point μ belongs to one of the component intervals of this set:

$$\mu \in \delta_{i'}^{(k')}.$$

However, this implies that the intervals $\delta_i^{(k)}$ and $\delta_{i'}^{(k')}$ have non-void intersection; this contradicts the condition

$$G_k G_{k'} = 0.$$

Thus each of the $\delta_i^{(k)}$ is a component interval of the set G. On the other hand, each point of G belongs to at least one $\delta_i^{(k)}$. Finally, all these intervals are distinct. Therefore the family of sets

$$\{\delta_i^{(k)}\} \qquad (i = 1, 2, \ldots; \; k = 1, 2, \ldots)$$

is the family of all component intervals of the sum G. Having established this, it is easy to complete the proof:

$$mG = \sum_{i,k} m\delta_i^{(k)} = \sum_k \left(\sum_i m\delta_i^{(k)} \right) = \sum_k mG_k,$$

as we wished to prove.

In order to carry over the theorem (with appropriate modifications) to the case of a sum of sets which intersect, we shall need two simple lemmas.

Lemma 2. *Let the closed interval $[P, Q]$ be covered by a finite system H of open intervals (λ, μ). Then*

$$Q - P < \sum_H m(\lambda, \mu).$$

Proof. We select a subsystem H^* of H, formed in the following manner. Designate

---

*The Russian expression means "complete additivity"; however, the term "countable additivity" is more accurate and is the term in common use in the English language at the present time.—E.H.

by (λ_1, μ_1) any of the intervals of the system H which contains the point P. That is,
$$\lambda_1 < P < \mu_1.$$
(At least one such interval exists.) If it happens that
$$\mu_1 > Q,$$
then the required system H^* consists of (λ_1, μ_1) alone. However, if $\mu_1 \leqslant Q$, then $\mu_1 \in [P, Q]$, and there is an interval (λ_2, μ_2) of the system H which contains the point μ_1; that is,
$$\lambda_2 < \mu_1 < \mu_2.$$
If
$$\mu_2 > Q,$$
the process is complete, and the intervals (λ_1, μ_1) and (λ_2, μ_2) comprise the system H^*.

If $\mu_2 \leqslant Q$, then $\mu_2 \in [P, Q]$, and there is an interval (λ_3, μ_3) of the system H, containing μ_2:
$$\lambda_3 < \mu_2 < \mu_3.$$
If $\mu_3 > Q$, the process is completed, and if $\mu_3 \leqslant Q$ we continue the process.

But the system H is finite by hypothesis, and our process demands that we choose a new interval from H at each step, because
$$\mu_1 < \mu_2 < \mu_3 < \cdots$$
Hence the process must of necessity terminate and the final step occurs when the point μ_k lies to the right of the point Q.

Suppose that
$$\mu_n > Q,$$
and that $\mu_{n-1} \leqslant Q$, i.e., that the process terminates with the n-th step. Then the intervals
$$(\lambda_1, \mu_1), \quad (\lambda_2, \mu_2), \ldots, (\lambda_n, \mu_n)$$
comprise the system H^*. Here we have
$$\lambda_{k+1} < \mu_k \qquad (k = 1, 2, \ldots n-1).$$
This implies that
$$\sum_{k=1}^{n} (\mu_k - \lambda_k) > \sum_{k=1}^{n-1} (\lambda_{k+1} - \lambda_k) + (\mu_n - \lambda_n) = \mu_n - \lambda_1;$$
and since
$$\mu_n - \lambda_1 > Q - P$$
we have
$$Q - P < \sum_{k=1}^{n} (\mu_k - \lambda_k),$$
which implies immediately that
$$Q - P < \sum_{H} (\mu - \lambda).$$

LEMMA 3. *Let the open interval Δ be the sum of a finite or denumerable family of open sets:*

1. The Measure of a Bounded Open Set

$$\Delta = \sum_k G_k.$$

Then
$$m\Delta \leq \sum_k mG_k.$$

Proof. Let $\Delta = (A, B)$ and let the component intervals of the set G_k be $\delta_i^{(k)}$ ($i = 1, 2, \ldots$; $k = 1, 2, \ldots$). Select a positive number ε $\left(0 < \varepsilon < \frac{B-A}{2}\right)$ and consider the closed interval $[A + \varepsilon, B - \varepsilon]$, which is contained in the interval Δ. This closed interval is covered by the family of open intervals $\delta_i^{(k)}$ ($i = 1, 2, \ldots$; $k = 1, 2, \ldots$). Applying Borel's theorem (Theorem 7, §2, Chapter II) to this family of intervals, we obtain a finite family of intervals

$$\delta_{i_s}^{(k_s)} \qquad (s = 1, 2, \ldots, n),$$

which covers the closed interval $[A + \varepsilon, B - \varepsilon]$. By the preceding lemma, we have

$$B - A - 2\varepsilon < \sum_{s=1} m\delta_{i_s}^{(k_s)},$$

from which it follows that

$$B - A - 2\varepsilon < \sum_{i,k} m\delta_i^{(k)} = \sum_k \left(\sum_i m\delta_i^{(k)}\right) = \sum_k mG_k.$$

Since the number ε is arbitrarily small, we infer that

$$B - A \leq \sum_k mG_k,$$

and the lemma is proved.

Theorem 3. *If a bounded open set G is the sum of a finite or denumerable number of open sets G_k,*

$$G = \sum_k G_k,$$

then

$$mG \leq \sum_k mG_k.$$

Proof. Let Δ_i ($i = 1, 2, \ldots$) be the component intervals of the open set G. Then

$$mG = \sum_i m\Delta_i.$$

However,

$$\Delta_i = \Delta_i \cdot \sum_k G_k = \sum_k (\Delta_i \cdot G_k),$$

whence, by Lemma 3,

$$m\Delta_i \leq \sum_k m(\Delta_i \cap G_k)$$

and we have

$$mG \leq \sum_i \left[\sum_k m(\Delta_i \cdot G_k)\right] = \sum_k \left[\sum_i m(\Delta_i \cdot G_k)\right]. \qquad (*)$$

On the other hand,

$$G_k = G_k \cdot \sum_i \Delta_i = \sum_i (\Delta_i \cdot G_k).$$

The individual terms of the right-hand side of the last equality are pairwise disjoint (since $\Delta_i \Delta_{i'} = 0$ for $i \ne i'$). Hence we are in a position to apply Theorem 2, and so

$$\sum_i m(\Delta_i G_k) = m G_k. \qquad (*_*)$$

Comparing (*) and ($*_*$), we have the present theorem.

§ 2. THE MEASURE OF A BOUNDED CLOSED SET

Let F be a non-void bounded closed set and let S be the smallest closed interval containing the set F. As we have shown (Theorem 5, §3, Chapter II), the set $C_S F$ is open and therefore has a definite measure $m[C_S F]$. This makes it possible to state the following definition.

DEFINITION 1. *The measure of a non-void bounded closed set F is the number*

$$mF = B - A - m[C_S F],$$

where $S = [A, B]$ is the smallest closed interval containing the set F.

It is unnecessary to define the measure of the void closed set since this set is open and we have already agreed that its measure is the number 0. Further, it is not difficult to prove, although we shall not take time to do so, that a non-void closed bounded set cannot be open, so that there is no need to study the connection between the definitions of measure of an open set and of a closed set.

We next consider some examples.

1. $F = [a, b]$. Here it is obvious that $S = [a, b]$ and $C_S F = 0$, so that

$$m[a, b] = b - a$$

i.e., the measure of a closed interval is equal to its length.

2. F is the sum of a finite number of pairwise disjoint closed intervals:

$$F = [a_1, b_1] + [a_2, b_2] + \ldots + [a_n, b_n].$$

We may consider the closed intervals as being enumerated in the order of increasing left end points; then, obviously,

$$b_k < a_{k+1} \qquad (k = 1, 2, \ldots, n),$$

It follows that

$$S = [a_1, b_n], \quad C_S F = (b_1, a_2) + (b_2, a_3) + \ldots + (b_{n-1}, a_n)$$

Thus we have

$$mF = b_n - a_1 - \sum_{k=1}^{n-1}(a_{k+1} - b_k) = \sum_{k=1}^{n}(b_k - a_k)$$

i.e., the measure of a sum of a finite number of pairwise disjoint closed intervals equals the sum of the length of these intervals.

3. Let $F = P_0$ (Cantor's perfect set). In this case, $S = [0, 1]$ and $C_S F = G_0$; therefore

$$m P_0 = 1 - 1 = 0$$

i.e., the Cantor perfect set P_0 has measure zero.

It is interesting to note that the power of P_0 is c.

THEOREM 1. *The measure of a bounded closed set F is non-negative.*

2. The Measure of a Bounded Closed Set

Proof. Using the notation of Definition 1, it is clear that
$$C_S F \subset (A, B)$$
and by Theorem 1, §1,
$$m[C_S F] \leqslant m(A, B) = B - A,$$
from which it follows that
$$mF \geqslant 0.$$

LEMMA. *Let F be a bounded closed set contained in the open interval Δ. Then*
$$mF = m\Delta - m[C_\Delta F].$$

Proof. The set $C_\Delta F$ is open, so that the preceding lemma is applicable. Let $\Delta = (A, B)$, and let $S = [a, b]$ be the smallest closed interval containing the set F (Fig. 8).

Fig. 8

Then it is easy to see that
$$C_\Delta F = C_\Delta S + C_S F.$$

Both terms of the right-hand member are open and are disjoint. This fact and the additive property of measure (Theorem 2, §1) imply that
$$m[C_\Delta F] = m[C_\Delta S] + m[C_S F].$$
Since it is obvious that $C_\Delta S = (A, a) + (b, B)$, we see that
$$m[C_\Delta S] = (a - A) + (B - b),$$
and consequently
$$m[C_\Delta F] = (B - A) - (b - a) + m[C_S F],$$
which proves the lemma.

THEOREM 2. *Let F_1 and F_2 be two bounded closed sets. If $F_1 \subset F_2$, then*
$$mF_1 \leqslant mF_2.$$

Proof. Let Δ be an interval containing the set F_2. It is easy to verify that
$$C_\Delta F_1 \supset C_\Delta F_2,$$
and we have
$$m[C_\Delta F_1] \geqslant m[C_\Delta F_2],$$
so that the preceding lemma can be applied.

COROLLARY. *The measure of a bounded closed set F is the least upper bound of the measures of all closed sets contained in F.*

THEOREM 3. *Let F be a closed set and let G be a bounded open set. If $F \subset G$, then*
$$mF \leqslant mG.$$

Proof. Let Δ be an interval containing the set G. It is easy to see that
$$\Delta = G + C_\Delta F;$$
and Theorem 3, §1 gives us the inequality $m\Delta \leqslant mG + m[C_\Delta F]$.

We then apply the preceding lemma.

THEOREM 4. *The measure of a bounded open set G is the least upper bound of the measures of all closed sets contained in G.*

Proof. By the preceding theorem, mG is an upper bound for the measures of closed sets $F \subset G$, and we must prove that the measure of a closed subset of G can be arbitrarily close to mG. Let the component intervals of the set G be (λ_k, μ_k) ($k = 1, 2, \ldots$) so that
$$mG = \sum (\mu_k - \lambda_k).$$
We take an arbitrary $\varepsilon > 0$ and find a natural number n so large that
$$\sum_{k=1}^{n} (\mu_k - \lambda_k) > mG - \frac{\varepsilon}{2}.$$
Then, for every k ($k = 1, 2, \ldots, n$) we choose a closed interval $[\alpha_k, \beta_k]$ so that
$$[\alpha_k, \beta_k] \subset (\lambda_k, \mu_k), \quad m[\alpha_k, \beta_k] > m(\lambda_k, \mu_k) - \frac{\varepsilon}{2n}$$
(to accomplish this, it suffices to take an η_k such that
$$0 < \eta_k < \min\left[\frac{\mu_k - \lambda_k}{2}, \frac{\varepsilon}{4n}\right]$$
and put $\alpha_k = \lambda_k + \eta_k$, $\beta_k = \mu_k - \eta_k$). Finally, we set
$$F_0 = \sum_{k=1}^{n} [\alpha_k, \beta_k].$$
It is then clear that $F_0 \subset G$, F_0 is closed, and
$$mF_0 = \sum_{k=1}^{n} (\beta_k - \alpha_k) > \sum_{k=1}^{n} (\mu_k - \lambda_k) - \frac{\varepsilon}{2} > mG - \varepsilon.$$
Since ε is arbitrarily small, the theorem is proved.

THEOREM 5. *The measure of a bounded closed set F is the greatest lower bound of the measures of all possible bounded open sets containing F.*

Proof. As above, we need only prove that it is possible to find a bounded open set which contains the set F and has measure arbitrarily close to mF.

For this purpose, we take an open interval Δ containing the set F and consider the open set $C_\Delta F$. For any $\varepsilon > 0$, we can, by Theorem 4, find a closed set Φ such that
$$\Phi \subset C_\Delta F, \quad m\Phi > m[C_\Delta F] - \varepsilon.$$
Set
$$G_0 = C_\Delta \Phi.$$
It is clear that G_0 is an open set containing F. Also,
$$mG_0 = m\Delta - m\Phi < m\Delta - m[C_\Delta F] + \varepsilon = mF + \varepsilon.$$
Since ε is arbitrary, this completes the proof.

THEOREM 6. *Let the bounded closed set F be the sum of a finite number of pairwise disjoint closed sets :*

$$F = \sum_{k=1}^{n} F_k \qquad (F_k F_{k'} = 0,\ k \neq k').$$

Then
$$mF = \sum_{k=1}^{n} mF_k.$$

Proof. It is clearly sufficient to consider the case of two sets :
$$F = F_1 + F_2 \qquad (F_1 F_2 = 0).$$
We take an arbitrary $\varepsilon > 0$ and choose two bounded open sets G_1 and G_2 such that
$$G_i \supset F_i,\quad mG_i < mF_i + \frac{\varepsilon}{2} \qquad (i = 1, 2)$$
which is possible in view of the preceding theorem. We put
$$G = G_1 + G_2.$$
Then G is a bounded open set containing the set F. This implies that
$$mF \leqslant mG \leqslant mG_1 + mG_2 < mF_1 + mF_2 + \varepsilon.$$
Since ε is arbitrary, it follows from this that
$$mF \leqslant mF_1 + mF_2. \tag{*}$$

On the other hand, by the separation theorem (Ch. II, §4, Theorem 2), there exist open sets B_1 and B_2 such that
$$B_i \supset F_i\ (i = 1, 2),\quad B_1 B_2 = 0$$
Noting this, we again take an arbitrary $\varepsilon > 0$ and select a bounded open set G such that
$$G \supset F,\quad mG < mF + \varepsilon.$$
Then the sets $B_1 G$ and $B_2 G$ are disjoint bounded open sets which contain the sets F_1 and F_2 respectively. It follows that
$$mF_1 + mF_2 \leqslant m(B_1 G) + m(B_2 G) = m[(B_1 G) + (B_2 G)]$$
(in view of the additivity of measure for open sets). Since $B_1 G + B_2 G \subset G$, we can now infer that
$$mF_1 + mF_2 \leqslant mG < mF + \varepsilon$$
and, since ε is arbitrary,
$$mF_1 + mF_2 \leqslant mF. \tag{*_*}$$

Comparing (*) and (*$_*$), we obtain
$$mF = mF_1 + mF_2,$$
as we wished to prove.

§ 3. THE OUTER AND INNER MEASURES OF A BOUNDED SET

DEFINITION 1. The outer measure m^*E of a bounded set E is the greatest lower bound of the measures of all bounded open sets containing the set E:
$$m^*E = \inf_{G \supset E} \{mG\}.$$

It is plain that the outer measure of every bounded set E is well-defined and that
$$0 \leqslant m^*E < +\infty.$$

DEFINITION 2. *The inner measure m_*E of a bounded set E is the least upper bound of the measures of all closed sets contained in the set E:*
$$m_*E = \sup_{F \subset E} \{mF\}.$$

It is again plain that the inner measure of every bounded set E is well-defined and that
$$0 \leqslant m_*E < +\infty$$

THEOREM 1. *If G is a bounded open set, then*
$$m^*G = m_*G = mG.$$

This theorem follows from the corollary of Theorem 1, §1, and from Theorem 4, §2.

THEOREM 2. *If F is a bounded closed set, then*
$$m^*F = m_*F = mF.$$

This theorem follows from the corollary to Theorem 2, §2, and from Theorem 5, §2.

THEOREM 3. *For every bounded set E,*
$$m_*E \leqslant m^*E.$$

Proof. Let G be a bounded open set containing the set E. For any closed subset F of the set E, we have $F \subset G$ and, by Theorem 3, §2, $mF \leqslant mG$. Hence
$$m_*E \leqslant mG.$$
Since this is true for every open bounded set G containing E, we see that
$$m_*E \leqslant m^*E,$$
as was to be proved.

THEOREM 4. *Let A and B be bounded sets. If $A \subset B$, then*
$$m_*A \leqslant m_*B, \quad m^*A \leqslant m^*B.$$

Proof. These inequalities have similar proofs. Let us prove the first of them. Let S be the set of numbers consisting of the measures of all closed subsets of the set A and let T be the analogous set for the set B. Then
$$m_*A = \sup S, \quad m_*B = \sup T.$$
If F is a closed subset of A, then F is necessarily a subset of B. From this, it follows that
$$S \subset T$$
and the present theorem follows from the well-known fact that the least upper bound of the subset of any set cannot exceed the least upper bound of the set itself.

THEOREM 5. *If a bounded set E is the sum of a finite number or a denumerable set of sets E_k,*
$$E = \sum_k E_k,$$
then
$$m^*E \leqslant \sum_k m^*E_k.$$

3. OUTER AND INNER MEASURES OF A BOUNDED SET

Proof. The theorem is trivial if the series $\Sigma m^* E_k$ diverges. Let us suppose that this series converges. Choosing an arbitrary $\varepsilon > 0$, we can find bounded open sets G_k such that

$$G_k \supset E_k, \quad mG_k < m^* E_k + \frac{\varepsilon}{2^k} \qquad (k = 1, 2, 3, \ldots).$$

Designate by Δ any open interval containing the set E. Then $E \subset \Delta \sum_k G_k$. Then, by Theorem 3, §1, we may write

$$m^* E \leqslant m \left[\Delta \sum_k G_k\right] = m \left[\sum_k \Delta G_k\right] \leqslant \sum_k m(\Delta G_k) \leqslant \sum_k mG_k \leqslant \sum_k m^* E_k + \varepsilon,$$

The theorem follows, since ε is arbitrary.

THEOREM 6. *If a bounded set E is the sum of a finite or denumerable number of pairwise disjoint sets E_k,*

$$E = \sum_k E_k \qquad (E_k E_{k'} = 0, \; k \neq k'),$$

then

$$m_* E \geqslant \sum_k m_* E_k.$$

Proof. Consider the first n sets $E_1, E_2, \ldots, E_n$. For an arbitrary $\varepsilon > 0$, there exist closed sets F_k such that

$$F_k \subset E_k, \quad mF_k > m_* E_k - \frac{\varepsilon}{n}, \qquad (k = 1, 2, \ldots, n).$$

The sets F_k are pairwise disjoint, and their sum $\sum_{k=1}^{n} F_k$ is closed. Hence, applying Theorem 6, §2, we obtain

$$m_* E \geqslant m \left[\sum_{k=1}^{n} F_k\right] = \sum_{k=1}^{n} mF_k > \sum_{k=1}^{n} m_* E_k - \varepsilon.$$

Since ε is arbitrary, it follows that

$$\sum_{k=1}^{n} m_* E_k \leqslant m_* E.$$

This proves the theorem for the case of a finite number of summands. If there is a denumerable number of summands, then, noting that the number n is arbitrary, we establish the covergence of the series $\Sigma m_* E_k$ and the inequality

$$\sum_{k=1}^{\infty} m_* E_k \leqslant m_* E.$$

It is easy to see that the theorem ceases to be valid if we omit the condition that the sets E_k have no common points. For example, if

$$E_1 = [0, 1], \quad E_2 = [0, 1], \quad E = E_1 + E_2,$$

then $m_* E = 1$, $m_* E_1 + m_* E_2 = 2$.

THEOREM 7. *Let E be a bounded set. If Δ is an open interval containing E, then*

$$m^* E + m_* [C_\Delta E] = m\Delta.$$

Proof. Take an arbitrary $\varepsilon > 0$ and select a closed set F such that

$$F \subset C_\Delta E, \quad mF > m_* [C_\Delta E] - \varepsilon.$$

The set $G = C_\Delta F$ is a bounded open set containing the set E. From this fact and the lemma of §2, we infer that

$$m^* E \leqslant mG = m\Delta - mF < m\Delta - m_* [C_\Delta E] + \varepsilon.$$

Since ε is arbitrary, the last inequalities show that
$$m^*E + m_*[C_\Delta E] \leq m\Delta.$$
In order to obtain the reverse inequality
$$m^*E + m_*[C_\Delta E] \geq m\Delta, \qquad (*)$$
it is necessary to reason more closely. Take $\varepsilon > 0$ and select a bounded open set G_0 such that
$$G_0 \supset E, \quad mG_0 < m^*E + \frac{\varepsilon}{3}.$$
Designate the end points of the interval Δ by A and B, and select an open interval (a, b) contained in Δ such that
$$A < a < A + \frac{\varepsilon}{3}, \quad B - \frac{\varepsilon}{3} < b < B.$$
Having done this, set
$$G = \Delta G_0 + (A, a) + (b, B).$$
The set G is bounded and open, contains E and has the property that
$$mG < m^*E + \varepsilon.$$
But besides this, we have the essential fact that the set
$$F = C_\Delta G$$
is *closed*, which follows from the easily verified identity
$$F = [a, b] \cdot CG.$$
Since $F \subset C_\Delta E$, we find that
$$m_*[C_\Delta E] \geq mF = m\Delta - mG > m\Delta - m^*E - \varepsilon.$$
Since ε is arbitrary, the inequality (*) follows and with it, the present theorem.

COROLLARY. *Using the notation of the last theorem, we have*
$$m^*[C_\Delta E] - m_*[C_\Delta E] = m^*E - m_*E.$$
In fact, if we interchange the roles of the sets E and $C_\Delta E$, we obtain
$$m^*[C_\Delta E] + m_*E = m\Delta,$$
and consequently
$$m^*[C_\Delta E] + m_*E = m^*E + m_*[C_\Delta E].$$
This is equivalent to the assertion to be proved.

§ 4. MEASURABLE SETS

DEFINITION. A bounded set E is said to be *measurable* if its outer and inner measures are equal:
$$m^*E = m_*E.$$
The common value of these two measures is called the *measure of the set E* and is designated by mE:

4. Measurable Sets

$$mE = m^*E = m_*E.$$

This method of defining the concept of measure is due to Lebesgue, so that a measurable set is sometimes called a set "measurable in the Lebesgue sense," or, more briefly, "measurable (L)."

If the set E is non-measurable, it is impossible to talk about its measure, and the symbol mE is meaningless. In particular, we consider all unbounded sets non-measurable.*

THEOREM 1. *A bounded open set is measurable and its measure, according to the new definition, coincides with its measure as introduced in §1.*

This result is an immediate consequence of Theorem 1, §3. In exactly the same way, the following theorem is deducible from Theorem 2, §3.

THEOREM 2. *A bounded closed set is measurable and its measure, according to the new definition, coincides with that introduced in §2.*

From the corollary to Theorem 7, §3 we infer the following fact.

THEOREM 3. *If E is a bounded set contained in the open interval Δ, then the sets E and $C_\Delta E$ are both measurable or are both non-measurable.*

By combining Theorems 5 and 6, §3, we obtain another result.

THEOREM 4. *If a bounded set E is the sum of a finite or denumerable number of measurable sets which are pairwise disjoint,*

$$E = \sum_k E_k \qquad (E_k E_{k'} = 0,\ k \neq k'),$$

then the set E is measurable and

$$mE = \sum_k mE_k.$$

The proof results from the following chain of inequalities:

$$\sum_k mE_k = \sum_k m_* E_k \leq m_* E \leq m^* E \leq \sum_k m^* E_k = \sum_k mE_k.$$

The property of measure just established is called *countable additivity*.

In the last theorem, it is essential that the individual terms be pairwise disjoint. For the moment, we remove this limitation only for the case of a finite number of summands.

THEOREM 5. *The sum of a finite number of measurable sets is a measurable set.*

Proof. Let

$$E = \sum_{k=1}^n E_k,$$

where the sets $E_k (k = 1, 2, \ldots, n)$ are measurable. Take an arbitrary $\varepsilon > 0$ and find a closed set F_k and a bounded open set G_k such that

$$F_k \subset E_k \subset G_k, \quad mG_k - mF_k < \frac{\varepsilon}{n}.$$

Having done this, set

$$F = \sum_{k=1}^n F_k, \quad G = \sum_{k=1}^n G_k.$$

---

* The author's convention that all unbounded sets are non-measurable is not in consonance with the point of view adopted by most modern writers on measure and integration theory. This convention is convenient for the author's purposes of exposition, but it introduces a serious restriction in the range of application of the theory of measure and integration. We describe the theory of measure for not necessarily bounded sets in an appendix, which is §9 of the present chapter.—E.H.

It is obvious that F is closed, that G is bounded and open, and that

$$F \subset E \subset G.$$

It follows that

$$mF \leq m_*E \leq m^*E \leq mG. \qquad (*)$$

The set $G - F$ is open (since it can be represented in the form $G \cdot CF$) and is bounded. Therefore the set $G - F$ is measurable. The set F is likewise measurable, and hence, inasmuch as

$$G = F + (G - F)$$

and the sets F and $G - F$ have void intersection, we can apply the preceding theorem, which yields

$$mG = mF + m(G - F).$$

Therefore

$$m(G - F) = mG - mF.$$

We establish in like manner the equalities

$$m(G_k - F_k) = mG_k - mF_k \qquad (k = 1, 2, \ldots, n)$$

We next note the easily verified inclusion

$$G - F \subset \sum_{k=1}^{n} (G_k - F_k).$$

All sets involved are bounded and open, so that, on the basis of the theorems of §1, we have

$$m(G - F) \leq \sum_{k=1}^{n} m(G_k - F_k)$$

or

$$mG - mF \leq \sum_{k=1}^{n} [mG_k - mF_k] < \varepsilon.$$

From this and (*), it follows that

$$m^*E - m_*E < \varepsilon,$$

and, since ε is arbitrarily small,

$$m^*E = m_*E.$$

THEOREM 6. *The intersection of a finite number of measurable sets is a measurable set.*

Proof. Let

$$E = \prod_{k=1}^{n} E_k,$$

where the sets E_k are measurable. Let Δ be any open interval containing all of the sets E_k. It is easy to verify that

$$C_\Delta E = \sum_{k=1}^{n} C_\Delta E_k.$$

The sets $C_\Delta E_k$ are measurable, because the sets E_k are measurable. Theorem 5 then implies that the set $C_\Delta E$ is measurable. Hence E is also measurable, as we wished to prove.

THEOREM 7. *The difference of two measurable sets is a measurable set.*

Proof. Let
$$E = E_1 - E_2,$$
where the sets E_1 and E_2 are measurable. Let Δ be any open interval containing both of the sets E_1 and E_2. Then
$$E = E_1 \cdot C_\Delta E_2,$$
and we may refer to the preceding theorem.

THEOREM 8. *If E_1 and E_2 are measurable sets such that $E_1 \supset E_2$, then*
$$mE = mE_1 - mE_2 \quad (E = E_1 - E_2).$$

Proof. We write
$$E_1 = E + E_2 \qquad (E E_2 = 0),$$
and apply Theorem 4 to write
$$mE_1 = mE + mE_2$$
which is equivalent to the present theorem.

THEOREM 9. *If a bounded set E is the sum of a denumerable number of measurable sets, then E is measurable.*

Proof. Let
$$E = \sum_{k=1}^{\infty} E_k.$$
We introduce sets A_k ($k = 1, 2, \ldots$), setting
$$A_1 = E_1, \quad A_2 = E_2 - E_1, \ldots, A_k = E_k - (E_1 + \ldots E_{k-1}), \ldots$$
It is obvious that
$$E = \sum_{k=1}^{\infty} A_k.$$
All of the sets A_k are measurable and are pairwise disjoint (the key to the proof lies in this fact), so that this is equivalent to Theorem 4.

THEOREM 10. *The intersection of a denumerable number of measurable sets is measurable.*

Proof. Let
$$E = \prod_{k=1}^{\infty} E_k,$$
where all the sets E_k are measurable. Since $E \subset E_1$, the set E is bounded. Let Δ be any open interval containing E and let
$$A_k = \Delta E_k \qquad (k = 1, 2, 3, \ldots).$$
Then
$$E = \Delta E = \Delta \prod_{k=1}^{\infty} E_k = \prod_{k=1}^{\infty} (\Delta E_k) = \prod_{k=1}^{\infty} A_k.$$

It is easy to verify that

$$C_\Delta E = \sum_{k=1}^{\infty} C_\Delta A_k,$$

and now Theorems 3 and 9 can be applied.

In conclusion, we establish two theorems which play an important rôle in the theory of functions.

THEOREM 11. *Let the sets $E_1, E_2, E_3, \ldots$ be measurable. If*

$$E_1 \subset E_2 \subset E_3 \subset \ldots,$$

and if the sum $E = \sum_{k=1}^{\infty} E_k$ is a bounded set, then

$$mE = \lim_{n \to \infty} [mE_n].$$

Proof. It is obvious that the set E can be represented in the form

$$E = E_1 + (E_2 - E_1) + (E_3 - E_2) + (E_4 - E_3) + \ldots$$

where the individual terms are pairwise disjoint. It follows from Theorems 4 and 8 that

$$mE = mE_1 + \sum_{k=1}^{\infty} m(E_{k+1} - E_k) = mE_1 + \sum_{k=1}^{\infty} [mE_{k+1} - mE_k].$$

In view of the definition of the sum of an infinite series, the last equality can be represented as follows:

$$mE = \lim_{n \to \infty} \{mE_1 + \sum_{k=1}^{n-1} [mE_{k+1} - mE_k]\}.$$

and this is equivalent to the present theorem, because

$$mE_1 + \sum_{k=1}^{n-1} [mE_{k+1} - mE_k] = mE_n.$$

THEOREM 12. *Let $E_1, E_2, E_3, \ldots$ be measurable sets and let $E = \prod_{k=1}^{\infty} E_k$. If*

$$E_1 \supset E_2 \supset E_3 \supset \ldots,$$

then

$$mE = \lim_{n \to \infty} [mE_n].$$

Proof. This theorem can easily be reduced to the preceding. In fact, letting Δ be any open interval containing the set E_1, we have

$$C_\Delta E_1 \subset C_\Delta E_2 \subset C_\Delta E_3 \subset \ldots,$$

$$C_\Delta E = \sum_{k=1}^{\infty} C_\Delta E_k.$$

Theorem 11 shows that

$$m(C_\Delta E) = \lim_{n \to \infty} [m(C_\Delta E_n)],$$

or, equivalently,

$$m\Delta - mE = \lim_{n \to \infty} [m\Delta - mE_n],$$

which leads at once to the present theorem.

§ 5. MEASURABILITY AND MEASURE AS INVARIANTS UNDER ISOMETRIES

Let there be given two sets A and B consisting of objects of an arbitrary nature. If a rule is specified which assigns one and only one element b of the set B to each element a of the set A, we say a *single-valued mapping* of the set A into the set B has been established. It is not assumed that every element of the set B corresponds to some element of A. If every element of B is the image of some $a \in A$, then we frequently say that the mapping is a mapping of A *onto* B. If this is not necessarily the case, we say that the mapping is a mapping of A *into* B. The concept of a general mapping as just defined is a direct generalization of the concept of a (real-valued) function. The element $b \in B$ which corresponds to the element $a \in A$ is frequently designated by $f(a)$, and we write

$$b = f(a).$$

If $b = f(a)$, the element b is said to be the image of the element a, and the element a is the inverse image of the element b. A single element b may have many inverse images.

Let A^* be a subset of the set A and let B^* be the set of images of all elements of A^* (in other words, if $a \in A^*$, then $f(a) \in B^*$, and if $b \in B^*$, then there exists at least one element $a \in A^*$ such that $f(a) = b$). In such a case, the set B^* is called the *image* of the set A^*, and we write

$$B^* = f(A^*).$$

The set A^* is called the inverse image of the set B^*.

Having set down these general concepts, we turn to the consideration of an important special form of mapping.

DEFINITION 1. A single-valued mapping $\varphi(x)$ of the set of real numbers Z into itself is called an isometry if the distance between the images of any two points of the line equals the distance between the points themselves:

$$|\varphi(x) - \varphi(y)| = |x - y|.$$

for all x and y. In other words, an isometry is a mapping of the set Z into the set Z which does not alter the distance between pairs of points of Z.

In the definition of an isometry, it is not specified that every point of Z be the image of some point or that distinct points of Z have distinct images. However, both of these properties obtain. For the time being, we verify only the latter property.

THEOREM 1. Let $\varphi(x)$ be an isometry. If $x \neq y$, then $\varphi(x) \neq \varphi(y)$.

We need merely note that

$$|\varphi(x) - \varphi(y)| = |x - y| \neq 0.$$

THEOREM 2. a) If $A \subset B$, then $\varphi(A) \subset \varphi(B)$.

b) $\varphi\left(\sum_\xi E_\xi\right) = \sum_\xi \varphi(E_\xi).$

c) $\varphi\left(\prod_\xi E_\xi\right) = \prod_\xi \varphi(E).$

d) If E_0 is the void set, then $\varphi(E_0) = E_0$.

The proof is left to the reader; we point out that Theorem 1 is used only in the proof of part c).

It is easy to verify that the following three mappings are isometries:

I. $\varphi(x) = x + d$ *(translation)*;
II. $\varphi(x) = -x$ *(reflection in the origin)*;
III. $\varphi(x) = -x + d$.

It is a very important fact that all possible isometries of Z are exhausted by these three types (properly speaking two, since II is a special case of III).

THEOREM 3. *If $\varphi(x)$ is an isometry, then either*

$$\varphi(x) = x + d$$

or

$$\varphi(x) = -x + d$$

for some $d \in Z$.

Proof. Set

$$\varphi(0) = d.$$

Then for every x, it is clear that

$$|\varphi(x) - d| = |x|,$$

and we have

$$\varphi(x) = (-1)^{\sigma(x)} x + d \qquad [\sigma(x) = 0, 1].$$

The function $\sigma(x)$ is defined for every $x \neq 0$. Our problem is to show that $\sigma(x)$ is a constant. Let x and y be two points such that $x \neq 0$, $y \neq 0$, $x \neq y$. Then

$$\varphi(x) - \varphi(y) = (-1)^{\sigma(x)} x - (-1)^{\sigma(y)} y$$

or,

$$\varphi(x) - \varphi(y) = (-1)^{\sigma(x)} [x - (-1)^\rho y],$$

where $\rho = \sigma(y) - \sigma(x)$ has one of the three values

$$\varrho = 1, 0, -1.$$

Taking absolute values in the last equality and using the definition of isometry, we see that

$$|x - (-1)^\varrho y| = |x - y|.$$

Therefore we must have either

$$x - (-1)^\varrho y = x - y$$

or

$$x - (-1)^\varrho y = -x + y.$$

But the second case is impossible, since it implies that

$$2x = y[1 + (-1)^\varrho].$$

Then, if $\rho = \pm 1$, we would have $x = 0$, and if $\rho = 0$, we would have $x = y$, and this contradicts our hypotheses on x and y.

Hence the first case is the only possibility; this yields $\rho = 0$, i.e.,

$$\sigma(x) = \sigma(y).$$

5. Measurability and Measure as Invariants under Isometries

Therefore the function $\sigma(x)$ has one and the same value for all $x \neq 0$:

$$\sigma(x) = \sigma \qquad (\sigma = 0, 1),$$

so that

$$\varphi(x) = (-1)^\sigma x + d.$$

This equality is obviously valid for $x = 0$, and the present theorem is proved.

COROLLARY. *Under an isometry φ, every point $y \in Z$ is the image of some point $x \in Z$, i.e., $\varphi(Z) = Z$.*

If $\varphi(x) = (-1)^\sigma x + d$, the inverse image of y is the point

$$x = (-1)^\sigma (y - d)$$

If $\varphi(x) = (-1)^\sigma x + d$ is any isometry, then the isometry

$$\varphi^{-1}(x) = (-1)^\sigma (x - d)$$

is called the *inverse* isometry. These two isometries are connected by the relation

$$\varphi[\varphi^{-1}(x)] = \varphi^{-1}[\varphi(x)] = x.$$

In other words, if the point x under the isometry φ has the point y as its image, then under the isometry φ^{-1}, the point y has the point x as its image. It is an important fact that every isometry has an inverse.

THEOREM 4. *Under an isometry, the following is true: a) every open interval is mapped into an open interval of the same measure, and the end points of the image interval are the images of the end points of the original interval.*

b) the image of a bounded set is a bounded set.

Proof. Let $\Delta = (a, b)$ be an open interval. Then under the isometry $\varphi(x) = x + d$, the image of the interval Δ is the interval $(a + d, b + d)$, and under the isometry $\varphi(x) = -x + d$, the image of Δ is the interval $(d - b, d - a)$. In both cases

$$m\varphi(\Delta) = b - a = m\Delta.$$

To prove b), let E be a bounded set. If Δ is an open interval containing the set E, then

$$\varphi(E) \subset \varphi(\Delta),$$

so that $\varphi(E)$ is bounded. We can also reason as follows: if we have $|x| < k$ for every x in E, then we have $|y| < k + |d|$ for every y in $\varphi(E)$.

THEOREM 5. *Under an isometry, we have the following: a) a closed set goes into a closed set; b) an open set goes into an open set.*

Proof. a) Let $\varphi(F)$ be the image of a closed set F. Designate by y_0 any limit point of the set $\varphi(F)$, and choose a sequence $\{y_n\}$ for which

$$\lim y_n = y_0, \qquad y_n \in \varphi(F).$$

Let

$$x_0 = \varphi^{-1}(y_0), \qquad x_n = \varphi^{-1}(y_n).$$

Then $x_n \in F$. But

$$|x_n - x_0| = |y_n - y_0|$$

so that

$$x_n \to x_0;$$

and, since F is closed, $x_0 \in F$; therefore
$$y_0 = \varphi(x_0) \in \varphi(F).$$
It follows that $\varphi(F)$ is a closed set.

b) Let G be an open set. Set
$$F = CG.$$
Then F is a closed set and
$$G+F = Z, \quad G \cdot F = 0.$$
Hence, by Theorem 2 and the Corollary of Theorem 3,
$$\varphi(G)+\varphi(F) = Z, \quad \varphi(G) \cdot \varphi(F) = 0,$$
i.e., $\varphi(G)$ is the complement of the closed set $\varphi(F)$ and is therefore open.

THEOREM 6. *The measure of a bounded open set is invariant under all isometries.*

Proof. Let G be a bounded open set. Then $\varphi(G)$ also is a bounded open set. Designate the component intervals of the set G by δ_k ($k = 1, 2, 3, \ldots$). Theorem 4 shows that every interval $\varphi(\delta_k)$ is a component interval of the set $\varphi(G)$, and it is easy to verify that all component intervals of the set $\varphi(G)$ are of this form. Therefore
$$m\varphi(G) = \sum_k m\varphi(\delta_k) = \sum_k m\delta_k = mG,$$
as we wished to prove.

THEOREM 7. *Isometries leave the outer and inner measures of a bounded set invariant.*

Proof. a) Let E be a bounded set. Taking an arbitrary $\varepsilon > 0$, we find a bounded open set G such that
$$G \supset E, \quad mG < m^*E + \varepsilon.$$
In this case, $\varphi(G)$ is a bounded open set containing the set $\varphi(E)$. We have accordingly,
$$m^*\varphi(E) \leq m\varphi(G) = mG < m^*E + \varepsilon.$$
Since the number ε is arbitrary, it follows that
$$m^*\varphi(E) \leq m^*E,$$
so that the outer measure of a bounded set does not increase under an isometry. Neither does it decrease because otherwise the inverse isometry would lead to an increase in outer measure. We have, therefore,
$$m^*\varphi(E) = m^*E.$$

b) Let Δ be any open interval containing the set E. Then $\varphi(\Delta)$ is an open interval containing the set $\varphi(E)$. Let
$$A = C_\Delta E.$$
The relations
$$E+A = \Delta, \quad E \cdot A = 0$$
give us
$$\varphi(E)+\varphi(A) = \varphi(\Delta), \quad \varphi(E) \cdot \varphi(A) = 0,$$
so that $\varphi(E)$ is the complement of the set $\varphi(A)$ with respect to the interval $\varphi(\Delta)$.

Hence, by Theorem 7, § 3,
$$m^*\varphi(A) + m_*\varphi(E) = m\varphi(\Delta)$$
and, on the basis of part a) of the present theorem and Theorem 4, we infer
$$m^*A + m_*\varphi(E) = m\Delta.$$
This implies that
$$m_*\varphi(E) = m\Delta - m^*(C_\Delta E),$$
and applying Theorem 7, § 3, again, we find that
$$m_*\varphi(E) = m_*E.$$

COROLLARY *Under an isometry, a measurable set goes into a measurable set having the same measure.*

DEFINITION 2. The sets A and B are called *congruent* if there exists an isometry under which one of them goes into the other.

Using this term the results just established can be expressed as follows:

THEOREM 8. *Congruent sets have the same exterior and interior measure. A set which is congruent with a measurable set is measurable and has the same measure.*

§ 6. THE CLASS OF MEASURABLE SETS

In §§ 4 and 5, we studied the properties of certain measurable sets; we shall now consider some properties of the entire class of measurable sets.

THEOREM 1. *Every bounded denumerable set is measurable and has measure zero.*

Proof. Let the bounded set E consist of the points
$$x_1, x_2, x_3, \ldots$$
Let E_k be the one-element set consisting of the point x_k. Obviously, E_k is a measurable set of measure zero, and the theorem follows from the equality
$$E = \sum_{k=1}^{\infty} E_k$$
and Theorem 4, § 4.

As the Cantor perfect set P_0 shows, the converse of the preceding theorem is false.

DEFINITION 1. If the set E is representable as the sum of a denumerable number of closed sets
$$E = \sum_{k=1}^{\infty} F_k,$$
then E is said to be a set of type F_σ.

DEFINITION 2. If the set E is representable as the intersection of a denumerable number of open sets,
$$E = \prod_{k=1}^{\infty} G_k,$$
then E is said to be a set of type G_δ.

By Theorems 9 and 10, § 4, we have

THEOREM 2. *Every bounded set of type F_σ or of type G_δ is measurable.*

Proof. This is obvious for sets of type F_σ because the boundedness of the sum of sets implies the boundedness of the components, and since the latter are closed, they are also measurable. If E is a bounded set of type G_δ, then designating an open interval containing the set E by Δ, we can represent E as an intersection of bounded open sets,

$$E = \prod_{k=1}^{\infty} (\Delta G_k)$$

from which the measurability of the set E is obvious.

DEFINITION 3. *If the set E can be obtained from closed and open sets by using a finite or a denumerable number of union and intersection operations, then the set E is called a Borel set. A bounded Borel set is said to be measurable (B).**

For example, sets of type F_δ and of type G_δ are Borel sets.

Arguing as in the proof of Theorem 2, we establish the following theorem.

THEOREM 3. *A set measurable (B) is measurable (L).*

The converse theorem is not true: there exist examples of sets measurable (L) which are not measurable (B). The first actual example of such a set was constructed by the Russian mathematician M. Ya. Suslin (1894-1919), who died at an early age. Suslin discovered a very important and extensive class of sets, the so-called A-sets, each of which (under the condition of boundedness) is measurable (L). This class contains the class of all Borel sets, but is much larger.

It is interesting to find out if there exist bounded sets which are not measurable (L). It is impossible to answer this question by a cardinal number argument, as the following theorem shows.

THEOREM 4. *The family M of all measurable sets has the same power as the family of all subsets of Z, i.e., 2^c.*

Proof. First of all, it is clear that

$$\bar{\bar{M}} \leqslant 2^c.$$

On the other hand, we take some measurable set E of measure zero and of power c (for example, the Cantor set P_0) and designate by S the family of all its subsets. Every subset of a set of measure zero also has outer measure zero, and is hence measurable. Therefore

$$S \subset M,$$

and inasmuch as $\bar{\bar{S}} = 2^c$, it is clear that

$$\bar{\bar{M}} \geqslant 2^c.$$

This proves the present theorem.

Nevertheless, we have the following theorem.

THEOREM 5. *Bounded non-measurable sets exist.*

To prove this, we introduce the following example.

Example of a non-measurable set. Subdivide all points of the closed interval $[-\frac{1}{2}, +\frac{1}{2}]$ into classes, putting two points x and y into the same class if and only if their difference

$$x - y$$

---

* Here again the author diverges from the usual convention. Most writers refer to all Borel sets as being measurable (B).—E.H.

6. THE CLASS OF MEASURABLE SETS

is a rational number. This construction can be made explicit in the following manner. For each point $x \in [-\frac{1}{2}, +\frac{1}{2}]$, let the class $K(x)$ consist of all those points of the segment $[-\frac{1}{2}, +\frac{1}{2}]$ which have the form

$$x + r,$$

where r is a rational number. In particular, $x \in K(x)$. We shall show that distinct [1] classes $K(x)$ and $K(y)$ are disjoint. In fact, assume that two such classes have non-void intersection and that $z \in K(x) K(y)$. Then

$$z = x + r_x = y + r_y,$$

where r_x and r_y are rational numbers; thus

$$y = x + r_x - r_y.$$

Now, if $t \in K(y)$,

$$t = y + r = x + (r_x - r_y + r) = x + r',$$

so that $t \in k(x)$ and $K(y) \subset K(x)$. We establish in the same way that $K(x) \subset K(y)$, and see that $K(x) = K(y)$; i.e., $K(x)$ and $K(y)$ are one and the same class, contrary to the hypothesis that they are distinct. The set of all classes obtained in this manner form a partition of $[-1, 1]$. We now choose one point from each class and designate by A the set of all points selected.

The set A is non-measurable.

To prove this, we enumerate all rational points of the closed interval $[-1, +1]$:

$$r_0 = 0, r_1, r_2, r_3, \ldots$$

and designate by A_k the set obtained from the set A by the translation

$$\varphi_k(x) = x + r_k.$$

(In other words, if $x \in A$, then $x + r_k \in A_k$, and if $x \in A_k$, then $x - r_k \in A$.) In particular, $A_0 = A$. All the sets A_k are congruent with one another, and hence (Theorem 8, § 5)

$$m_* A_k = m_* A = \alpha$$
$$m^* A_k = m^* A = \beta \qquad (k = 0, 1, 2, \ldots).$$

We will show that

$$\beta > 0. \tag{1}$$

To do this, we notice that

$$\left[-\frac{1}{2}, +\frac{1}{2}\right] \subset \sum_{k=0}^{\infty} A_k. \tag{2}$$

In fact, if $x \in [-\frac{1}{2}, +\frac{1}{2}]$, then x lies in one of the classes of the partition formed above. If x_0 is the representative of this class in the set A, the difference $x - x_0$ is a rational number which obviously belongs to the closed interval $[-1, +1]$. That is,

$$x - x_0 = r_k$$

and $x \in A_k$. This proves (2).

---

[1] "Distinct" in the sense of the theory of sets, i.e., such that $K(x) \neq K(y)$. On the other hand, it is quite possible that $K(x) = K(y)$ although $x \neq y$, and then these classes are *not* distinct.

We now apply Theorem 5, § 3 and write

$$1 = m^*\left[-\frac{1}{2}, +\frac{1}{2}\right] \leqslant m^*\left[\sum_{k=0}^{\infty} A_k\right] \leqslant \sum_{k=0}^{\infty} m^* A_k,$$

i.e.,

$$1 \leq \beta + \beta + \beta + \ldots,$$

and from this, (1) follows.

On the other hand, it is easy to show that

$$\alpha = 0. \tag{3}$$

To do this, we show first that for $n \neq m$,

$$A_n A_m = 0. \tag{4}$$

In fact, if the point z were in $A_n A_m$, the points

$$x_n = z - r_n, \quad x_m = z - r_m$$

would be (obviously distinct) points of the set A, i.e., representatives of two distinct classes, which is impossible, because their difference

$$x_n - x_m = r_m - r_n$$

is a rational number. This proves (4). It is also easy to see that for arbitrary k

$$A_k \subset \left[-\frac{3}{2}, +\frac{3}{2}\right]$$

(because if $x \in A_k$, then $x = x_0 + r_k$, where $|x_0| < \frac{1}{2}$, $|r_k| < 1$), so that

$$\sum_{k=0}^{\infty} A_k \subset \left[-\frac{3}{2}, +\frac{3}{2}\right]. \tag{5}$$

It follows from (5) and (4) and Theorem 6, § 3 that

$$3 = m_*\left[-\frac{3}{2}, +\frac{3}{2}\right] \geqslant m_*\left[\sum_{k=0}^{\infty} A_k\right] \geqslant \sum_{k=0}^{\infty} m_* A_k,$$

whence

$$\alpha + \alpha + \alpha + \ldots \leqslant 3$$

and $\alpha = 0$.

Comparing (1) and (3) we obtain

$$m_* A < m^* A,$$

which proves the non-measurability of the set A.

Remark. If initially we had partitioned not the segment $[-\frac{1}{2}, +\frac{1}{2}]$ into classes, but rather an arbitrary measurable set E of positive measure, then repeating literally the argument set forth above, we would have arrived at a non-measurable set $A \subset E$. Therefore every measurable set of positive measure contains a non-measurable subset.

§ 7. GENERAL REMARKS ON THE PROBLEM OF MEASURE

The negative result obtained at the end of § 6 suggests that Lebesgue's definition of measure is faulty and leads naturally to the question whether it is possible to improve somewhat on Lebesgue's definition. To give an answer to this question, it is necessary first of all to formulate exactly the problem which we wish to solve. The problem of measure for point sets can be posed in two ways.

I. *The difficult problem of the theory of measure.*[2]
It is required to assign to every bounded set E a non-negative number μE, its measure, so that the following requirements are satisfied:
  1. If $E = [0, 1]$, then $\mu E = 1$.
  2. If the sets A and B are congruent, then $\mu A = \mu B$.
  3. If the set E is the sum of a finite or a denumerable number of pairwise disjoint sets E_k ($k = 1, 2, 3, \ldots$) then
$$\mu E = \sum_k \mu E_k \text{ (countable additivity)}.$$

We have formulated the problem for the case of linear sets, *i.e.*, for the one-dimensional space $Z = R_1$. It could be also posed for the plane R_2 and, in general, for n-dimensional space R_n, except that in requirement 1, it would be necessary to refer not to the closed interval $[0, 1]$, but to the square $[0, 1\, ;\, 0, 1]$ or, in general, to the n-dimensional unit cube.

However, it is easy to establish the following theorem.

THEOREM 1. *The difficult problem of the theory of measure has no solution in the space R_1.*

Proof. In the example of a non-measurable set presented in §6, we constructed pairwise disjoint sets
$$A_0, A_1, A_2, \ldots.$$
congruent in pairs, such that
$$\left[-\frac{1}{2}, +\frac{1}{2}\right] \subset \sum_{k=0}^{\infty} A_k \subset \left[-\frac{3}{2}, +\frac{3}{2}\right].$$

If the difficult problem of the theory of measure were solvable, then it would be the case that
$$\mu\left[-\frac{1}{2}, +\frac{1}{2}\right] \leqslant \sum_{k=0}^{\infty} \mu A_k \leqslant \mu\left[-\frac{3}{2}, +\frac{3}{2}\right].$$

But the segment $[-\frac{1}{2}, +\frac{1}{2}]$ is congruent with the segment $[0, 1]$; besides (for arbitrary k)
$$\mu A_k = \mu A = \sigma,$$
and, finally, the set
$[-\frac{3}{2}, +\frac{3}{2}]$ is bounded; consequently
$$1 \leqslant \sigma + \sigma + \sigma + \ldots < +\infty$$
which is impossible, either for $\sigma > 0$ or for $\sigma = 0$. This completes the proof.

---

[2] The terms "difficult", "easy" in reference to the problem of measure are not generally accepted. We introduce them here for brevity in the formulation of statements, etc., although we do not consider them very felicitous.

We also pose another problem in the theory of measure.

II. *The easy problem of the theory of measure*, which is formulated almost like problem I, with the single difference, that requirement 3 is set down only for the case of a *finite* number of summands, i.e., instead of countable additivity of our measure we require only finite additivity.

In connection with the easy problem of the theory of measure, we state the following results, without proof.[3]

THEOREM 2 (S. BANACH). *The easy problem of the theory of measure is solvable for the spaces R_1 and R_2, but not uniquely.*

THEOREM 3 (F. HAUSDORFF). *For the space R_n, where $n > 2$, the easy problem of the theory of measure is unsolvable.*

The difference between these results is explained by the fact that the concept of congruence, involved in the formulation of the problem, is essentially connected with the concept of isometry. Since in a space with a large number of dimensions, the group of isometries is extremely large, it is quite natural that it is more difficult to find measures invariant under such a group.

In conclusion, we present a discussion which, to a certain extent, justifies Lebesgue's definition of measure.

Suppose that we have a solution of the easy problem of the theory of measure. Then it follows from the relation

$$A \subset B,$$

that

$$\mu A \leqslant \mu B \text{ (Principle of Monotonicity)},$$

since $\mu B = \mu A + \mu(B-A)$. It follows immediately that the measure μE of every set E which consists of only one point equals zero, for on the closed interval $[0, 1]$ we can choose as large a number of sets congruent to E as we please. It follows from this that the measure μ of an arbitrary finite set equals zero and also that

$$\mu(a, b) = \mu(a, b] = \mu[a, b) = \mu[a, b].$$

Furthermore, the relation

$$[0, 1] = \left[0, \frac{1}{n}\right] + \left(\frac{1}{n}, \frac{2}{n}\right] + \ldots + \left(\frac{n-1}{n}, 1\right],$$

shows that

$$\mu\left[0, \frac{1}{n}\right] = \frac{1}{n}$$

which makes it clear that the measure of the closed interval $[a, b]$ with rational length equals $b - a$. Using the principle of monotonicity, one shows easily that the equality

$$\mu[a, b] = b - a$$

holds for all closed intervals $[a, b]$. It is then clear that for of an arbitrary open bounded set G having a finite number of component intervals,

$$\mu G = mG,$$

---

[3] The proof of Banach's theorem requires acquaintance with functional analysis and is beyond the scope of this book.

and if the component intervals form a denumerable set, then

$$\mu G \geqslant mG.$$

The most natural solutions of the easy problem of the theory of measure are those solutions for which the measure μ of a bounded open set G is the sum of the lengths of its component intervals (it is clear from what is stated above that it is sufficient to require only that $\mu G \leqslant \sum_k \mu \delta_k$ and this only for the case of an infinite set of intervals).

It can be seen from the proof of Banach's theorem that such solutions actually exist (we accept this without proof). If we call such solutions *regular*, then

THEOREM 4. *For every regular solution of the easy problem of the theory of measure, the measure μE of a measurable set E equals its Lebesgue measure mE.*

Proof. The definition of a regular solution implies that the measure μG of a bounded open set G equals its Lebesgue measure mG. For every bounded closed set F,

$$\mu F = mF.$$

Now, if a bounded set E contains a closed set F and is contained in a bounded open set G, then, by the principle of monotonicity,

$$mF \leqslant \mu E \leqslant mG.$$

Therefore we have

$$m_* E \leqslant \mu E \leqslant m^* E$$

and the present theorem is proved.

§ 8. VITALI'S THEOREM

DEFINITION. Let E be a point set and let M be a family of closed intervals (none consisting of a single point). If for every point $x \in E$ and every $\varepsilon > 0$, there exists a closed interval $d \in M$ such that

$$x \in d, \quad md < \varepsilon$$

we shall say the set E is covered by the family M in the sense of Vitali.

In other words, the set E is covered by the family M in the sense of Vitali if every point of the set E is contained in arbitrarily small closed intervals belonging to the family M.

The following theorem has many applications in the theory of functions.

THEOREM 1 (VITALI). *If a bounded set E is covered by a family M of closed intervals in the sense of Vitali, then it is possible to find a finite or denumerable family of closed intervals $\{d_k\}$ in M such that*

$$d_k d_i = 0 \ (k \neq i), \quad m^*[E - \sum_k d_k] = 0.$$

In other words, the closed intervals d_k are pairwise disjoint and cover the set E, up to a set of measure zero.

We give Banach's proof of this remarkable theorem.

Proof. Select any open interval Δ containing the set E (E is bounded), and remove from M the intervals which are not contained completely in Δ. It is clear that the system M_0, consisting of the remaining closed intervals (those intervals of the initial system M which are contained entirely in Δ) also covers the set E in the sense of Vitali. Having noted this, we consider any closed interval $d_1 \in M_0$. If $E \subset d_1$, the construction is complete. In the contrary case, we choose one interval after another from M_0, according to the following rule. Suppose that the intervals

$$d_1, d_2, \ldots, d_n \tag{1}$$

have already been chosen and that they are pairwise disjoint. If

$$E \subset \sum_{k=1}^{n} d_k,$$

the construction is complete and the theorem is proved. If, however,

$$E - \sum_{k=1}^{n} d_k \neq 0, \tag{2}$$

we set

$$F_n = \sum_{k=1}^{n} d_k, \qquad G_n = \Delta - F_n$$

and consider all those intervals of the system M_0 which are contained in the open set G_n. By (2), such intervals exist, and their lengths are bounded (for example, by the number $m\Delta$). Let k_n denote the least upper bound of the lengths of these intervals and let d_{n+1} be any one of them for which [4]

$$m d_{n+1} > \frac{1}{2} k_n. \tag{3}$$

It is clear that the interval d_{n+1} is disjoint from all of the intervals (1). If the process of constructing the intervals $d_1, d_2, d_3, \ldots$ does not terminate after a finite number of steps (in which case the proof is complete), it yields a sequence

$$d_1, d_2, d_3, \ldots \tag{4}$$

of intervals, which are pairwise disjoint. We shall show that this is the sequence sought, i.e., that

$$m^*(E - S) = 0 \tag{5}$$

where

$$S = \sum_{k=1}^{\infty} d_k.$$

For this purpose, we construct, for every k, a closed interval D_k having the same midpoint as the closed interval d_k but having five times greater length: $mD_k = 5\,md_k$. It is easy to see that

$$\sum_{k=1}^{\infty} mD_k < +\infty \tag{6}$$

In fact, the segments d_k are disjoint and are all contained in Δ, so that

$$\sum_{k=1}^{\infty} md_k \leqslant m\Delta \tag{7}$$

---

[4] Since no interval of the family M reduces to a single point, we necessarily have $k_n > 0$.

whence (6) follows. To establish the relation (5), it is thus sufficient to show that
$$E - S \subset \sum_{k=i}^{\infty} D_k \qquad (8)$$
for all i. Let $x \in E - S$. Then $x \in G_i$ for all i, since G_i is open there exists an interval d of the system M_0 such that
$$x \in d \subset G_i.$$
However, it is impossible that
$$d \subset G_n \qquad (9)$$
for all n, since it would follow from this that
$$md \leqslant k_n < 2 m d_{n+1}$$
for all n, and this is impossible, since by (7), $md_n \to 0$. This implies that for some n, the relation (9) is not fulfilled and then the relation
$$d \cdot F_n \neq 0. \qquad (10)$$
holds. Now let n be the *smallest* integer for which (10) holds. Since
$$d \cdot F_i = 0$$
and $F_1 \subset F_2 \subset F_3 \subset \ldots$, it is clear that
$$n > i.$$
From the definition of n, it follows that
$$d \cdot F_{n-1} = 0;$$
We can now infer two facts. First,
$$d \cdot d_n \neq 0 \qquad (11)$$
and second, $d \subset G_{n-1}$ and
$$md \leqslant k_{n-1} < 2 m d_n. \qquad (12)$$
From (11) and (12), it obviously follows that
$$d \subset D_n$$
and consequently
$$d \subset \sum_{k=i}^{\infty} D_k.$$
This implies that
$$x \in \sum_{k=i}^{\infty} D_k,$$
whence (8) follows. This completes the present proof.

A modified form of Vitali's theorem is sometimes useful in applications.

THEOREM 2 (VITALI). *Under the hypotheses of Theorem 1, for every $\varepsilon > 0$, there exists a finite system $d_1, d_2, \ldots, d_n$ of pairwise disjoint closed intervals of the system M for which*
$$m^*\left(E - \sum_{k=1}^{n} d_k\right) < \varepsilon.$$

Proof. Let Δ be any open interval containing the set E, and eliminate from M all closed intervals which are not contained in the interval Δ. The remaining family of closed intervals, M_0 then obviously covers the set E in the sense of Vitali. Apply Theorem 1 to the system M_0, and construct thereby a system $\{d_k\}$ of pairwise disjoint intervals of the family M_0, for which

$$m^*\left[E - \sum_k d_k\right] = 0.$$

If the family $\{d_k\}$ is finite, the construction is complete. If this family is infinite, then we have obviously

$$\sum_{k=1}^{\infty} m d_k \leqslant m\Delta,$$

and it is possible to find an n so large that

$$\sum_{k=n+1}^{\infty} m d_k < \varepsilon.$$

It is easy to see that

$$E - \sum_{k=1}^{n} d_k \subseteq \left[E - \sum_{k=1}^{\infty} d_k\right] + \sum_{k=n+1}^{\infty} d_k \tag{13}$$

and accordingly

$$m^*\left[E - \sum_{k=1}^{n} d_k\right] < \varepsilon$$

because the outer measure of the first term in the right member of (13) equals zero.

§ 9. EDITOR'S APPENDIX TO CHAPTER III.

We here extend the definition and basic properties of measurable sets from the case of bounded sets, treated in the present chapter, to arbitrary sets on the real line.

DEFINITION 1. Let E be an arbitrary subset of the real line Z. Then E is said to be measurable if all of the sets $E[-n, n]$ are measurable in the sense of the definition of § 4.

It is easy to see that the foregoing definition is consistent with the definition of § 4; that is, a bounded set E is measurable according to the new definition if and only if it is measurable according to the old definition.

DEFINITION 2. Let E be any measurable set. Then the measure mE of E is defined by the relation

$$mE = \lim_{n \to \infty} m(E[-n, n]).$$

Since all of the sets $E[-n, n]$ are measurable, each term on the right-hand side is defined, and since the sets $E[-n, n]$ form an increasing sequence of sets, we see that the sequence of numbers $m(E[-n, n])$ is increasing, and therefore has a limit, which is either finite or $+\infty$. Hence the number mE is well defined for every measurable set. We make the conventions that $+\infty + \infty = +\infty$, that $a + \infty = +\infty$ for all real numbers a, and that $\lim_{n\to\infty} b_n = +\infty$ if the b_n's are an increasing sequence one of which is $+\infty$.

9. Editor's Appendix

Examples.
1. $mZ = +\infty$, since $m(Z[-n, n]) = 2n$, and $\lim\limits_{n\to\infty} 2n = +\infty$.
2. Similarly, $m([0, \infty)) = m((-\infty, 0]) = \infty$.
3. The measure of any bounded measurable set is the same as its measure according to §4, since from a certain point on, all of the sets $E[-n, n]$ are equal to E.
4. All finite or denumerable subsets B of Z are measurable and have measure 0. This follows at once from the fact that $B[-n, n]$ is finite or denumerable for all n, if B is finite or denumerable.
5. Let $P = \sum\limits_{n=-\infty}^{+\infty} (n + P_0)$. Then it is easy to see that $mP = 0$.
6. Let $E = \sum\limits_{n=2}^{\infty} (n - \frac{1}{n}, n + \frac{1}{n})$. Then $mE = \lim\limits_{k\to\infty} \sum\limits_{n=2}^{k} \frac{2}{n} = +\infty$.
7. Let $E = \sum\limits_{n=1}^{+\infty} (n - 2^{-n}, n + 2^{-n})$. Then $mE = \sum\limits_{n=1}^{+\infty} 2^{-n+1} = 2$.

It is obvious that there exist non-measurable unbounded sets. If $E = A + [1, \infty)$, where A is a non-measurable subset of $(0, 1)$, it is clear that E is non-measurable. Of course much more elaborate examples can be constructed.

We now discuss the theorems of §§4, 5, and 6, extending them where possible to unbounded sets.

THEOREM 1. *Every open set and every closed set are measurable, and the measures of such sets can be any number greater than or equal to zero and less than or equal to $+\infty$.*

We need verify only that for every open set G and every natural number n, the set $G[-n, n]$ is measurable in the sense of §4. Since

$$G[-n, n] = \prod_{k=1}^{\infty} G \cap (-n - \tfrac{1}{k}, -n + \tfrac{1}{k})$$

and all of the sets appearing on the right-hand side are open, we can apply Theorems 1 and 10, §4, to see that $G[-n, n]$ is measurable.

THEOREM 2. *The sum of a finite or denumerable family of measurable sets is measurable.*

Proof. Let $E_1, E_2, E_3, \ldots, E_k, \ldots$ be a denumerable family of measurable sets. Then, for all natural numbers n and k, we have that

$$E_k[-n, n]$$

is a measurable set. Since

$$\left(\sum_{k=1}^{\infty} E_k\right)[-n, n] = \sum_{k=1}^{\infty} E_k[-n, n],$$

we have only to apply Theorem 9, §4, to prove the present theorem.

THEOREM 3. *The intersection of a finite or denumerable family of measurable sets is a measurable set.*

Proof. Let the notation be as in the preceding theorem. Then

$$\left(\prod_{k=1}^{\infty} E_k\right)[-n, n] = \prod_{k=1}^{\infty} E_k[-n, n],$$

and a reference to Theorem 10, §4, completes the proof.

In §4, it was impossible to prove that the complement of a measurable set is measurable, since the complement of a bounded set is necessarily unbounded. However, under our present definition of measurability, complements of measurable sets are always measurable.

THEOREM 4. *Let E be a measurable set. Then CE is also measurable.*
Proof. By hypothesis, all of the sets $E[-n, n]$ are measurable according to the definition of §4. The present theorem follows from this fact, the relation
$$(CE)[-n, n] = C(E[-n, n])[-n, n],$$
and Theorem 7, § 4.

THEOREM 5. *Every Borel set is measurable.*
Proof. This follows immediately from Theorems 1, 2, and 4.

THEOREM 6. *If E and F are measurable sets and $E \subset F$, then $mE \leq mF$.*
Proof. This follows at once from Definitions 1 and 2 and Theorem 8, § 4.

THEOREM 7 (COUNTABLE ADDITIVITY). *Let $E_1, E_2, E_3, \ldots$ be a finite or denumerable sequence of measurable sets which are pairwise disjoint. Then*
$$m\left(\sum_k E_k\right) = \sum_k m E_k.$$
Proof. For all natural numbers n and p, we have
$$m\left(\left(\sum_{k=1}^p E_k\right)[-n, n]\right) = m\left(\sum_{k=1}^p E_k[-n, n]\right) = \sum_{k=1}^p m(E_k[-n, n]).$$
Taking the limit of both sides of this equality as $n \to \infty$, we have
$$m\left(\sum_{k=1}^p E_k\right) = \lim_{n\to\infty} m\left(\left(\sum_{k=1}^p E_k\right)[-n, n]\right) = \lim_{n\to\infty}\left(\sum_{k=1}^p m(E_k[-n, n])\right) =$$
$$\sum_{k=1}^p \lim_{n\to\infty} m(E_k[-n, n]) = \sum_{k=1}^p m(E_k).$$
This establishes the present theorem for a finite number of summands E_k. For a denumerable number, we argue as follows. The relations
$$m\left(\sum_{k=1}^\infty E_k\right) = \lim_{n\to\infty} m\left(\left(\sum_{k=1}^\infty E_k\right)[-n, n]\right) = \lim_{n\to\infty} m\left(\sum_{k=1}^\infty E_k[-n, n]\right) =$$
$$\lim_{n\to\infty} \sum_{k=1}^\infty m(E_k[-n, n]) \leq \sum_{k=1}^\infty m(E_k)$$
are easily verified. Hence $m\left(\sum_{k=1}^\infty E_k\right) \leq \sum_{k=1}^\infty m(E_k)$. To establish the reverse inequality, we use the finite part of the theorem already proved. That is, we write
$$m\sum_{k=1}^\infty (E_k) = \sum_{k=1}^p m(E_k) + m\left(\sum_{k=p+1}^\infty E_k\right) = \sum_{k=1}^\infty m(E_k) + \lim_{p\to\infty} m\left(\sum_{k=p+1}^\infty E_k\right) \geq \sum_{k=1}^\infty m(E_k).$$
This completes the proof.

Remark. The above proof shows in particular that if $m(\sum_k E_k)$ is infinite, then the series $\sum_k mE_k$ diverges or has at least one term equal to ∞.

THEOREM 8. *If $E_1, E_2, E_3, \ldots, E_n, \ldots$ are measurable and if $E_1 \subset E_2 \subset E_3 \ldots \subset E_n \subset \ldots$, then*
$$m\left(\sum_k E_k\right) = \lim_{n\to\infty} mE_n.$$
Proof. We write
$$\sum_k E_k = E_1 + (E_2 - E_1) + (E_3 - E_2) + \ldots + (E_n - E_{n-1}) + \ldots,$$
and apply Theorem 7, using the same argument as in the proof of Theorem 11, § 4.

6. Editor's Appendix

THEOREM 9. *If $E_1, E_2, E_3, \ldots, E_n, \ldots$ are measurable sets, if $E_1 \supset E_2 \supset \ldots \supset E_n \supset \ldots$, and $m(E_1)$ is finite, then*
$$m\left(\prod_{n=1}^{\infty} E_n\right) = \lim_{n \to \infty} m(E_n)$$

Proof. Write $\prod_{n=1}^{\infty} E_n$ as A, and write $E_n - E_{n+1}$ as B_n ($n = 1, 2, 3, \ldots$). Then the sets A and $B_1, B_2, B_3, \ldots$ are pairwise disjoint. We have in addition
$$E_1 = A + B_1 + B_2 + B_3 + \ldots + B_n + \ldots,$$
and
$$E_p = A + B_p + B_{p+1} + \ldots \qquad (p = 1, 2, 3, \ldots).$$
Furthermore,
$$mE_1 = mA + \sum_{n=1}^{\infty} mB_n,$$
by Theorem 7. Since mE_1 is finite, the infinite series on the right-hand side converges, and thus
$$0 = \lim_{p \to \infty} \sum_{n=p}^{\infty} mB_n = \lim_{p \to \infty} m\left(\sum_{n=p}^{\infty} B_n\right).$$
Since
$$m(E_p) = mA + m\left(\sum_{n=p}^{\infty} B_n\right),$$
we may take the limit as $p \to \infty$ on both sides of this expression and find
$$\lim_{p \to \infty} m(E_p) = mA + \lim_{p \to \infty} m\left(\sum_{n=p}^{\infty} B_n\right) = mA.$$

The extension of the theorems of §4 to the case of not necessarily bounded sets is thus complete. Theorems 6, 7, and 8 of §5 are obviously true for unbounded as well as bounded sets. We omit the details. We also omit a discussion of Vitali's theorem for unbounded sets, since we shall not need it in the sequel. We close the present section with two theorems which are in a sense analogues of Definitions 1 and 2, §3, and the definition of §4.

THEOREM 10. *Let E be any measurable set. If $mE < \infty$ and ε is any number > 0, then there exist an open set G and a bounded closed set F such that $F \subset E \subset G$ and $mG - m(E) < \varepsilon$, $m(E) - m(F) < \varepsilon$. If $mE = +\infty$, then for every positive number K, there exists a bounded closed set $F \subset E$ such that $mF > K$.*

Proof. Let $E_n = E[-n, n]$ ($n = 1, 2, 3, \ldots$). There exists a closed set $F_n \subset E_n$ such that $m(F_n) > m(E_n) - \frac{1}{n}$. Let $C_n = \sum_{k=1}^{n} F_n$. Then we obviously have
$$mE_n \geq m(C_n) > m(E_n) - \frac{1}{n} \qquad (n = 1, 2, 3, \ldots).$$
Taking the limits as $n \to \infty$ of all the members of these inequalities, we obtain
$$mE = \lim_{n \to \infty} mC_n.$$
This shows that the bounded closed set F required exists both for $mE < +\infty$ and $mE = +\infty$.

To show the existence of an open $G \supset E$ with the required property, in case $mE < +\infty$, we argue as follows. Let G_n be an open set such that $E_n \subset G_n$ and $mE_n > mG_n - \frac{\varepsilon}{2^n}$ ($n = 1, 2, 3, \ldots$). Let $H_n = \sum_{k=1}^{n} G_n$. We wish to show that

$mH_n - mE_n < \varepsilon$. We note first that $m(G_n - E_n) < \frac{\varepsilon}{2^n}$ for all n, and that

$$H_n - E_n = (\sum_{k=1}^{n} G_k) - E_n = \sum_{k=1}^{n}(G_k - E_n) \subset \sum_{k=1}^{n} G_k - E_k.$$

Hence we may write

$$m(H_n - E_n) = mH_n - mE_n \leqslant m(\sum_{k=1}^{n} G_k - E_k) \leqslant$$

$$\sum_{k=1}^{n} m(G_k - E_k) = \sum_{k=1}^{n} mG_k - mE_k < \sum_{k=1}^{n} \frac{\varepsilon}{2^k} < \varepsilon.$$

Since $H_1 \subset H_2 \subset H_3 \subset \ldots$, we apply Theorem 8 to write

$$\lim_{n \to \infty} mH_n = m(\sum_{n=1}^{\infty} H_n).$$

Combining our results, we have
$$mE_n \leqslant mH_n < mE_n + \varepsilon.$$

Taking the limits as $n \to \infty$ of all members of these inequalities, we see that the set $G = \sum_{n=1}^{\infty} H_n$ satisfies the requirements of the present theorem.

Exercises for Chapter III

Prove the following assertions.

1. Every perfect set contains a perfect subset of measure zero.

2. If A is a measurable set of positive measure, then there exist points x and y in A such that $|x - y|$ is rational.

3. A necessary and sufficient condition that the bounded set E be measurable is that for every $\varepsilon > 0$ there exist a closed set $F \subset E$ such that $m^*(E - F) < \varepsilon$ (de la Vallée-Poussin's test).

4. For every bounded set E, it is possible to find sets A and B such that $A \subset E \subset B$, A is of type F_σ, B is of type G_σ, and
$$mA = m_*E, \quad mB = m^*E.$$

5. If A and B are two measurable sets without common points, then for an arbitrary set E,
$$m^*[E \cdot (A+B)] = m^*(E \cdot A) + m^*(E \cdot B), \quad m_*[E \cdot (A+B)] = m_*(E \cdot A) + m_*(E \cdot B).$$

6. A necessary and sufficient condition that the bounded set E be measurable is that for an arbitrary bounded set A,
$$m^*A = m^*(A \cdot E) + m^*(A \cdot CE)$$
(Carathéodory's test).

7. The set E is said to be nowhere dense if every open interval contains points of the set CE'. Construct a bounded perfect set which is nowhere dense and has positive measure.

8. Construct a measurable set E contained in $U = [0, 1]$ such that for an arbitrary open interval $\Delta \subset U$,
$$m(\Delta \cdot E) > 0, \quad m(\Delta \cdot CE) > 0.$$

9. If $E = \sum_{k=1}^{\infty} E_k$, $E_1 \subset E_2 \subset E_3 \subset \ldots$ and E is bounded, then for $n \to \infty$,
$$m^*E_n \to m^*E.$$

10. For every method of solving the easy problem of the theory of measure, the measure of a bounded denumerable set equals zero.

11. Show by an example that Theorem 9, § 9, may fail to be true if $mE_1 = \infty$.

CHAPTER IV

MEASURABLE FUNCTIONS

§ 1. THE DEFINITION AND THE SIMPLEST PROPERTIES OF MEASURABLE FUNCTIONS

If for every x in a set E, there is defined a number $f(x)$, then we say that the function $f(x)$ is defined on the set E. We allow the function to have infinite values, provided that they have a definite sign; this compels us to introduce the ideal numbers $-\infty$ and $+\infty$. These numbers are related to each other and to an arbitrary finite number a by the inequalities

$$-\infty < a < +\infty$$

We establish the following laws of operation for $+\infty$ and $-\infty$:

$$+\infty \pm a = +\infty, \quad +\infty + (+\infty) = +\infty, \quad +\infty - (-\infty) = +\infty,$$
$$-\infty \pm a = -\infty, \quad -\infty + (-\infty) = -\infty, \quad -\infty - (+\infty) = -\infty,$$
$$|+\infty| = |-\infty| = +\infty,$$
$$+\infty \cdot a = a \cdot (+\infty) = +\infty, \quad -\infty \cdot a = a \cdot (-\infty) = -\infty, \quad \text{if} \quad a > 0,$$
$$+\infty \cdot a = a \cdot (+\infty) = -\infty, \quad -\infty \cdot a = a \cdot (-\infty) = +\infty, \quad \text{if} \quad a < 0,$$
$$0 \cdot (\pm\infty) = (\pm\infty) \cdot 0 = 0,$$
$$(+\infty) \cdot (+\infty) = (-\infty) \cdot (-\infty) = +\infty, \quad (+\infty) \cdot (-\infty) = (-\infty) \cdot (+\infty) = -\infty,$$
$$\frac{a}{\pm\infty} = 0.$$

Here a designates an arbitrary finite real number. The expressions

$$+\infty - (+\infty), \quad -\infty - (-\infty), \quad +\infty + (-\infty), \quad -\infty + (+\infty), \quad \frac{\pm\infty}{\pm\infty}, \quad \frac{a}{0}$$

are not defined at all.[1]

In dealing with a function $f(x)$ defined on the set E, we shall designate by the symbol

$$E(f > a)$$

the set of those x in the set E for which the inequality

$$f(x) > a$$

holds. We define the symbols $E(f \geq a)$, $E(f = a)$, $E(f \leq a)$, $E(a < f \leq b)$, etc., in

---

[1] Usually the symbols $0 \cdot (\pm\infty)$ and $(\pm\infty) \cdot 0$ are also taken as meaningless. It is more convenient for us to consider them as being equal to zero.

the same way. If the set on which the function $f(x)$ is defined is designated by a letter other than E, for example A or B, we shall write

$$A(f > a), \quad B(f > a)$$

and so on.

DEFINITION 1. The function $f(x)$ defined on the set E is said to be *measurable* if the set E is measurable and if the set

$$E(f > a)$$

is measurable for all a.*

Since we are concerned here with sets which are measurable in Lebesgue's sense, we frequently speak of functions measurable (L).

THEOREM 1. *Every function defined on a set of measure zero is measurable.*

This assertion is obvious.

THEOREM 2. *Let $f(x)$ be a measurable function defined on the set E. If A is a measurable subset of E, then $f(x)$, considered only for $x \in A$, is measurable.*[2]

In fact,

$$A(f > a) = A \cdot E(f > a).$$

THEOREM 3. *Let $f(x)$ be defined on the measurable set E, which is the sum of a finite or denumerable number of measurable sets E_k,*

$$E = \sum_k E_k.$$

If $f(x)$ is measurable on each of the sets E_k, then it is also measurable on E.

We need only note that

$$E(f > a) = \sum_k E_k(f > a).$$

DEFINITION 2. Two functions $f(x)$ and $g(x)$, defined on the same set E, are said to be *equivalent* if

$$mE(f \neq g) = 0.$$

It is usual to indicate that the functions $f(x)$ and $g(x)$ are equivalent by the following expression:

$$f(x) \sim g(x).$$

DEFINITION 3. Let some property S hold for all the points of a set E, except for the points of a subset E_0 of the set E. If $mE_0 = 0$, we say that S holds *almost everywhere* on the set E, or *for almost all* points of E.

In particular, the set of points E_0 excluded may be the void set.

We can say, for example, that two functions defined on the set E are equivalent if they are equal almost everywhere on E.

---

[2] Instead of using this rather cumbersome expression, we will say that $f(x)$ is measurable on the set A.

*Since only bounded sets can be measurable, according to the definition given by the author in Chapter III, a function defined on an unbounded set cannot be measurable. In §6 of the present chapter, we sketch the theory of measurable functions under the extended notion of measure presented in §9 of Chapter III.—E. H.

1. Definition and Simplest Properties of Measurable Functions

Theorem 4. *If $f(x)$ is a measurable function defined on the set E, and if $g(x) \sim f(x)$, then $g(x)$ is also measurable.*

Proof. Let
$$A = E(f \neq g), \qquad B = E - A.$$
Then $mA = 0$, so that B is measurable. This implies that the function $f(x)$ is measurable on the set B. But the functions $f(x)$ and $g(x)$ are identical on the set B, so that $g(x)$ also is measurable on B. Inasmuch as $g(x)$ is also measurable on A (since $mA = 0$), it is measurable on $E = A + B$.

Theorem 5. *If $f(x) = c$ for all points of a measurable set E, then the function $f(x)$ is measurable.*

In fact,
$$E(f > a) = \begin{cases} E & \text{for } a < c, \\ 0 & \text{for } a \geqslant c. \end{cases}$$

We note that the c of the preceding theorem may be infinite.

A function $f(x)$ defined on the closed interval $[a, b]$ is said to be a *step function* if we can subdivide $[a, b]$ by the points
$$c_0 = a < c_1 < c_2 < \ldots < c_n = b$$
into a finite number of sub-intervals in the interior of which (*i.e.*, in the open intervals (c_k, c_{k+1}) $(k = 0, 1, \ldots, n-1)$) $f(x)$ is *constant*. Theorems 3 and 5 have the following corollary.

Corollary. *A step function is measurable.*

Theorem 6. *If $f(x)$ is a measurable function defined on the set E, then the sets*
$$E(f \geqslant a), \quad E(f = a), \quad E(f \leqslant a), \quad E(f < a)$$
are measurable for all a.

Proof. It is easy to verify that
$$E(f \geqslant a) = \prod_{n=1}^{\infty} E\left(f > a - \frac{1}{n}\right)$$
from which the measurability of the set $E(f \geqslant a)$ follows. The measurability of the other sets mentioned follows from the relations
$$E(f = a) = E(f \geqslant a) - E(f > a), \qquad E(f \leqslant a) = E - E(f > a),$$
$$E(f < a) = E - E(f \geqslant a).$$

Remark. It can easily be proved that if at least one of the sets
$$E(f \geqslant a), \quad E(f \leqslant a), \quad E(f < a) \tag{1}$$
is measurable for every a, the function $f(x)$ is measurable on the set E (E is assumed to be measurable). For example, the identity
$$E(f > a) = \sum_{n=1}^{\infty} E\left(f \geqslant a + \frac{1}{n}\right)$$
shows that $f(x)$ is measurable if all the sets $E(f \geqslant a)$ are measurable. The remaining assertions are established in a similar manner. So, in the definition of a measurable function, we can replace the set $E(f > a)$ by any one of the sets (1).

IV. MEASURABLE FUNCTIONS

THEOREM 7. *If the function $f(x)$ defined on the set E is measurable and if k is a finite number, then the functions* 1) $f(x) + k$, 2) $kf(x)$, 3) $|f(x)|$, 4) $f^2(x)$, *and* * 5) $\frac{1}{f(x)}$ *(if $f(x) \neq 0$) are also measurable.*

Proof. 1) The measurability of the function $f(x) + k$ follows from the identity

$$E(f + k > a) = E(f > a - k).$$

2) The measurability of the function $kf(x)$ for $k = 0$ follows from Theorem 5. For other k, measurability of $kf(x)$ follows from the obvious relations

$$E(kf > a) = \begin{cases} E(f > \frac{a}{k}), & \text{if } k > 0 \\ E(f < \frac{a}{k}), & \text{if } k < 0. \end{cases}$$

3) The function $|f(x)|$ is measurable because

$$E(|f| > a) = \begin{cases} E & \text{if } a < 0 \\ E(f > a) + E(f > -a), & \text{if } a \geq 0. \end{cases}$$

4) Analogously, from the fact that

$$E(f^2 > a) = \begin{cases} E & \text{if } a > 0, \\ E(|f| > \sqrt{a}), & \text{if } a \geq 0, \end{cases}$$

it follows that the function $f^2(x)$ is measurable.

5) Finally, for $f(x) \neq 0$, we have

$$E(\tfrac{1}{f} > a) = \begin{cases} E(f > 0), & \text{if } a = 0, \\ E(f > 0) \cdot E(f < \tfrac{1}{a}), & \text{if } a > 0, \\ E(f > 0) + E(f < 0) \cdot E(f < \tfrac{1}{a}), & \text{if } a < 0, \end{cases}$$

which shows that $\frac{1}{f(x)}$ is measurable.

THEOREM 8. *Every function $f(x)$ which is defined and continuous on the closed interval $E = [a, b]$ is measurable.*

Proof. We establish first that the set

$$F = E(f \leq a)$$

is closed. In fact, if x_0 is a limit point of F and

$$x_n \to x_0 \qquad\qquad (x_n \in F),$$

then $f(x_n) \leq a$ for all n and we have

$$f(x_0) \leq a$$

since $f(x)$ is continuous. Hence $x_0 \in F$, and this establishes the fact that F is closed. Consequently the set

$$E(f > a) = E - E(f \leq a)$$

is measurable, and the present theorem is proved.

From the very definition of a measurable function it follows that a function

---

* The expression $f(x) \neq 0$ must be interpreted as meaning that $f(x)$ is equal to zero nowhere on E, since $\frac{1}{0}$ is undefined, and since functions must be defined everywhere. If we were to allow functions to be undefined on a set of measure zero, then $\frac{1}{f(x)}$ could be defined and would be measurable provided that $f(x)$ is different from zero for almost all $x \in E$.—E. H.

1. DEFINITION AND SIMPLEST PROPERTIES OF MEASURABLE FUNCTIONS

defined on a non-measurable set is non-measurable. However, it is easy to show the existence of non-measurable functions defined on measurable sets.

DEFINITION 4. *Let M be a subset of the closed interval $E = [A, B]$. The function $\varphi_M(x)$, equal to unity on the set M and to zero on the set $E - M$, is called the characteristic function of the set M.*

THEOREM 9. *The set M and its characteristic function $\varphi_M(x)$ are both measurable or both non-measurable.*

Proof. If the function $\varphi_M(x)$ is measurable, the measurability of the set M follows from the relation
$$M = E(\varphi_M > 0).$$
Conversely, if M is a measurable set, the relations
$$E(\varphi_M > a) = \begin{cases} 0, & \text{if } a \geq 1 \\ M, & \text{if } 0 \leq a < 1 \\ E, & \text{if } a < 0 \end{cases}$$
establish the measurability of the function $\varphi_M(x)$.

Among other things, examples of discontinuous measurable functions are obtained very simply from the foregoing.

§ 2. FURTHER PROPERTIES OF MEASURABLE FUNCTIONS

Lemma. If two measurable functions $f(x)$ and $g(x)$ are defined on a set E, then the set
$$E(f > g)$$
is measurable.

If we enumerate all rational numbers
$$r_1, r_2, r_3, \ldots,$$
then we easily verify the validity of the relation
$$E(f > g) = \sum_{k=1}^{\infty} E(f > r_k) \cdot E(g < r_k)$$
and the lemma follows immediately from this equality.

THEOREM 1. *Let $f(x)$ and $g(x)$ be finite measurable functions defined on the set E. Then each of the functions* 1) $f(x) - g(x)$, 2) $f(x) + g(x)$, 3) $f(x) \cdot g(x)$, *and if* $g(x) \neq 0$,* 4) $\frac{f(x)}{g(x)}$ *is measurable.*

Proof. 1) The function $a + g(x)$ is measurable for arbitrary a. This fact and the preceding lemma imply that the set $E(f > a + g)$ is measurable, and since
$$E(f - g > a) = E(f > a + g)$$
the function $f(x) - g(x)$ is measurable.

2) The measurability of the sum $f(x) + g(x)$ follows from the fact that
$$f(x) + g(x) = f(x) - [-g(x)].$$

3) The measurability of the product $f(x) g(x)$ follows from the identity
$$f(x) g(x) = \frac{1}{4} \{[f(x) + g(x)]^2 - [f(x) - g(x)]^2\}$$
and Theorem 7, §1.

---

* See editor's footnote on page 92.

4) Finally, the measurability of the quotient $\frac{f(x)}{g(x)}$ is a consequence of the identity

$$\frac{f(x)}{g(x)} = f(x) \cdot \frac{1}{g(x)}.$$

This theorem shows that arithmetic operations applied to measurable functions do not take us outside of this class of functions. The following theorem establishes a similar result with respect to taking the limit, which is not an arithmetic operation.

THEOREM 2. *Let a sequence of measurable functions $f_1(x), f_2(x),\ldots$ be defined on the set E. If the limit (finite or infinite)*

$$F(x) = \lim_{n \to \infty} f_n(x),$$

exists for every point $x \in E$, then the function $F(x)$ is measurable.

Proof. Take an arbitrary fixed a and consider the sets

$$A_m^{(k)} = E\left(f_k > a + \frac{1}{m}\right), \quad B_m^{(n)} = \prod_{k=n}^{\infty} A_m^{(k)}.$$

These sets obviously are measurable. To prove the theorem, it is sufficient to verify that

$$E(F > a) = \sum_{n,\,m} B_m^{(n)}.$$

We proceed to do this. If $x_0 \in E(F > a)$, then $F(x_0) > a$, and we can find a natural number m such that $F(x_0) > a + \frac{1}{m}$. Since $f_k(x_0) \to F(x_0)$, we can find an n such that for $k \geqslant n$, we have

$$f_k(x_0) > a + \frac{1}{m}.$$

In other words, $x_0 \in A_m^{(k)}$ for all $k \geqslant n$; then $x_0 \in B_m^{(n)}$ and $x_0 \in \sum_{n,\,m} B_m^{(n)}$ necessarily. We have thus shown that

$$E(F > a) \subset \sum_{n,\,m} B_m^{(n)}.$$

The reverse inclusion

$$\sum_{n,\,m} B_m^{(n)} \subset E(F > a) \tag{*}$$

remains to be established. Let $x_0 \in \sum_{n,\,m} B_m^{(n)}$. Then $x_0 \in B_m^{(n)}$ for some fixed n and m. That is, $x_0 \in A_m^{(k)}$ for all $k \geqslant n$; in other words, we have

$$f_k(x_0) > a + \frac{1}{m}$$

for $k \geqslant n$. Taking the limit as k goes to infinity in the last inequality, we obtain

$$F(x_0) \geqslant a + \frac{1}{m}.$$

It is clear from this that $F(x_0) > a$, i.e., $x_0 \in E(F > a)$. This proves the inclusion (*) and hence the present theorem.

The theorem just proved allows the following generalization.

THEOREM 3. *Let the measurable functions $f_1(x), f_2(x),\ldots$ and some function $F(x)$ be defined on the set E. If the relation*

$$\lim_{n \to \infty} f_n(x) = F(x) \tag{α}$$

is satisfied almost everywhere on E, then the function $F(x)$ is measurable.

Proof. Let A denote the set of all those points $x \in E$ for which the relation (α) does not hold (at such points, the limit $\lim f_n(x)$ need not exist at all). By hypothesis, $mA = 0$ and $F(x)$ is measurable on the set A. By Theorem 2, it is also measurable on the set $E - A$, and consequently it is measurable on the entire set E.

§ 3. SEQUENCES OF MEASURABLE FUNCTIONS. CONVERGENCE IN MEASURE

In this section, we consider sets of the form
$$E(|f - g| \geqslant \sigma), \qquad E(|f - g| < \sigma)$$
where $f(x)$ and $g(x)$ are functions defined on the set E and σ is a positive number. Points at which both functions $f(x)$ and $g(x)$ take on infinite values of the same sign are not included in either of these sets, strictly speaking, inasmuch as at these points the difference $f(x) - g(x)$ is devoid of meaning. Since the indicated situation represents known inconveniences we agree once and for all to put these points into the set $E(|f - g| \geqslant \sigma)$.[3] Under this convention, we obviously have
$$E = E(|f - g| \geqslant \sigma) + E(|f - g| < \sigma)$$
and the sets on the right-hand side are disjoint.

THEOREM 1 (LEBESGUE). *Let there be given a sequence $f_1(x), f_2(x), f_3(x), \ldots$ of measurable functions on a set E, all of which are finite almost everywhere. Suppose that $f_n(x) \to f(x)$ as $n \to \infty$ almost everywhere on E and that $f(x)$ is finite almost everywhere. Then*
$$\lim_{n \to \infty} [mE(|f_n - f| \geqslant \sigma)] = 0.$$
for all $\sigma > 0$.

Proof. First of all, note that by Theorem 3, §2, the limit function $f(x)$ is measurable and the sets under consideration are measurable. Set
$$A = E(|f| = +\infty), \quad A_n = E(|f_n| = +\infty), \quad B = E(f \text{ non} \to f),$$
$$Q = A + \sum_{n=1}^{\infty} A_n + B.$$

It is clear that
$$mQ = 0. \tag{1}$$

Furthermore, let
$$E_k(\sigma) = E(|f_k - f| \geqslant \sigma), \quad R_n(\sigma) = \sum_{k=n}^{\infty} E_k(\sigma),$$
$$M = \prod_{n=1}^{\infty} R_n(\sigma).$$
All of these sets are measurable. Since
$$R_1(\sigma) \supset R_2(\sigma) \supset R_3(\sigma) \supset \ldots,$$

---

[3] This convention is of course artificial. In the sequel, however, we consider only functions which are finite almost everywhere, and the sets $E(|f - g| \geqslant \sigma)$ concern us only from the point of view of their measure. In the final analysis, it is therefore immaterial how we treat the set because its measure is zero.

we infer from Theorem 12, §4, Chapt. III, that
$$m R_n(\sigma) \to m M \tag{2}$$
as $n \to \infty$. We shall show that
$$M \subset Q. \tag{3}$$
In fact, if $x_0 \bar\in Q$, then
$$\lim_{k \to \infty} f_k(x_0) = f(x_0)$$
where all the numbers $f_1(x_0), f_2(x_0), \ldots$, and their limit $f(x_0)$, are finite. Thus we can find an n such that
$$|f_k(x_0) - f(x_0)| < \sigma$$
for $k \geq n$. In other words,
$$x_0 \bar\in E_k(\sigma) \qquad (k \geq n).$$
therefore $x_0 \bar\in R_n(\sigma)$ and $x_0 \bar\in M$. The inclusion (3) follows. From (3) and (1), we see that $m M = 0$, and (2) may be rewritten as
$$m R_n(\sigma) \underset{n \to \infty}{\to} 0. \tag{4}$$
This proves the present theorem, since
$$E_n(\sigma) \subset R_n(\sigma).$$

Remark. The result (4) established above is stronger than what we set out to prove. Below, in the proof of D. F. Egorov's theorem, we shall find it convenient to use exactly this stronger result.

The preceding theorem makes it natural to introduce the following definition

DEFINITION. Let the sequence of measurable functions $f_1(x), f_2(x), f_3(x), \ldots$ (*) be defined and finite almost everywhere on a measurable set E. Let $f(x)$ be a measurable function which is finite almost everywhere. If
$$\lim_{n \to \infty} [m E(|f_n - f| \geq \sigma)] = 0$$
for all positive numbers σ, then the *sequence* (*) *is said to converge in measure to the function* $f(x)$.

Following G. M. Fichtenholz, we designate convergence in measure by the symbol
$$f_n(x) \Rightarrow f(x).$$

With the aid of the notion of convergence in measure, we can formulate Lebesgue's Theorem as follows:

THEOREM 1*. *If a sequence of functions converges almost everywhere, it converges in measure to the same limit function.*

The following example shows that the converse of Theorem 1* is false.

Example. On the half-open interval $[0, 1)$ define the k functions
$$f_1^{(k)}(x), \ f_2^{(k)}(x), \ \ldots, \ f_k^{(k)}(x),$$
for every natural number k, according to the following definition:
$$f_i^{(k)}(x) = \begin{cases} 1 & \text{for } x \in \left[\dfrac{i-1}{k}, \dfrac{i}{k}\right) \\ 0 & \text{for } x \bar\in \left[\dfrac{i-1}{k}, \dfrac{i}{k}\right) \end{cases}$$

3. Sequences of Measurable Functions; Convergence in Measure

[In particular, $f_1^{(1)}(x) \equiv 1$ on $[0, 1)$.] Numbering all these functions with the indices 1, 2, 3,..., we obtain the sequence

$$\varphi_1(x) = f_1^{(1)}(x), \quad \varphi_2(x) = f_1^{(2)}(x), \quad \varphi_3(x) = f_2^{(2)}(x), \quad \varphi_4(x) = f_1^{(3)}(x), \ldots,$$

It is easy to see that the sequence of functions $\varphi(x)$ converges in measure to zero. In fact, if $\varphi_n(x) = f_i^{(k)}(x)$, we have

$$E(|\varphi_n| \geqslant \sigma) = \left[\frac{i-1}{k}, \frac{i}{k}\right)$$

for all σ such that $0 < \sigma \leqslant 1$ and the measure of this set, which is equal to $\frac{1}{k}$, tends to zero as $n \to \infty$.[4] At the same time, the relation

$$\varphi_n(x) \to 0$$

is fulfilled for no point x of the interval $[0, 1)$. In fact, if $x_0 \in [0, 1)$, we can find an i for every k, such that

$$x_0 \in \left[\frac{i-1}{k}, \frac{i}{k}\right),$$

so that $f_i^{(k)}(x_0) = 1$. In other words, no matter how far out we go in the sequence of numbers

$$\varphi_1(x_0), \varphi_2(x_0), \varphi_3(x_0), \ldots,$$

we shall always encounter in this sequence numbers equal to 1, which proves our assertion.

Thus, the concept of convergence in measure is a broader concept than the concepts of convergence almost everywhere and of convergence everywhere.

It is natural to ask to what degree the relation

$$f_n(x) \Rightarrow f(x)$$

determines the function $f(x)$, i.e., whether the limit function is unique for convergence in measure.

Theorems 2 and 3 give us an answer to this question.

THEOREM 2. *If the sequence of functions $f_n(x)$ converges in measure to the function $f(x)$, then the same sequence converges in measure to every function $g(x)$ which is equivalent to the function $f(x)$.*

Proof. For all $\sigma > 0$,

$$E(|f_n - g| \geqslant \sigma) \subset E(f \neq g) + E(|f_n - f| \geqslant \sigma),$$

and since $mE(f \neq g) = 0$, we infer that

$$mE(|f_n - g| \geqslant \sigma) \leqslant mE(|f_n - f| \geqslant \sigma).$$

This proves the theorem.

THEOREM 3. *If the sequence of functions $f_n(x)$ converges in measure to two functions $f(x)$ and $g(x)$, then these limit functions are equivalent.*

Proof. It is easy to verify that

$$E(|f - g| \geqslant \sigma) \subset E\left(|f_n - f| \geqslant \frac{\sigma}{2}\right) + E\left(|f_n - g| \geqslant \frac{\sigma}{2}\right) \quad (*)$$

---

[4] We may suppose that $\sigma \leqslant 1$, because otherwise the set $E(|\varphi_n| \geqslant \sigma)$ is void, and the statement becomes trivial.

for all $\sigma > 0$, because a point not in the set on the right-hand side of this inclusion cannot possibly be in the set on the left-hand side. The relations
$$f_n \Rightarrow f, \quad f_n \Rightarrow g$$
show that the measure of the right-hand member of (*) tends to zero as $n \to \infty$, from which it is clear that
$$mE(|f-g| \geq \sigma) = 0.$$
Since
$$E(f \neq g) \subset \sum_{n=1}^{\infty} E\left(|f-g| \geq \frac{1}{n}\right),[5]$$
we see that $f \sim g$, as was to be proved.

Theorems 2 and 3 show that if we wish to prove the property of uniqueness of the limit function for convergence in measure, we must agree to consider equivalent functions as being identical. As a matter of fact, this is usually done in dealing with questions in the theory of functions which are of a *metric* nature, *i.e.*, questions in which properties of a function are studied with the aid of the measure of the sets on which the function possesses or does not possess a given property. In the integral calculus, we find many examples of this approach.

Although convergence in measure is more general than convergence almost everywhere, the following theorem nevertheless holds.

THEOREM 4 (F. RIESZ). *Let $\{f_n(x)\}$ be a sequence of functions which converges in measure to the function $f(x)$. Then there exists a subsequence*
$$f_{n_1}(x), \; f_{n_2}(x), \; f_{n_3}(x), \; \ldots \qquad (n_1 < n_2 < n_3 < \ldots),$$
which converges to the function $f(x)$ almost everywhere. [6]

Proof. Consider a sequence of positive numbers
$$\sigma_1 > \sigma_2 > \sigma_3 > \ldots,$$
for which
$$\lim \sigma_k = 0.$$
Further, let
$$\eta_1 + \eta_2 + \eta_3 + \ldots \qquad (\eta_k > 0)$$
be a convergent series of positive terms. We select the required sequence of indices
$$n_1 < n_2 < n_3 < \ldots \qquad (*)$$
in the following way. Let n_1 be a natural number for which
$$mE(|f_{n_1} - f| \geq \sigma_1) < \eta_1.$$
Such a number must necessarily exist, because
$$mE(|f_n - f| \geq \sigma_1) \to 0$$

---

[5] It would not be valid to use = here instead of the symbol ⊂. For example, those points at which $f(x) = g(x) = +\infty$ are not in the left-hand member, but we have agreed to put them into *all* sets of the form $E(|f-g| \geq \sigma)$.

[6] In formulating this theorem, we tacitly assume all conditions stated in the definition of convergence in measure hold, as: the measurability of the set E on which the functions are defined, the measurability of the functions themselves, and so on.

as $n\to\infty$. Let n_2 be a natural number for which
$$mE(|f_{n_2}-f|\geqslant \sigma_2)<\eta_2, \qquad n_2>n_1.$$
In general, we choose the number n_k so that
$$mE(|f_{n_k}-f|\geqslant \sigma_k)<\eta_k, \qquad n_k>n_{k-1}.$$
Thus the sequence (*) is defined.

We now establish that
$$\lim_{k\to\infty} f_{n_k}(x)=f(x). \qquad (*_*)$$
almost everywhere on the set E. Let
$$R_i=\sum_{k=i}^{\infty} E(|f_{n_k}-f|\geqslant \sigma_k), \quad Q=\prod_{i=1}^{\infty} R_i.$$
Since
$$R_1\supset R_2\supset R_3\supset \ldots,$$
we apply Theorem 12, §4, Chapt. III, to assert that
$$mR_i \to mQ.$$
On the other hand, it is clear that
$$mR_i<\sum_{k=i}^{\infty}\eta_k,$$
so that $mR_i\to 0$, and hence
$$mQ=0.$$
It remains to verify that the relation $(*_*)$ holds for all x in the set $E-Q$. Let $x_0\in E-Q$. Then $x_0\bar\in R_{i_0}$ for some natural number i_0. In other words,
$$x_0\bar\in E(|f_{n_k}-f|\geqslant \sigma_k),$$
for all $k\geqslant i_0$, and in consequence
$$|f_{n_k}(x_0)-f(x_0)|<\sigma_k \qquad (k\geqslant i_0)$$
Since $\sigma_k\to 0$, it is clear that
$$f_{n_k}(x_0)\to f(x_0).$$
This completes the proof.

Lebesgue's Theorem (Theorem 5) motivated our introduction of convergence in measure. On the other hand, with the aid of this theorem, we can establish the following fundamental theorem due to D. F. Egorov.

THEOREM 5 (D. F. EGOROV). *Let a sequence of measurable and almost everywhere finite functions*
$$f_1(x),\ f_2(x),\ f_3(x),\ldots$$
be defined and finite almost everywhere on a measurable set E. Suppose that $f_n(x)$ converges almost everywhere to the measurable function $f(x)$ which is finite almost everywhere:
$$\lim_{n\to\infty} f_n(x)=f(x). \qquad (*)$$

Then, for every $\delta > 0$, there exists a measurable set $E_\delta \subset E$ such that:
1) $mE_\delta > mE - \delta$;
2) *The limit* (*) *is uniform on the set* E_δ.

Proof. In the proof of Lebesgue's Theorem, we showed that
$$mR_n(\sigma) \underset{n \to \infty}{\to} 0, \tag{1}$$
as $n \to \infty$ for all $\sigma > 0$, where
$$R_n(\sigma) = \sum_{k=n}^{\infty} E(|f_k - f| \geqslant \sigma).$$
With this in mind we consider a convergent series of positive terms
$$\eta_1 + \eta_2 + \eta_3 + \ldots \qquad (\eta_i > 0)$$
and a sequence of positive numbers tending to zero:
$$\sigma_1 > \sigma_2 > \sigma_3 > \ldots, \quad \lim \sigma_i = 0.$$
In view of (1), we can find for every natural number i another natural number n_i such that
$$mR_{n_i}(\sigma_i) < \eta_i.$$
Having done this, we select an i_0 such that
$$\sum_{i=i_0}^{\infty} \eta_i < \delta$$
(δ is the number appearing in the statement of the theorem), and we set
$$e = \sum_{i=i_0}^{\infty} R_{n_i}(\sigma_i).$$
It is obvious that
$$me < \delta.$$
Let
$$E_\delta = E - e.$$
We now show that E_δ is the required set. The inequality
$$mE_\delta > mE - \delta$$
is clear, so we need only verify that the limit
$$f_n(x) \to f(x)$$
is uniform on the set E_δ.

Choose an arbitrary $\varepsilon > 0$. We find an i such that
$$i \geqslant i_0, \quad \sigma_i < \varepsilon,$$
and show that for $k \geqslant n_i$ and for all $x \in E_\delta$,
$$|f_k(x) - f(x)| < \varepsilon.$$
If $x \in E_\delta$, then $x \bar{\in} e$. This implies in particular that
$$x \bar{\in} R_{n_i}(\sigma_i).$$

In other words,
$$x \bar{\in} E(|f_k - f| \geq \sigma_i),$$
for $k \geq n_i$, so that
$$|f_k(x) - f(x)| < \sigma_i \qquad (k \geq n_i),$$
and hence,
$$|f_k(x) - f(x)| < \varepsilon \qquad (k \geq n_i).$$
The theorem is proved, because n_i depends only on ε and not on x.

§ 4. THE STRUCTURE OF MEASURABLE FUNCTIONS

In studying any functions, the problem of representing the function exactly or approximately with the aid of functions of a simpler nature arises quite naturally. Such, for example, are the algebraic questions of decomposing a polynomial into linear factors or a rational function into simple rational functions. The problems of expanding a continuous function in a power series or trigonometric series are also of this type.

In this section, we establish various theorems on approximating measurable functions by means of continuous functions, *i.e.*, we solve an approximation problem for measurable functions. These theorems lead us to the basic structural property of measurable functions, which is stated in Theorem 4.

THEOREM 1. *Let a measurable function $f(x)$ be defined and finite almost everywhere on the set E. For any $\varepsilon > 0$, there exists a measurable bounded function $g(x)$ such that*
$$mE(f \neq g) < \varepsilon.$$

Proof. Let
$$A_k = E(|f| > k), \quad Q = E(|f| = +\infty).$$
By hypothesis, $mQ = 0$. In view of the obvious relations
$$A_1 \supset A_2 \supset A_3 \supset \ldots$$
$$Q = \prod_{k=1}^{\infty} A_k$$

Theorem 12, § 4, Chapt. III shows that
$$mA_k \to mQ = 0$$
as $k \to \infty$. Hence a natural number k_0 can be found such that
$$mA_{k_0} < \varepsilon.$$
Define on the set E a function $g(x)$, setting
$$g(x) = \begin{cases} f(x) & \text{for } x \in E - A_{k_0} \\ 0 & \text{for } x \in A_{k_0}. \end{cases}$$
This function is measurable and is bounded, inasmuch as
$$|g(x)| \leq k_0.$$

Finally,
$$E(f \neq g) = A_{k_0},$$
this proves the theorem.

The preceding theorem shows that every measurable function which is finite almost everywhere becomes *bounded* when we disregard a set of arbitrarily small measure.

DEFINITION. Let the function $f(x)$ be defined on the set E, let $x_0 \in E$, and let $f(x_0)$ be finite. We say the *function $f(x)$ is continuous at the point x_0* in two cases: 1) if x_0 is an isolated point of E; 2) if $x_0 \in E'$ and the relations

$$x_n \to x_0, \quad x_n \in E$$

imply that

$$f(x_n) \to f(x_0).$$

If $f(x)$ is continuous at every point of the set E, it is said to be continuous on the set E.

Lemma 1. *Let the sets $F_1, F_2, \ldots, F_n$ be closed and pairwise disjoint. If the function $\varphi(x)$, defined on the set*

$$F = \sum_{k=1}^{n} F_k,$$

is constant on each of the sets F_k, it is continuous on the set F.

Proof. Let $x_0 \in F'$ and suppose that

$$x_i \to x_0, \quad x_i \in F.$$

Since the set F is closed, the point x_0 belongs to this set and, hence, we can find an m such that

$$x_0 \in F_m.$$

But the sets F_k are disjoint in pairs. This implies that $x_0 \bar{\in} F_k$ if $k \neq m$. Since the set F_k is closed, the point x_0 cannot be a limit point of the set F_k. It follows that in the sequence $\{x_i\}$, there can be only a finite number of points belonging to the set F_k for $k \neq m$. Consider all of the members of the sequence which are in one of the sets

$$F_1, \ldots F_{m-1}, F_{m+1}, \ldots, F_n,$$

and let x_{i_0} be the last of them. Then, for $i > i_0$, we necessarily have

$$x_i \in F_m,$$

i.e., for $i > i_0$,

$$\varphi(x_i) = \varphi(x_0),$$

and this proves the lemma.

Lemma 2. *Let F be a closed set contained in the closed interval $[a, b]$. If the function $\varphi(x)$ is defined and continuous on the set F, then it is possible to define a function on $[a, b]$ with the following properties:*

1) $\psi(x)$ *is continuous*;
2) *if* $x \in F$, *then* $\psi(x) = \varphi(x)$;
3) $\max |\psi(x)| = \max |\varphi(x)|$.

Proof. Let $[\alpha, \beta]$ be the smallest closed interval containing the set F. If the required function $\psi(x)$ were already defined throughout $[\alpha, \beta]$, it would be sufficient to complete its definition by setting

$$\psi(x) = \begin{cases} \varphi(a), & \text{if } x \in [a, \alpha) \\ \varphi(\beta), & \text{if } x \in (\beta, b] \end{cases}$$

in order to obtain the required function on the entire closed interval $[a, b]$. We may therefore suppose without loss of generality that $[a, b]$ is the smallest closed interval containing the set F. If $F = [a, b]$, the theorem is trivial. Suppose then that $F \neq [a, b]$. The set $[a, b] - F$ consists of a finite or a denumerable family of pairwise disjoint open intervals, the endpoints of which belong to F (these are the complementary intervals of the set F). We define the function $\psi(x)$ by setting it equal to $\varphi(x)$ at all points of the set F and making it linear on all complementary intervals of F. Let us verify that the function $\psi(x)$ is continuous on $[a, b]$. Its continuity at every point of the set $[a, b] - F$ is obvious. Let x_0 be a point of the set F. We will show that $\psi(x)$ is continuous from the left at this point. If the point x_0 is the right endpoint of any complementary interval, continuity from the left of the function $\psi(x)$ at this point is obvious. Suppose then that x_0 is not the right endpoint of any complementary interval of F and let

$$x_1 < x_2 < x_3 < \cdots$$

be a sequence of points tending to x_0. If

$$x_n \in F \qquad (n = 1, 2, 3, \ldots),$$

then, using the fact that the function $\varphi(x)$ is continuous on the set F, we have

$$\psi(x_n) = \varphi(x_n) \to \varphi(x_0) = \psi(x_0).$$

We therefore need to consider only the case

$$x_n \overline{\in} F \qquad (n = 1, 2, 3, \ldots).$$

Under this hypothesis, the point x_1 lies in some complementary interval (λ_1, μ_1), where $\mu_1 < x_0$ (because x_0 is not the right endpoint of a complementary interval). Let n_1 be the natural number such that

$$\lambda_1 < x_k < \mu_1 \quad (k = 1, 2, \ldots, n_1), \qquad x_{n_1+1} > \mu_1.$$

Then x_{n_1+1} lies in another complementary interval (λ_2, μ_2), where again $\mu_2 < x_0$. Continuing this construction, we produce a sequence

$$(\lambda_1, \mu_1), \quad (\lambda_2, \mu_2), \quad (\lambda_3, \mu_3), \ldots$$

of complementary intervals with the property that

$$x_k \in (\lambda_i, \mu_i) \qquad (k = n_{i-1}+1, \ldots, n_i).$$

The relation

$$x_{n_i} < \mu_i < x_0$$

shows that

$$\mu_i \to x_0,$$

and from the inequalities

$$\mu_{i-1} \leq \lambda_i < x_0,$$

it is clear that

$$\lambda_i \to x_0.$$

also. However, λ_i and μ_i are in F, so that

$$\lim \psi(\lambda_i) = \lim \psi(\mu_i) = \psi(x_0).$$

By virtue of the fact that the values of a linear function in any interval lie between its values at the endpoints of this interval, it is clear that

$$\lim \psi(x_n) = \psi(x_0).$$

Thus left continuity of the function $\psi(x)$ is proved. Continuity on the right is established in precisely the same way. From the very way in which it was constructed, it is clear that $\psi(x)$ coincides with $\varphi(x)$ on the set F.

The function $|\psi(x)|$ is continuous on $[a, b]$, and hence, by a famous theorem of Weierstrass, this function is bounded on $[a, b]$ and attains its least upper bound at some point y. It is clear that $y \in F$, because $\psi(x)$ is linear on the complementary intervals of F. It follows that

$$\max |\psi(x)| = \max |\varphi(x)|.$$

This completes the proof.

THEOREM 2 (E. BOREL). *Let a measurable function $f(x)$ be defined and be finite almost everywhere on the closed interval $[a, b]$. For all numbers $\sigma > 0$ and $\varepsilon > 0$, there exists a function $\psi(x)$ continuous on $[a, b]$ for which*

$$mE(|f-\psi| \geqslant \sigma) < \varepsilon.$$

If $|f(x)| \leqslant K$, we can choose $\psi(x)$ so that

$$|\psi(x)| \leqslant K.$$

Proof. Suppose first that

$$|f(x)| \leqslant K,$$

i.e., that the function $f(x)$ is bounded. Fixing on arbitrary numbers $\sigma > 0$ and $\varepsilon > 0$, we choose a natural number m so large that

$$\frac{K}{m} < \sigma,$$

and construct the sets

$$E_i = E\left(\frac{i-1}{m} K \leqslant f < \frac{i}{m} K\right) \qquad (i = 1-m, \; 2-m, \; \ldots, \; m-1)$$

$$E_m = E\left(\frac{m-1}{m} K \leqslant f \leqslant K\right).$$

These sets are measurable, are pairwise disjoint, and have the obvious property that

$$[a, b] = \sum_{i=1-m}^{m} E_i.$$

For every i, we will choose a closed set $F_i \subset E_i$ such that

$$mF_i > mE_i - \frac{\varepsilon}{2m}$$

and set

$$F = \sum_{i=1-m}^{m} F_i.$$

4. The Structure of Measurable Functions

It is clear that $[a, b] - F = \Sigma (E_i - F_i)$, and therefore
$$m[a, b] - mF < \varepsilon.$$

We now define a function $\varphi(x)$ on the set F by setting
$$\varphi(x) = \frac{i}{m} K \text{ for } x \in F_i \qquad (i = 1 - m, \ldots, m).$$

By Lemma 1, this function is continuous on the set F; we also have $|\varphi(x)| \leq K$; finally, for $x \in F$,
$$|f(x) - \varphi(x)| < \sigma.$$

It remains to apply Lemma 2. Doing this, we find a continuous function $\psi(x)$ which coincides with the function $\varphi(x)$ on the set F, and which has the property that
$$|\psi(x)| \leq K.$$

Inasmuch as
$$E(|f - \psi| \geq \sigma) \subset [a, b] - F,$$

it is clear the function $\psi(x)$ satisfies the requirements of the theorem. The theorem is thus proved, for a bounded function $f(x)$.

Suppose now that $f(x)$ is unbounded. Then, using Theorem 1, we can find a bounded function $g(x)$ such that
$$mE(f \neq g) < \frac{\varepsilon}{2}.$$

Applying the present theorem to the bounded function $g(x)$, we can find a continuous function $\psi(x)$ such that
$$mE(|g - \psi| \geq \sigma) < \frac{\varepsilon}{2}.$$

It is easy to see that
$$E(|f - \psi| \geq \sigma) \subset E(f \neq g) + E(|g - \psi| \geq \sigma),$$

so that the function $\psi(x)$ satisfies all requirements.

COROLLARY. *For every measurable function $f(x)$ which is defined and is finite almost everywhere on the closed interval $[a, b]$, there exists a sequence of continuous functions $\psi_n(x)$ converging in measure to the function $f(x)$.*

Proof. Taking two decreasing sequences of positive numbers which tend to zero,
$$\sigma_1 > \sigma_2 > \sigma_3 > \ldots, \qquad \sigma_n \to 0,$$
$$\varepsilon_1 > \varepsilon_2 > \varepsilon_3 > \ldots, \qquad \varepsilon_n \to 0,$$

we construct for every n a continuous function $\psi_n(x)$ such that
$$mE(|f - \psi_n| \geq \sigma_n) < \varepsilon_n.$$

It is easy to see that
$$\psi_n(x) \Rightarrow f(x).$$

In fact, if σ is any positive number, there is a natural number n such that $\sigma_n < \sigma$ for $n \geq n_0$, and for these n's,
$$E(|f - \psi_n| \geq \sigma) \subset E(|f - \psi_n| \geq \sigma_n),$$

Our assertion follows at once.

Applying F. Riesz's Theorem (Theorem 4, § 3) to the sequence $\{\psi_n(x)\}$, we produce a sequence of continuous functions, $\{\psi_{n_k}(x)\}$, which converges to the function $f(x)$ almost everywhere.

In other words, we have established the following assertion.

THEOREM 3 (M. FRÉCHET). *For every measurable function $f(x)$ which is defined and is finite almost everywhere on the closed interval $[a, b]$, there exists a sequence of continuous functions converging to $f(x)$ almost everywhere.*

With the aid of this theorem we can easily establish a most remarkable and very important theorem due to Luzin.

THEOREM 4 (N. N. LUZIN). *Let $f(x)$ be a measurable function defined on $[a, b]$ which is finite almost everywhere. For every $\delta > 0$, there exists a continuous function $\varphi(x)$ such that*

$$mE(f \neq \varphi) < \delta.$$

If $|f(x)| \leq K$, then

$$|\varphi(x)| \leq K$$

also.

Proof. Let

$$\varphi_1(x),\ \varphi_2(x),\ \varphi_3(x),\ldots$$

be a sequence of continuous functions having the property described in Fréchet's theorem (Theorem 3). Using Egorov's Theorem, we can find a set E_δ such that

$$mE_\delta > b - a - \frac{\delta}{2},$$

and such that

$$\varphi_n(x) \to f(x)$$

uniformly with respect to x on the set E_δ. A basic theorem of analysis states that the function $f(x)$ is continuous on the set E_δ[7] (in the sense of the definition given at the beginning of the present section. It does not follow that the function $f(x)$, considered on $[a, b]$, is continuous at every point of the set E_δ. It is necessary by all means to avoid considering $f(x)$ outside the set E_δ). We find a closed subset F of the set E_δ such that

$$mF > mE_\delta - \frac{\delta}{2}.$$

If the function $f(x)$ is considered only on the set F, it is obviously continuous on this set. Applying Lemma 2, we produce a continuous function $\varphi(x)$, defined on $[a, b]$ and coinciding with $f(x)$ on the set F. Thus

$$E(f \neq \varphi) \subset [a, b] - F,$$

and the measure of this set is $< \delta$, so that $\varphi(x)$ is a function meeting all requirements of the present theorem. If $|f(x)| \leq K$, this inequality is valid also for x in F, and then, by virtue of Lemma 2, we have

$$|\varphi(x)| \leq K.$$

This completes the proof.

---

[7] In elementary analysis courses, the theorem is ordinarily proved for the case of functions defined on a closed interval, but the proof is valid for functions defined on an arbitrary set.

Luzin's theorem can also be formulated as follows. A measurable and almost everywhere finite function becomes continuous if we remove a set of arbitrarily small measure. Some authors[8] take this important property for the very definition of the concept of measurable function. It is not difficult to establish the equivalence of both definitions; the second is less formal and shows at once that the concept of measurable function is closely connected with the concept of continuous function.

§ 5. TWO THEOREMS OF WEIERSTRASS

In the preceding section, we proved a series of theorems on the approximation of measurable functions by means of continuous functions. We can go even further, replacing continuous functions by polynomials.

For this purpose, a theorem of Weierstrass, which incidentally is important of itself, will be necessary. We give S. N. Bernstein's proof.

Lemma 1. For every x,

$$\sum_{k=0}^{n} C_n^k x^k (1-x)^{n-k} = 1. \qquad (1)$$

$\left(\text{Here } C_n^k \text{ is the binomial coefficient } \dfrac{n!}{k!\,(n-k)!}\right).$

In fact, setting $a = x$ and $b = 1 - x$ in Newton's binomial formula,

$$(a+b)^n = \sum_{k=0}^{n} C_n^k a^k b^{n-k}$$

we obtain formula (1).

Lemma 2. For all real x,

$$\sum_{k=0}^{n} C_n^k (k-nx)^2 x^k (1-x)^{n-k} \leqslant \frac{n}{4}. \qquad (2)$$

Proof. Differentiate the identity

$$\sum_{k=0}^{n} C_n^k z^k = (1+z)^n \qquad (3)$$

with respect to z, and multiply the result by z:

$$\sum_{k=0}^{n} k C_n^k z^k = nz(1+z)^{n-1}. \qquad (4)$$

Differentiating (4) and multiplying the result by z, we obtain

$$\sum_{k=0}^{n} k^2 C_n^k z^k = nz(1+nz)(1+z)^{n-2}. \qquad (5)$$

---

[8] See, for example, P. S. Aleksandrov and A. N. Kolmogorov, "Introduction to the Theory of Functions of a Real Variable."

Set
$$z = \frac{x}{1-x}$$
in (3), (4) and (5), and multiply the identities thus obtained by $(1-x)^n$. This gives us

$$\sum_{k=0}^{n} C_n^k x^k (1-x)^{n-k} = 1, \tag{6}$$

$$\sum_{k=0}^{n} k C_n^k x^k (1-x)^{n-k} = nx, \tag{7}$$

and

$$\sum_{k=0}^{n} k^2 C_n^k x^k (1-x)^{n-k} = nx(1-x+nx). \tag{8}$$

Multiply (6) by $n^2 x^2$, (7) by $-2nx$, (8) by 1 and add together the resulting equalities. We obtain

$$\sum_{k=0}^{n} (k-nx)^2 C_n^k x^k (1-x)^{n-k} = nx(1-x).$$

To prove the lemma, it is sufficient to note that for all real x,
$$x(1-x) \leqslant \frac{1}{4}.\ ^9$$

DEFINITION 1. Let $f(x)$ be a finite function defined on the closed interval $[0, 1]$. The polynomial
$$B_n(x) = \sum_{k=0}^{n} f\left(\frac{k}{n}\right) C_n^k x^k (1-x)^{n-k} \tag{9}$$
is called the *Bernstein polynomial* of degree n for the function $f(x)$.

THEOREM 1 (S. N. BERNSTEIN). *If the function $f(x)$ is continuous on the segment $[0, 1]$, then*
$$B_n(x) \to f(x) \tag{10}$$
uniformly with respect to x, as $n \to \infty$.

Proof. Let
$$M = \max |f(x)|.$$
Let ε be any number > 0. Since f is continuous on the bounded closed interval $[0, 1]$, there exists a $\delta > 0$ such that
$$|f(x'') - f(x')| < \varepsilon,$$
whenever $|x'' - x'| < \delta$. Now select an arbitrary $x \in [0, 1]$.

It follows from (1) that
$$f(x) = \sum_{k=0}^{n} f(x) C_n^k x^k (1-x)^{n-k},$$
and therefore
$$|B_n(x) - f(x)| \leqslant \sum_{k=0}^{n} \left| f\left(\frac{k}{n}\right) - f(x) \right| C_n^k x^k (1-x)^{n-k}. \tag{11}$$

---

[9] In fact, $4x^2 - 4x + 1 = (2x-1)^2 \geqslant 0$. Hence $1 \geqslant 4x(1-x)$.

5. Two Theorems of Weierstrass

Divide all the numbers $k = 0, 1, 2, \ldots, n$ into two categories A and B according to the rules

$$k \in A \text{ if } |\tfrac{k}{n} - x| < \delta,$$
$$k \in B \text{ if } |\tfrac{k}{n} - x| \geq \delta.$$

If $k \in A$, then

$$\left|f\left(\tfrac{k}{n}\right) - f(x)\right| < \varepsilon,$$

and, on the basis of Lemma 1,

$$\sum_A \left|f\left(\tfrac{k}{n}\right) - f(x)\right| C_n^k x^k (1-x)^{n-k} \leq \varepsilon \sum_A C_n^k x^k (1-x)^{n-k} \leq$$

$$\leq \varepsilon \sum_{k=0}^n C_n^k x^k (1-x)^{n-k} = \varepsilon. \quad (12)$$

If $k \in B$, then

$$\frac{(k-nx)^2}{n^2\delta^2} \geq 1,$$

and, by Lemma 2,

$$\sum_B \left|f\left(\tfrac{k}{n}\right) - f(x)\right| \cdot C_n^k x^k (1-x)^{n-k} \leq \frac{2M}{n^2\delta^2} \sum_B (k-nx)^2 C_n^k x^k (1-x)^{n-k} \leq$$

$$\leq \frac{2M}{n^2\delta^2} \sum_{k=0}^n (k-nx)^2 C_n^k x^k (1-x)^{n-k} < \frac{M}{2n\delta^2}. \quad (13)$$

Assembling (11), (12) and (13), we find that for all $x \in [0, 1]$.

$$|B_n(x) - f(x)| < \varepsilon + \frac{M}{2n\delta^2}.$$

Consequently, if

$$n > \frac{M}{2\varepsilon\delta^2}$$

we have

$$|B_n(x) - f(x)| < 2\varepsilon,$$

and this completes the proof.

THEOREM 2 (WEIERSTRASS). *Let $f(x)$ be a continuous function defined on the closed interval $[a, b]$. For every $\varepsilon > 0$, there exists a polynominal $P(x)$ such that*

$$|f(x) - P(x)| < \varepsilon$$

for all $x \in [a, b]$.

Proof. If $[a, b] = [0, 1]$, the theorem follows directly from Bernstein's theorem. Suppose that $[a, b] \neq [0, 1]$. Consider the following function of the argument y:

$$f[a + y(b - a)].$$

This function is defined and continuous on the segment $[0, 1]$. We can find a polynomial $Q(y)$ such that for all $y \in [0, 1]$,

$$|f[a + y(b - a)] - Q(y)| < \varepsilon.$$

If $x \in [a, b]$, then $\frac{x-a}{b-a} \in [0, 1]$ and hence

$$\left|f(x) - Q\left(\frac{x-a}{b-a}\right)\right| < \varepsilon.$$

The polynomial $P(x) = Q\left(\frac{x-a}{b-a}\right)$ therefore satisfies our requirements.

With the aid of Weierstrass's theorem, we can reformulate the theorems of Borel and Fréchet (but not the theorem of Luzin). For example, Fréchet's theorem can be stated in the following form.

THEOREM 3 (M. FRÉCHET). *For every measurable function $f(x)$ which is defined on $[a, b]$ and is finite almost everywhere, there exists a sequence of polynomials converging to $f(x)$ almost everywhere.*

Proof. Let $\{\varphi_n(x)\}$ be a sequence of continuous functions converging almost everywhere to $f(x)$. If $P_n(x)$ is a polynomial such that

$$|P_n(x) - \varphi_n(x)| < \frac{1}{n},$$

for all $x \in [a, b]$, the sequence $\{P_n(x)\}$ converges to $f(x)$ at every point x for which

$$\varphi_n(x) \to f(x).$$

This proves the theorem.

It is suggested the reader recast Borel's theorem in like manner (preserving the inequalities $|P_n(x)| \leq \sup |f(x)|$).

Weierstrass's theorem on uniform approximation of continuous functions by polynomials (Theorem 2) has a close analogue, also proved by Weierstrass, on the approximation of continuous periodic functions by trigonometric polynomials. We proceed to prove this theorem.

DEFINITION 2. Any function of the form

$$T(x) = A + \sum_{k=1}^{n} (a_k \cos kx + b_k \sin kx)$$

is called a *trigonometric polynomial of the n-th order.*

If $b_1 = \ldots = b_n = 0$, the trigonometric polynomial $T(x)$ is said to be *even*.

LEMMA 3. a) *The function $\cos^k x$ can be written as an even trigonometric polynomial.*

b) *If $T(x)$ is a trigonometric polynominal, then $T(x) \sin x$ is also a trigonometric polynomial.*

c) *If $T(x)$ is a trigonometric polynominal, then $T(x + a)$ also is a trigonometric polynomial.*

We leave the proof to the reader.

LEMMA 4. *If the function $f(x)$ is defined and continuous on the closed interval $[0, \pi]$, then for every $\varepsilon > 0$, there exists an even trigonometric polynomial $T(x)$ such that*

$$|f(x) - T(x)| < \varepsilon$$

for all $x \in [0, \pi]$.

Proof. Consider the function

$$f(\arccos y).$$

This function is defined and continuous on the segment $[-1, +1]$. Therefore, we

5. Two Theorems of Weierstrass

can produce a polynomial $\sum_{k=0}^{n} a_k y^k$ such that for all $y \in [-1, +1]$,

$$\left| f(\arccos y) - \sum_{k=0}^{n} a_k y^k \right| < \varepsilon.$$

Now let $x \in [0, \pi]$. Then $\cos x \in [-1, +1]$ and, hence,

$$\left| f(x) - \sum_{k=0}^{\infty} a_k \cos^k x \right| < \varepsilon.$$

It remains to note that in view of Lemma 3, the function $\sum_{k=0}^{n} a_k \cos^k x$ is an even trigonometric polynomial.

COROLLARY. *If the even function $f(x)$ is defined for all x, has period 2π, and is continuous everywhere, then for every $\varepsilon > 0$, we can find a trigonometric polynomial such that*

$$|f(x) - T(x)| < \varepsilon$$

for all real x.

In fact, on the segment $[0, \pi]$, this inequality can be satisfied by an *even* trigonometric polynomial $T(x)$ so that the inequality is automatically satisfied for $x \in [-\pi, 0]$, and, then, since the difference $f(x) - T(x)$ has period 2π, the inequality is satisfied everywhere.

THEOREM 4 (WEIERSTRASS). *Let $f(x)$ be a continuous periodic function of period 2π. For every $\varepsilon > 0$, there exists a trigonometric polynomial $T(x)$ such that*

$$|f(x) - T(x)| < \varepsilon$$

for all x.

Proof. By the corollary to Lemma 4, for the *even* functions

$$f(x) + f(-x), \quad [f(x) - f(-x)] \sin x$$

there exist trigonometric polynomials $T_1(x)$ and $T_2(x)$ such that

$$f(x) + f(-x) = T_1(x) + \alpha_1(x), \quad [f(x) - f(-x)] \sin x = T_2(x) + \alpha_2(x),$$

where

$$|\alpha_1(x)| < \frac{\varepsilon}{2}, \quad |\alpha_2(x)| < \frac{\varepsilon}{2}.$$

Multiplying the first of these equalities by $\sin^2 x$ and the second by $\sin x$, then adding them and dividing by 2, we obtain

$$f(x) \sin^2 x = T_3(x) + \beta(x) \qquad \left(|\beta(x)| < \frac{\varepsilon}{2}\right),$$

where $T_3(x)$ is again some trigonometric polynomial. Here, the function $f(x)$ is an arbitrary continuous periodic function. Therefore, a similar equality is true of the function $f(x - \frac{\pi}{2})$:

$$f\left(x - \frac{\pi}{2}\right) \sin^2 x = T_4(x) + \gamma(x) \qquad \left(|\gamma(x)| < \frac{\varepsilon}{2}\right).$$

Replacing x by $x + \frac{\pi}{2}$, we obtain

$$f(x) \cos^2 x = T_5(x) + \delta(x) \qquad \left(|\delta(x)| < \frac{\varepsilon}{2}\right),$$

whence

$$f(x) = T_3(x) + T_5(x) + \beta(x) + \delta(x),$$

i.e., the polynomial $T_3(x) + T_5(x)$ satisfies all of our requirements.

We will not now study the connections between this theorem and the theory of measurable functions, but it will be very useful in the sequel.

§ 6. EDITOR'S APPENDIX TO CHAPTER IV

In this section, we discuss the basic properties of measurable functions defined on sets which are not necessarily bounded. We use the notion of measurability which was introduced in § 9, Chapter III, so that a large class of unbounded sets are measurable. We use the same definition of measurable functions as in Definition 1, § 1. The only difference is that the sets admitted as measurable may now be unbounded, always of course under the conditions imposed by Definition 1, § 9, Chapter III. Theorems 1, 2 and 3 of § 1 of the present Chapter are immediately seen to be true for this new class of measurable functions. Definitions 2 and 3 of § 1 of the present chapter are repeated in the present context without any change. Theorems 4 and 5 of § 1 of the present chapter are also true in the present context.

We give an extended definition of step function. A function defined on the entire line Z is said to be a step function if there exists a doubly infinite sequence of points $\{c_n\}_{n=-\infty}^{+\infty}$ such that

$$\ldots < c_{-n} < c_{-n+1} < \ldots < c_{-1} < c_0 < c_1 < \ldots < c_n < c_{n+1} < \ldots,$$

$\lim_{n \to \infty} c_n = +\infty$, $\lim_{n \to -\infty} c_n = -\infty$, and $f(x)$ is constant on every open interval (c_k, c_{k+1}). It is then easy to see that every step function is measurable.

Theorems 6 and 7, § 1 of the present chapter are all true under our extended definition without change. Theorem 8, § 1, remains true if $[A, B]$ is replaced by any interval (a, b) with $a \geqslant -\infty$ and $b \leqslant +\infty$ or any interval $[a, b]$ with $a > -\infty$ and $b \leqslant +\infty$; similarly for intervals $(a, b]$.

We alter Definition 4 § 1, as follows.

DEFINITION 1. *Let M be any subset of the subset E of the line Z. Then the function $\theta_M(x)$, which is equal to 1 for $x \in M$ and 0 for $x \in E - M$, and which is defined only on E, is said to be the characteristic function of the set M (with respect to the set E).*

THEOREM 1. *Let E be a measurable set and let M be a subset of E. The set M and its characteristic function $\theta_M(x)$ are both measurable or both non-measurable.*

All of the theorems of § 2 of the present chapter are true under our extended definition of measurable functions. We leave the details to the reader.

The situation alters radically as we go to the theorems of § 3 of the present chapter. Lebesgue's theorem (Theorem 1, § 3) fails to be true for functions defined on sets of infinite measure. For example, let E be the entire line Z, and let $f_n(x) = 0$ for $x \leqslant n$, $f_n(x) = 1$ for $x > n$ ($n = 1, 2, 3, \ldots$). Then $\lim_{n \to \infty} f_n(x) = 0$ for all $x \in Z$, but $m(E(f_n - 0 \geqslant 1) = +\infty$ for all n. Using the notation of Theorem 1, § 3, we can state the following sufficient condition for this theorem to hold even for functions defined on a set of infinite measure. If $mR_n(\sigma) < +\infty$ for some n, then this theorem holds. One has only to repeat the argument in the text, together with Theorem 9, § 9, Chapter III.

Convergence in measure is defined for functions defined on a set of infinite measure just as for functions defined on a set of finite measure. With this definition, Theorems 2 and 3, § 3, are obviously true. F. Riesz's theorem (Theorem 4, § 3) is also true in this more general context, and the proof given in the text can be repeated word for word.

Egorov's theorem (Theorem 5, § 3) is no longer true if the set E on which the functions are defined has infinite measure. The example given above in connection with Lebesgue's theorem shows that Egorov's theorem cannot be true for sets of infinite measure. However, an analogue of Egorov's theorem, due to Luzin, can be stated for arbitrary measurable subsets of the line Z.

THEOREM 2. (LUZIN). *Let E be any measurable subset of the real line Z. Let $f_1(x), f_2(x), f_3(x), \ldots$ be a sequence of measurable functions defined on E and finite almost everywhere, which converges almost everywhere on E to a function $f(x)$ which is finite almost everywhere. Then E can be written as the sum*
$$A + B_1 + B_2 + B_3 + \ldots,$$
where $mA = 0$ and $f_n(x)$ converges uniformly to $f(x)$ on each of the sets B_k ($K = 1, 2, 3, \ldots$).

Proof. It is obvious that E is the sum of a finite or denumerable number of sets of finite measure, since $E[-n, n]$ has measure $\leq 2n$. We can therefore consider only the case in which E has finite measure, since we can add up the sets obtained by applying the theorem to each of the (pairwise disjoint) sets $E_n = E([-n, -n+1) + (n-1, n])$. Then, using Theorem 5, § 3, we find a sequence of measurable subsets $B_1, B_2, B_3, \ldots$ of E such that $m(E - \sum_{k=1}^{n} B_k) \leq \frac{1}{n}$ for $n = 1, 2, 3, \ldots$, and such that $f_n(x)$ converges uniformly to $f(x)$ on each of the sets B_k. We need only set $A = E - \sum_{k=1}^{\infty} B_k$ to obtain the present theorem.

The results of § 4 of the present chapter are not true in their entirety for functions defined on sets of infinite measure. For example, Theorem 1, § 4 fails completely in this wider context. Let $f(x) = x$ for all real numbers x. Then $f(x)$ is measurable, since $E(f > a) = (a, +\infty)$, and this open interval is an open set, hence measurable. However, if $g(x)$ is any bounded function, say $|g(x)| \leq K$, then the set $E(f \neq g)$ contains at least the two intervals $(-\infty, -K)$ and $(K, +\infty)$. If $g(x)$ is measurable, the set $E(f \neq g)$ must therefore have infinite measure.

In spite of the above example, the remaining theorems of § 4 are all true for arbitrary measurable functions. We proceed to state and prove them.

THEOREM 3. *Let $f(x)$ be defined on the entire real line and be measurable and finite almost everywhere. Then, for every $\sigma > 0$ and $\varepsilon > 0$, there exists a continuous function $\theta(x)$ defined on the entire line such that*
$$mZ(|f - \theta| \geq \sigma) < \varepsilon.$$

Proof. For every integer n, $n = 0, \pm 1, \pm 2, \pm 3, \ldots$, let F_n be the closed interval $\left[n + \frac{\varepsilon}{2^{|n|+4}}, n+1 - \frac{\varepsilon}{2^{|n|+4}}\right]$ for $n \neq 0$, and let $F_0 = \left[\frac{\varepsilon}{8}, \frac{7\varepsilon}{8}\right]$. It is easy to show that $m(Z - \sum_{n=-}^{\infty} F_n) = \frac{\varepsilon}{2}$. If we consider the function $f(x)$ only on the set F_n, we can apply Borel's theorem (Theorem 2, § 4) to assert the existence of a continuous function $\theta_n(x)$ on F_n such that
$$mF_n(|\theta - f| \geq \sigma) < \delta_n,$$
where $\delta_0 = \frac{\varepsilon}{4}$ and $\delta_n = \frac{\varepsilon}{2^{|n|+3}}$ for $n = \pm 1, \pm 2, \ldots$.
A simple generalization of Lemma 2, § 4, may now be applied to assert that there exists a function $\theta(x)$ defined on the entire line such that $\theta(x)$ is continuous and such that $\theta(x) = \theta_n(x)$ for all $x \in F_n$ ($n = 0, \pm 1, \pm 2, \ldots$). Upon adding up the sets where $|\theta - f| \geq \sigma$, we see at once that their total measure is less than ε, and this proves the theorem.

THEOREM 4. *Let $f(x)$ be as in the preceding theorem. Then there exists a sequence of continuous functions defined on the entire line which converges almost everywhere to $f(x)$.*

Proof. From Theorem 3, we infer that there is a sequence of continuous functions converging in measure to $f(x)$. By F. Riesz's theorem, which is true for functions defined on sets of infinite measure, we select a subsequence converging almost everywhere to $f(x)$.

THEOREM 5. (LUZIN). *Let $f(x)$ be as in Theorem 3 above. Let ε be any positive number. Then there exists a continuous function θ on Z such that*

$$mZ(f \neq \theta) < \varepsilon.$$

Proof. We argue from the form of Luzin's theorem given in Theorem 4, § 4, by considering the closed intervals F_n used in the proof of Theorem 3 above. We then find continuous functions θ_n on F_n such that

$$mF_n(f \neq \theta_n) < \delta_n,$$

again just as in the proof of Theorem 3 above. The rest of the proof is virtually obvious, and we leave the details to the reader.

It is clear that not all of the results of §5 are true for functions defined on unbounded sets. Weierstrass's theorem on polynomial approximation fails, of course, for continuous functions on the entire line, as the function e^x, for example, shows. For any polynomial $P(x)$, we have $\lim_{x \to \infty} [e^x - P(x)] = +\infty$. However, Theorem 3, § 5 (Fréchet's theorem) is true for functions defined on the entire line. Let $f(x)$ be a function defined on the entire line, measurable, and finite almost everywhere. Then, by Theorem 4 above, there exists a sequence $\theta_n(x)$ of continuous functions on the line converging almost everywhere to $f(x)$. By Weierstrass's theorem, there exists for every natural number n a polynomial $P_n(x)$ such that $|\theta_n(x) - P_n(x)| < \frac{1}{n}$ for all $x \in [-n, n]$. Now let x be any point for which $\theta_n(x) \to f(x)$. We have $m - 1 < |x| \leq m$ for some natural number m. Then, for $n \geq m$, we have $|P_n(x) - f(x)| \leq |P_n(x) - \theta_n(x)| + |\theta_n(x) - f(x)| < \frac{1}{n} + |\theta_n(x) - f(x)|$. This shows that $p_n(x) \to f(x)$ for all x such that $\theta_n(x) \to f(x)$, i.e., for almost all x.

Exercises for Chapter IV

1. Prove that if $f_n(x) \Longrightarrow f(x)$, $g(x) \Longrightarrow g(x)$, then $f_n(x) + g_n(x) \Longrightarrow f(x) + g(x)$.

2. Prove that if $f_n(x) \Longrightarrow f(x)$ and $g(x)$ is measurable and almost everywhere finite, then $f_n(x) g_n(x) \Longrightarrow f(x) g(x)$.

3. Generalize Egorov's theorem to the case of a sequence of functions which has limit $+\infty$ at every point of the set E.

4. Prove that there exists a series of polynomials $p_1(x) + p_2(x) + p_3(x) + \ldots$, possessing the following property. For any function $f(x)$ continuous on an arbitrary closed interval $[a, b]$, we can group the terms of the series (without changing their order) so that the series $\sum_{k=1}^{\infty} [p_{n_k+1}(x) + \ldots + p_{n_{k+1}}(x)]$ converges to $f(x)$ uniformly on $[a, b]$.

5. Prove the following assertion. A necessary and sufficient condition that the sequence of measurable and almost everywhere finite functions $f_1(x), f_2(x), \ldots$ converge in measure, is that to every pair of numbers $\sigma > 0$ and $\varepsilon > 0$ there correspond an N such that for $n > N$ and $m > N$,

$$mE(|f_n - f_m| \geq \sigma) < \varepsilon$$

(F. Riesz).

6. In Borel's and Fréchet's theorems we can use trigonometric polynomials (if the closed interval is $[-\pi, +\pi]$), instead of continuous functions. Establish this fact.

7. Prove that the least upper bound of a finite or denumerable set of measurable functions is a measurable function.

8. Prove the following. If for arbitrary fixed n,
$$f_k^{(n)}(x) \Longrightarrow f^{(n)}(x),$$
as $k \to \infty$, and if
$$f^{(n)}(x) \Longrightarrow f(x),$$
as $n \to \infty$, then we can extract a sequence from the set $\{f_k^{(n)}(x)\}$ which converges in measure to $f(x)$.

9. Show that the preceding result is not true if we substitute convergence in the usual sense for convergence in measure.

10. Let $f(t)$ be a measurable and almost everywhere finite function, defined on the closed interval $E = [a, b]$. Prove the existence of a *decreasing* function $g(t)$, defined on $[a, b]$, which satisfies the relation $mE(g > x) = mE(f > x)$ for all real x.

11. Let $f(t)$ be a measurable and almost everywhere finite function defined on the closed interval $[a, b]$. Prove the existence and uniqueness of a number h such that
$$mE(f \geqslant h) \geqslant \frac{b-a}{2} \quad \text{and} \quad mE(f \geqslant H) < \frac{b-a}{2} \text{ if } H > h.$$

(L. V. Kantorovič).

CHAPTER V

THE LEBESGUE INTEGRAL OF A BOUNDED FUNCTION

§ 1. DEFINITION OF THE LEBESGUE INTEGRAL

The classical definition of an integral, given first by Cauchy and developed by Riemann, runs as follows. We consider a finite function $f(x)$ defined on the closed interval $[a, b]$; this interval is divided into subintervals by the points

$$x_0 = a < x_1 < x_2 < \ldots < x_n = b.$$

In each subinterval $[x_k, x_{k+1}]$, we select a point ξ_k and form the Riemann sum

$$\sigma = \sum_{k=0}^{n-1} f(\xi_k)(x_{k+1} - x_k).$$

If the sum σ tends to a finite limit I as the number $\lambda = \max(x_{k+1} - x_k)$ approaches zero, independent of the method of subdividing $[a, b]$ and of the choice of the points ξ_n, this limit I is called the Riemann integral of the function $f(x)$ and is designated by the symbol

$$\int_a^b f(x)\, dx.$$

When we wish to emphasize that we are dealing with the Riemann integral, we write

$$(R) \int_a^b f(x)\, dx.$$

Functions for which the Riemann integral exists are said to be integrable in the Riemann sense, or, more briefly, integrable (R). In order that the function $f(x)$ be integrable (R), it is necessary that it be bounded.

Cauchy himself proved that every continuous function is integrable (R). There also exist discontinuous functions which are integrable (R). For example, all monotonic functions are integrable (R). However, it is easy to construct a bounded function which is not integrable (R). We consider, for example, *Dirichlet's function* $\psi(x)$, which is defined on the closed interval $[0, 1]$ in the following way.

$$\psi(x) = \begin{cases} 1 & \text{if } x \text{ is rational,} \\ 0 & \text{if } x \text{ is irrational.} \end{cases}$$

It is easy to see that this function is not integrable (R), because the sum σ is 0 if all the points ξ_k are irrational and $\sigma = 1$ if all ξ_k are rational. It thus appears that Riemann's definition of an integral suffers from essential shortcomings—even very simple functions turn out to be non-integrable.

It is not difficult to penetrate to the fundamental cause of this phenomenon. The situation is actually as follows. In defining the Riemann sum σ, we subdivide the segment $[a, b]$ into small subintervals $[x_0, x_1], [x_1, x_2], \ldots, [x_{n-1}, x_n]$ (designate these

1. Definition of the Lebesgue Integral

subintervals by $e_0, e_1, \ldots, e_{n-1}$); we take a point ξ_k in each subinterval e_k and, upon writing the sum

$$\sigma = \sum_{k=0}^{n-1} f(\xi_k) \, me_k,$$

we require that it have a limit independent of the choice of the points ξ_k in the sets e_k. In other words, any point x of the set e_k can be taken for ξ_k, and varying this point must have no appreciable influence on the value of the sum σ. This is possible only if variation of the points ξ_k produces little change in the magnitude of $f(\xi_k)$. But what connects the different points x of the set e_k? They are connected by the fact they are near one another, since e_k is a small segment $[x_k, x_{k+1}]$. If the function $f(x)$ is continuous, a sufficient proximity of different values of x implies arbitrary proximity of the corresponding values of the function, and we are right to expect that changing the points ξ_k within the limits of the set e_k has little influence on the magnitude of the sum σ. For a discontinuous function, on the other hand, this need not be the case at all.

In other words, we can say that the sets e_k are chosen so that the value $f(\xi_k)$ can be considered a normal representation of all values of the function on e_k only for continuous functions. It thus appears that the definition of the Riemann integral can be considered as completely justifiable only for continuous functions. For all other functions, it appears to be highly artificial. We shall see below that for integrability (R), it is *necessary* that the function considered be not too discontinuous.

Seeking to extend the concept of the integral to a wider class of functions, Lebesgue proposed another process of integration in which the points x are combined into the sets e_k not by the accidental rule of proximity on the axis Ox, but rather by the rule of sufficient proximity of the corresponding values of the function. For this purpose, rather than subdividing the closed interval $[a, b]$ lying on the axis of abscissas, Lebesgue subdivides a closed interval $[A, B]$ lying on the axis of ordinates and including all values of the function $f(x)$:

$$A = y_0 < y_1 < \ldots < y_n = B.$$

If we construct the sets e_k by the definition

$$e_k = E(y_k \leq f < y_{k+1}),$$

it is clear that, in fact, to all points $x \in e_k$, there correspond values of the function lying close together, although, in contrast to the Riemann process, the points x themselves can be very far from one another.

The number y_k, for example, can serve as a representative of the values of the function on the set e_k, so that it is natural to set down the sum

$$\sum_{k=0}^{n-1} y_k me_k.$$

as the basis of the concept of integral.

We turn now to a precise discussion of the problem. Let a measurable bounded function $f(x)$ be defined on the measurable set E, such that

$$A < f(x) < B. \tag{1}$$

Subdivide the closed interval $[A, B]$ by means of the points

$$y_0 = A < y_1 < y_2 < \ldots < y_n = B$$

and define the set

$$e_k = E(y_k \leq f < y_{k+1}) \qquad (k = 0, 1, \ldots, n-1)$$

for every half-open interval $[y_k, y_{k+1})$. It is easy to verify the following four properties of the sets e_k.

1) The sets e_k are pairwise disjoint: $e_k e_{k'} = 0$ $(k \neq k')$.
2) The sets e_k are measurable.
3) $E = \sum_{k=0}^{n-1} e_k.$
4) $mE = \sum_{k=0}^{n-1} me_k.$

We now introduce the lower and upper Lebesgue sums s and S:

If we set
$$s = \sum_{k=0}^{n-1} y_k me_k, \quad S = \sum_{k=0}^{n-1} y_{k+1} me_k.$$

$$\lambda = \max(y_{k+1} - y_k),$$

we have

$$0 \leqslant S - s \leqslant \lambda mE. \tag{2}$$

The following lemma gives us the fundamental property of Lebesgue sums.

LEMMA. *Let the Lebesgue sums s_0 and S_0 correspond to some method of subdividing the closed interval $[A, B]$. If we add a new point of subdivision, $\bar{y}$, and again compute the Lebesgue sums s and S, then we have*

$$s_0 \leqslant s, \quad S \leqslant S_0.$$

In other words, the lower sum does not decrease and the upper sum does not increase when new subdivision points are added.

Proof. Suppose that

$$y_i < \bar{y} < y_{i+1}. \tag{3}$$

Then for $k \neq i$, the half-open intervals $[y_{k1}\, y_{k+1})$ and the corresponding sets e_k enter into the new method of subdivision as well as the old. In the new method of subdivision, the half-open interval $[y_i, y_{i+1})$ is replaced by the two half-open intervals

$$[y_i, \bar{y}), \quad [\bar{y}, y_{i+1}).$$

The set e_i is divided into two sets

$$e_i' = E(y_i \leqslant f < \bar{y}), \quad e_i'' = E(\bar{y} \leqslant f < y_{i+1}).$$

It is obvious that

$$e_i = e_i' + e_i'', \quad e_i' e_i'' = 0,$$

so that

$$me_i = me_i' + me_i''. \tag{4}$$

It is now clear that the sum s is obtained from the sum s_0 by replacing the term

$$y_i me_i$$

by the two terms

$$y_i me_i' + \bar{y} me_i'',$$

1. Definition of the Lebesgue Integral

from this observation, together with (3) and (4), it follows that

$$s \geqslant s_0.$$

The reasoning is analogous for the upper sums.

COROLLARY. *None of the lower sums s is greater than any of the upper sums S.*

Proof. Consider any two methods, I and II, of subdivision of the segment $[A, B]$. Let the lower sums s_1 and s_2 and upper sums S_1 and S_2 correspond, respectively, to these two methods. Set up a third method, method III, of subdivision of $[A, B]$ in which the points of division are the points of division of both methods I and II. If the sums s_3 and S_3 correspond to method III, then, by the lemma,

$$s_1 \leqslant s_3, \quad S_3 \leqslant S_2,$$

Since

$$s_3 \leqslant S_3,$$

it is clear that

$$s_1 \leqslant S_2,$$

and this is what we wished to prove.

Choose any fixed upper sum S_0. Since for every lower sum s, we have

$$s \leqslant S_0,$$

the set $\{s\}$ of all lower Lebesgue sums is bounded above. Let U be its least upper bound:

$$U = \sup \{s\}.$$

It is obvious that

$$U \leqslant S_0.$$

Since the sum S_0 is arbitrary, the last inequality proves that the set $\{S\}$ of all upper Lebesgue sums is bounded below. Designate by V its greatest lower bound:

$$V = \inf \{S\}.$$

For every method of subdivision, we clearly have

$$s \leqslant U \leqslant V \leqslant S.$$

But, as noted above, $S - s \leqslant \lambda \, m \, E$, and therefore

$$0 \leqslant V - U \leqslant \lambda m E$$

Since λ can be made arbitrarily small, we infer that

$$U = V.$$

DEFINITION. The common value of the numbers U and V is called the *Lebesgue integral* of the function $f(x)$ on the set E and is designated by the symbol

$$(L) \int_E f(x) \, dx.$$

Where no confusion with other definitions of the integral can arise, we write simply

$$\int_E f(x) \, dx.$$

In particular, if E is the closed interval $[a, b]$, the symbols

$$(L)\int_a^b f(x)\,dx, \quad \int_a^b f(x)\,dx$$

are used.

It follows from the foregoing that *every bounded measurable function is integrable in the Lebesgue sense*, or, more briefly, integrable (L). It can already be seen from this remark that the process of integration (L) is applicable to a much wider class of functions than the process of integration (R). In particular, all questions connected with conditions ensuring integrability, which for integrals (R) have a relatively complicated character, vanish completely for Lebesgue integrals.*

THEOREM 1. *As $\lambda \to 0$, the Lebesgue sums s and S approach the integral $\int_E f(x)\,dx$.*

This theorem follows directly from the inequalities

$$s \leqslant \int_E f(x)\,dx \leqslant S,$$

$$S - s \leqslant \lambda \cdot mE.$$

It follows among other things from Theorem 1 that the value of the Lebesgue integral, which by virtue of its definition depends on the numbers A and B, turns out to be actually independent of them.

In fact, suppose that

$$A < f(x) < B, \quad A < f(x) < B^*,$$

with $B^* < B$. Divide the closed interval $[A, B]$ into subintervals by the division points

$$A = y_0 < y_1 < \ldots < y_n = B,$$

where we include the point B^* as a division point,

$$B^* = y_m.$$

Upon defining the sets e_k, it is easy to see that

$$e_k = 0 \qquad (k \geqslant m).$$

This implies that

$$s = \sum_{k=0}^{n-1} y_k m e_k = \sum_{k=0}^{m-1} y_k m e_k = s^*,$$

where s^* is the lower Lebesgue sum constructed on the closed interval $[A, B^*]$ with the division points $y_0, y_1, \ldots, y_m$. Increasing the number of division points and taking the limit, we find that

$$I = I^*,$$

where I and I^* are the values of the Lebesgue integrals corresponding to the closed

---

*Lebesgue sums can be defined only for bounded measurable functions; so that the Lebesgue integral as defined here exists for all functions for which the approximating sums can be defined. This is to be contrasted with the situation obtaining for Riemann integrals, where the approximating sums can be defined for *all* bounded functions.—E. H.

intervals $[A, B]$ and $[A, B^*]$. Thus changing the number B to B^* has no effect on the value of the integral, provided only that $|f(x)| < B^*$ for all x. The corresponding fact is true of the number A. This observation is highly essential, because only now do we see that the definition of the integral is independent of the bounds A and B.

§2. FUNDAMENTAL PROPERTIES OF THE INTEGRAL

In the present section, we establish a number of facts concerning the integral of a bounded measurable function.

THEOREM 1. *If the measurable function $f(x)$ satisfies the inequalities*
$$a \leqslant f(x) \leqslant b,$$
on the measurable set E, then
$$a \cdot mE \leqslant \int_E f(x)\,dx \leqslant b \cdot mE.$$

Usually this theorem is called the *first law of the mean for integrals*.

Proof. Let n be a natural number. If we set
$$A = a - \frac{1}{n}, \qquad B = b + \frac{1}{n},$$
then it is obvious that
$$A < f(x) < B,$$
and Lebesgue sums can be defined by dividing the closed interval $[A, B]$. But if
$$A \leqslant y_k \leqslant B,$$
then
$$A \sum_{k=0}^{n-1} me_k \leqslant \sum_{k=0}^{n-1} y_k me_k \leqslant B \sum_{k=0}^{n-1} me_k$$
or, equivalently,
$$A \cdot mE \leqslant s \leqslant B \cdot mE.$$

We have therefore in the limit
$$\left(a - \frac{1}{n}\right) mE \leqslant \int_E f(x)\,dx \leqslant \left(b + \frac{1}{n}\right) mE.$$

Since the number n is arbitrary, the theorem is proved.

Several simple corollaries follow from this theorem.

COROLLARY 1. *If the function $f(x)$ is constant on the measurable set E, $f(x) = c$, then*
$$\int_E f(x)\,dx = c \cdot mE.$$

COROLLARY 2. *If the function $f(x)$ is non-negative (non-positive), then its integral is non-negative (non-positive).*

COROLLARY 3. *If $mE = 0$, we have*
$$\int_E f(x)\,dx = 0$$
for all bounded functions $f(x)$ defined on E.

THEOREM 2. *Let a bounded measurable function $f(x)$ be defined on a measurable*

V. THE LEBESGUE INTEGRAL OF A BOUNDED FUNCTION

set E. If the set E is the sum of a finite or denumerable number of pairwise disjoint measurable sets,
$$E = \sum_k E_k \qquad (E_k E_{k'} = 0,\ k \neq k'),$$
then
$$\int_E f(x)\,dx = \sum_k \int_{E_k} f(x)\,dx.$$

The property of the integral expressed by this theorem is called *countable additivity*.

Proof. Consider first the simple case in which there are only two summands:
$$E = E' + E'' \qquad (E'E'' = 0).$$

Suppose that
$$A < f(x) < B$$
on the set E. Having subdivided the closed interval $[A, B]$ by means of the points $y_0, y_1, \ldots, y_n$, we define the sets
$$e_k = E(y_k \leqslant f < y_{k+1}), \quad e'_k = E'(y_k \leqslant f < y_{k+1}), \quad e''_k = E''(y_k \leqslant f < y_{k+1})$$
then we obviously have
$$e_k = e'_k + e''_k \qquad (e'_k e''_k = 0).$$

It follows that
$$\sum_{k=0}^{n-1} y_k m e_k = \sum_{k=0}^{n-1} y_k m e'_k + \sum_{k=0}^{n-1} y_k m e''_k$$
and in the limit as $\lambda \to 0$, we obtain
$$\int_E f(x)\,dx = \int_{E'} f(x)\,dx + \int_{E''} f(x)\,dx.$$

Thus, the theorem is proved for the case of two summands. Applying the technique of mathematical induction, we easily generalize the theorem to the case of an arbitrary finite number of summands.

The case in which
$$E = \sum_{k=1}^\infty E_k$$
remains to be considered. Here
$$\sum_{k=1}^\infty m E_k = mE,$$
so that for $n \to \infty$, we have
$$\sum_{k=n+1}^\infty m E_k \to 0. \tag{*}$$

We find it convenient to write
$$\sum_{k=n+1}^\infty E_k = R_n.$$

Since the theorem is already proved for a finite number of component terms, we may write
$$\int_E f\,dx = \sum_{k=1}^n \int_{E_k} f\,dx + \int_{R_n} f\,dx.$$

2. Fundamental Properties of the Integral

By virtue of the theorem of the mean,
$$A \cdot mR_n \leq \int_{R_n} f dx \leq B \cdot mR_n,$$
and by (*), the measure $m R_n$ of the set R_n tends to zero as $n \to \infty$. We see then that
$$\int_{R_n} f dx \to 0$$
as $n \to \infty$.

We infer that
$$\int_E f dx = \sum_{k=1}^{\infty} \int_{E_k} f dx.$$

A series of corollaries follow from this theorem.

COROLLARY 1. *If the bounded measurable functions $f(x)$ and $g(x)$, both defined on the set E, are equivalent, then*
$$\int_E f(x) dx = \int g(x) dx.$$

In fact, if
$$A = E(f \neq g), \quad B = E(f = g),$$
then $mA = 0$, and
$$\int_A f dx = \int_A g dx = 0.$$

On the set B, the functions are identical and
$$\int_B f dx = \int_B g dx.$$

One then combines this equality with the preceding one and applies Theorem 2.

In particular, *the integral of a function which is equivalent to zero, equals zero.*

Naturally, the converse of the last assertion is false. For example, if $f(x)$ is defined on the closed interval $[-1, +1]$ by
$$f(x) = \begin{cases} 1 & \text{for } x \geq 0, \\ -1 & \text{for } x < 0, \end{cases}$$
then [1]
$$\int_{-1}^{+1} f(x) dx = \int_{-1}^{0} f(x) dx + \int_{0}^{1} f(x) dx = -1 + 1 = 0,$$
even though the function $f(x)$ is not equivalent to zero.

However the following statement is true.

---

[1] Since removing one point from the set E does not change the integral $\int_E f dx$, we are justified in designating the integral over an arbitrary one of the intervals $[a, b)$, $(a, b]$, (a, b) by the symbol $\int_a^b f(x) dx$, exactly as over the closed interval $[a, b]$.

V. THE LEBESGUE INTEGRAL OF A BOUNDED FUNCTION

COROLLARY 2. *If the integral of a non-negative bounded measurable function $f(x)$ equals zero,*

$$\int_E f(x)\,dx \qquad (f(x) \geq 0)$$

then $f(x)$ is equivalent to zero.

In fact, it is easy to see that

$$E(f>0) = \sum_{n=1}^{\infty} E\left(f > \frac{1}{n}\right).$$

If $f(x)$ were not equivalent to zero, then we could necessarily find an n_0 such that

$$mE\left(f > \frac{1}{n_0}\right) = \sigma > 0.$$

Setting

$$A = E\left(f > \frac{1}{n_0}\right), \quad B = E - A,$$

we would have

$$\int_A f(x)\,dx \geq \frac{1}{n_0}\sigma, \quad \int_B f(x)\,dx \geq 0,$$

and combining these inequalities, we would obtain

$$\int_E f(x)\,dx \geq \frac{1}{n_0}\sigma,$$

which contradicts the hypothesis.

THEOREM 3. *If two measurable bounded functions $f(x)$ and $F(x)$ are defined on a measurable set Q, then*

$$\int_Q [f(x) + F(x)]\,dx = \int_Q f(x)\,dx + \int_Q F(x)\,dx.$$

Proof. Let

$$a < f(x) < b, \quad A < F(x) < B.$$

Subdivide the segments $[a, b]$ and $[A, B]$ by the points

$$a = y_0 < y_1 < \ldots < y_n = b, \quad A = Y_0 < Y_1 < \ldots < Y_N = B,$$

respectively, and introduce the sets

$$e_k = Q(y_k \leq f < y_{k+1}), \quad E_i = Q(Y_i \leq F < Y_{i+1})$$

$$T_{i,k} = E_i e_k \quad (i = 0, 1, \ldots, N-1;\ k = 0, 1, \ldots, n-1).$$

Obviously,

$$Q = \sum_{i,k} T_{i,k},$$

and the sets $T_{i,k}$ are pairwise disjoint. Hence

$$\int_Q (f+F)\,dx = \sum_{i,k} \int_{T_{i,k}} (f+F)\,dx.$$

2. Fundamental Properties of the Integral

On the set $T_{i,k}$, we have
$$y_k + Y_i \leqslant f(x) + F(x) < y_{k+1} + Y_{i+1},$$
The first law of the mean implies that
$$(y_k + Y_i) mT_{i,k} \leqslant \int_{T_{i,k}} (f+F)\,dx \leqslant (y_{k+1} + Y_{i+1}) mT_{i,k}.$$
Combining all these inequalities, we obtain
$$\sum_{i,k} (y_k + Y_i) mT_{i,k} \leqslant \int_Q (f+F)\,dx \leqslant \sum_{i,k} (y_{k+1} + Y_{i+1}) mT_{i,k}. \tag{1}$$
We evaluate the sum
$$\sum_{i,k} y_k mT_{i,k}. \tag{2}$$
by itself. This sum can be written in the form
$$\sum_{k=0}^{n-1} y_k \left(\sum_{i=0}^{N-1} mT_{i,k} \right).$$
But
$$\sum_{i=0}^{N-1} mT_{i,k} = m\left[\sum_{i=0}^{N-1} T_{i,k} \right] = m\left[\sum_{i=0}^{N-1} E_i e_k \right] = m\left[e_k \sum_{i=0}^{N-1} E_i \right] = m(e_k Q) = me_k,$$
so that the sum (2) can also be written as
$$\sum_{k=0}^{n-1} y_k me_k.$$
In other words, the sum (2) is a lower Lebesgue sum s_f of the function $f(x)$. The other sums in the inequality (1) are evaluated analogously, so that the inequality can be written in the form
$$s_f + s_F \leqslant \int_Q (f+F)\,dx \leqslant S_f + S_F, \tag{3}$$
the notation employed being self-evident. Increasing the number of points of subdivision of the closed intervals $[a, b]$ and $[A, B]$ and taking the limit in the inequalities (3), we obtain the present theorem.

THEOREM 4. *If $f(x)$ is a bounded measurable function defined on the measurable set E and c is a finite constant, then*
$$\int_E cf(x)\,dx = c\int_E f(x)\,dx.$$

Proof. If $c = 0$, the theorem is trivial. Consider the case $c > 0$. Let
$$A < f(x) < B.$$
Subdividing the segment $[A, B]$ by means of points y_k and introducing the sets e_k in the usual way, we have
$$\int_E cf(x)\,dx = \sum_{k=0}^{n-1} \int_{e_k} cf(x)\,dx.$$

On the set e_k, the inequalities
$$cy_k \leqslant cf(x) < cy_{k+1},$$
hold, so that by the first law of the mean,
$$cy_k me_k \leqslant \int_{e_k} cf(x)\,dx \leqslant cy_{k+1} me_k.$$
Combining all of these inequalities, we obtain
$$cs \leqslant \int_E cf(x)\,dx \leqslant cS,$$
where s and S are Lebesgue sums for the function $f(x)$. The theorem is obtained from the last inequality by taking $S-s$ arbitrarily small. Finally, consider the case $c < 0$. Here
$$0 = \int_E [cf(x) + (-c)f(x)]\,dx = \int_E cf(x)\,dx + (-c)\int_E f(x)\,dx,$$
and the theorem follows.

COROLLARY. *If $f(x)$ and $F(x)$ are bounded and measurable on the set E, then*
$$\int_E [F(x) - f(x)]\,dx = \int_E F(x)\,dx - \int_E f(x)\,dx.$$

THEOREM 5. *Let $f(x)$ and $F(x)$ be bounded and measurable on the measurable set E. If*
$$f(x) \leqslant F(x),$$
then
$$\int_E f(x)\,dx \leqslant \int_E F(x)\,dx.$$

The function $F(x) - f(x)$ is non-negative, so that
$$\int_E F\,dx - \int_E f\,dx = \int_E (F-f)\,dx \geqslant 0.$$

THEOREM 6. *If the function $f(x)$ is bounded and measurable on the measurable set E, then*
$$\left| \int_E f(x)\,dx \right| \leqslant \int_E |f(x)|\,dx.$$

Proof. Let
$$P = E(f \geqslant 0), \qquad N = E(f < 0).$$
Then
$$\int_E f\,dx = \int_P f\,dx + \int_N f\,dx = \int_P |f|\,dx - \int_N |f|\,dx,$$
$$\int_E |f|\,dx = \int_P |f|\,dx + \int_N |f|\,dx,$$
and the present proof is reduced to establishing the elementary inequality
$$|a - b| \leqslant a + b \qquad\qquad (a \geqslant 0,\ b \geqslant 0).$$

§ 3. PASSAGE TO THE LIMIT UNDER THE INTEGRAL SIGN

We now consider the following problem. On the measurable set E, let there be given a sequence of bounded measurable functions

$$f_1(x), f_2(x), f_3(x), \ldots, f_n(x), \ldots$$

which in some sense (everywhere, almost everywhere, in measure) converges to a measurable bounded function $F(x)$. The question is whether the relation

$$\lim_{n \to \infty} \int_E f_n(x)\, dx = \int_E F(x)\, dx \tag{1}$$

is valid. If (1) is valid, then we say that it is permissible to pass to the limit under the integral sign.

It is easy to see that one cannot always pass to the limit under the integral sign. For example, if functions $f_n(x)$ are defined on the segment $[0, 1]$ as follows,

$$f_n(x) = \begin{cases} n & \text{for } x \in \left(0, \dfrac{1}{n}\right), \\ 0 & \text{for } x \bar{\in} \left(0, \dfrac{1}{n}\right), \end{cases}$$

then for every $x \in [0, 1]$ we have

$$\lim_{n \to \infty} f_n(x) = 0,$$

but

$$\int_0^1 f_n(x)\, dx = 1,$$

and this integral does not go to zero as $n \to \infty$.

For this reason, it is natural to inquire into the additional limitations which must be placed on the function $f_n(x)$ so that the equality (1) may hold. We limit ourselves for the present to a proof of the following theorem.

THEOREM (LEBESGUE). *Let a sequence $f_1(x), f_2(x), f_3(x), \ldots$ of bounded measurable functions, converging in measure to the bounded measurable function $F(x)$,*

$$f_n(x) \Rightarrow F(x)$$

be defined on the measurable set E. If there exists a constant K such that for all n and for all x,

$$|f_n(x)| < K,$$

then

$$\lim_{n \to \infty} \int_E f_n(x)\, dx = \int_E F(x)\, dx. \tag{1}$$

Proof. First of all, we note that for almost $x \in E$,

$$|F(x)| \leqslant K. \tag{2}$$

For, on the basis of F. Riesz's theorem (Theorem 4, § 3, Ch. IV), it is possible to extract a subsequence $\{f_{n_k}(x)\}$ from the sequence $\{f_n(x)\}$ which converges to $F(x)$ almost everywhere. At all points where

$$f_{n_k}(x) \to F(x),$$

it is possible to pass to the limit in the inequality $|f_{n_k}(x)| < K$, which leads to (2). Now let σ be a positive number. Set

$$A_n(\sigma) = E(|f_n - F| \geq \sigma), \quad B_n(\sigma) = E(|f_n - F| < \sigma).$$

Then

$$\left| \int_E f_n \, dx - \int_E F \, dx \right| \leq \int_E |f_n - F| \, dx = \int_{A_n(\sigma)} |f_n - F| \, dx + \int_{B_n(\sigma)} |f_n - F| \, dx.$$

By the inequality $|f_n(x) - F(x)| \leq |f_n(x)| + |F(x)|$, we have

$$|f_n(x) - F(x)| < 2K$$

for almost all x of the set $A_n(\sigma)$. The first law of the mean implies that

$$\int_{A_n(\sigma)} |f_n - F| \, dx \leq 2K \cdot mA_n(\sigma) . \tag{3}$$

The possibility that the inequality $|f_n - F| < 2K$ may fail on a set of measure 0 has no effect on our argument. It is possible, for example, to change the function $|f_n(x) - F(x)|$ on this set by setting it equal to zero; then inequality (2′) will be satisfied at all points of E. Since changing the function on a set of measure 0 has no effect on the value of the integral, (3) is true also without changing $|f_n(x) - F(x)|$. On the other hand, again by the first law of the mean,

$$\int_{B_n(\sigma)} |f_n - F| \, dx \leq \sigma m B_n(\sigma) \leq \sigma m E.$$

Combining this inequality with (3), we find that

$$\left| \int_E f_n \, dx - \int_E F \, dx \right| \leq 2K \cdot mA_n(\sigma) + \sigma m E. \tag{4}$$

Now take an arbitrary $\varepsilon > 0$, and select a $\sigma > 0$ so small that

$$\sigma \cdot mE < \frac{\varepsilon}{2}.$$

Having fixed this σ, the definition of convergence in measure ensures that we will have

$$mA_n(\sigma) \to 0$$

as $n \to \infty$ and therefore

$$2K \cdot mA_n(\sigma) < \frac{\varepsilon}{2}$$

for $n > N$. For such n, inequality (4) assumes the form

$$\left| \int_E f_n \, dx - \int_E F \, dx \right| < \varepsilon,$$

this proves the theorem.

It is clear that the preceding theorem retains its validity if the inequality

$$|f_n(x)| < K$$

is satisfied only almost everywhere on the set E. The proof remains the same.

Further, since convergence in measure is more general than ordinary convergence, the theorem is true if

$$f_n(x) \to F(x)$$

almost everywhere (or everywhere).

§ 4. COMPARISON OF RIEMANN AND LEBESGUE INTEGRALS

Let the function $f(x)$ (not necessarily finite) be defined on the closed interval $[a, b]$. Let $x_0 \in [a, b]$ and $\delta > 0$. Designate by $m_\delta(x_0)$ and $M_\delta(x_0)$, respectively, the greatest lower and the least upper bounds of the function $f(x)$ in the open interval $(x_0 - \delta < x < x_0 + \delta)$,

$$m_\delta(x_0) = \inf\{f(x)\}, \quad M_\delta(x_0) = \sup\{f(x)\} \quad (x_0 - \delta < x < x_0 + \delta).$$

(It is evident that we should consider only those points of the open interval $(x_0 - \delta, x_0 + \delta)$ which lie in the closed interval $[a, b]$.)
Obviously,

$$m_\delta(x_0) \leqslant f(x_0) \leqslant M_\delta(x_0).$$

As δ decreases, $m_\delta(x_0)$ does not decrease and $M_\delta(x_0)$ does not increase. For this reason, the limits

$$m(x_0) = \lim_{\delta \to +0} m_\delta(x_0), \quad M(x_0) = \lim_{\delta \to +0} M_\delta(x_0)$$

exist; we have, clearly enough,

$$m_\delta(x_0) \leqslant m(x_0) \leqslant f(x_0) \leqslant M(x_0) \leqslant M_\delta(x_0).$$

DEFINITION. The functions $m(x)$ and $M(x)$ are called the lower and upper Baire functions, respectively, of the function $f(x)$.

THEOREM 1. *Let the function $f(x)$ be finite at the point x_0. A necessary and sufficient condition that the function $f(x)$ be continuous at this point is that*

$$m(x_0) = M(x_0). \tag{*}$$

Proof. Suppose that $f(x)$ is continuous at the point x_0. Taking an arbitrary $\varepsilon > 0$, we can find a $\delta > 0$ such that

$$|x - x_0| < \delta,$$

whenever

$$|f(x) - f(x_0)| < \varepsilon.$$

In other words, for all $x \in (x_0 - \delta, x_0 + \delta)$, we have

$$f(x_0) - \varepsilon < f(x) < f(x_0) + \varepsilon.$$

But this implies that

$$f(x_0) - \varepsilon \leqslant m_\delta(x_0) \leqslant M_\delta(x_0) \leqslant f(x_0) + \varepsilon$$

and we have

$$f(x_0) - \varepsilon \leqslant m(x_0) \leqslant M(x_0) \leqslant f(x_0) + \varepsilon.$$

These inequalities imply (*), since ε is arbitrary. Thus the necessity of condition (*) is proved.

Suppose, conversely, that (*) is satisfied. Then
$$m(x_0) = M(x_0) = f(x_0)$$
and the common value of the Baire functions at the point x_0 is finite. Take an arbitrary $\varepsilon > 0$, and find a $\delta > 0$ so small that
$$m(x_0) - \varepsilon < m_\delta(x_0) \leqslant m(x_0), \quad M(x_0) \leqslant M_\delta(x_0) < M(x_0) + \varepsilon.$$
These inequalities imply that
$$f(x_0) - \varepsilon < m_\delta(x_0), \quad M_\delta(x_0) < f(x_0) + \varepsilon.$$
Now, if $x \in (x_0 - \delta, x_0 + \delta)$, the value of the function at x lies between $m_\delta(x_0)$ and $M_\delta(x_0)$, so that
$$f(x_0) - \varepsilon < f(x) < f(x_0) + \varepsilon.$$
In other words, $|x - x_0| < \delta$ implies that
$$|f(x) - f(x_0)| < \varepsilon,$$
i.e., the function $f(x)$ is continuous at the point x_0.

FUNDAMENTAL LEMMA. *Consider a sequence of subdivisions of the closed interval $[a, b]$*
$$a = x_0^{(1)} < x_1^{(1)} < \ldots < x_{n_1}^{(1)} = b$$
$$\cdots\cdots\cdots\cdots\cdots\cdots\cdots\cdots$$
$$a = x_0^{(i)} < x_1^{(i)} < \ldots < x_{n_i}^{(i)} = b$$
$$\cdots\cdots\cdots\cdots\cdots\cdots\cdots\cdots$$
where
$$\lambda_i = \max [x_{k+1}^{(i)} - x_k^{(i)}] \to 0$$
as $i \to \infty$. Let $m_k^{(i)}$ be the greatest lower bound of the values of the function $f(x)$ on the closed interval $[x_k^{(i)}, x_{k+1}^{(i)}]$. Define the function $\varphi_i(x)$ by setting
$$\varphi_i(x) = m_k^{(i)} \quad \text{for} \quad x \in (x_k^{(i)}, x_{k+1}^{(i)})$$
$$\varphi_i(x) = 0 \quad \text{for} \quad x = x_0^{(i)}, x_1^{(i)}, \ldots, x_{n_i}^{(i)}.$$
If x_0 does not coincide with any of the points $x_k^{(i)}$ ($i = 1, 2, 3, \ldots$; $k = 0, 1, 2, \ldots, n_i$), then
$$\lim_{i \to \infty} \varphi_i(x_0) = m(x_0).$$

Proof. Choose an arbitrary but fixed i, and let $[x_{k_0}^{(i)}, x_{k_0+1}^{(i)}]$ be the interval of the i-th method of subdivision which contains the point x_0. Since x_0 does not coincide with any of the division points, we have
$$x_{k_0}^{(i)} < x_0 < x_{k_0+1}^{(i)}$$
and, consequently, for sufficiently small $\delta > 0$, we have
$$(x_0 - \delta, x_0 + \delta) \subset [x_{k_0}^{(i)}, x_{k_0+1}^{(i)}].$$

4. Comparison of Riemann and Lebesgue Integrals

Therefore
$$m_{k_0}^{(i)} \leq m_\delta(x_0)$$
or, equivalently,
$$\varphi_i(x_0) \leq m_\delta(x_0).$$

Letting δ go to zero and taking the limit, we find that for arbitrary i
$$\varphi_i(x_0) \leq m(x_0).$$
This proves the lemma for the case $m(x_0) = -\infty$. Now suppose that $m(x_0) > -\infty$, and let
$$h < m(x_0).$$
We can then find a $\delta > 0$ such that
$$m_\delta(x_0) > h.$$
Fixing this δ, we find an i_0 so large that for $i > i_0$ we have
$$[x_{k_0}^{(i)}, x_{k_0+1}^{(i)}] \subset (x_0 - \delta, x_0 + \delta),$$
where $[x_{k_0}^{(i)}, x_{k_0+1}^{(i)}]$ is, as defined above, a closed interval containing the point x_0. The existence of such an i_0 follows from the condition
$$\lambda_i \to 0.$$
For all $i > i_0$, we have
$$m_{k_0}^{(i)} \geq m_\delta(x_0) > h,$$
or, equivalently,
$$\varphi_i(x_0) > h.$$
Thus, for every $h < m(x_0)$, we can find an i_0 such that for $i > i_0$
$$h < \varphi_i(x_0) \leq m(x_0),$$
and this implies that $\varphi_i(x_0) \to m(x_0)$. The proof is thus complete.

COROLLARY 1. *The Baire functions $m(x)$ and $M(x)$ are measurable.*

In fact, the set of division points $\{x_k^{(i)}\}$ is denumerable and therefore has measure zero. Hence the lemma implies that
$$\varphi_i(x) \to m(x)$$
almost everywhere.

But $\varphi_i(x)$ is measurable, since it is a step function; this implies that the function $m(x)$ is also measurable. Similar reasoning proves that the upper Baire function $M(x)$ is measurable.

COROLLARY 2. *Let the conditions of the fundamental lemma be satisfied. If the initial function $f(x)$ is bounded, then*
$$(L)\int_a^b \varphi_i(x)\,dx \to (L)\int_a^b m(x)\,dx.$$
In fact, if $|f(x)| \leq K$, then it is clear that
$$|\varphi_i(x)| \leq K, \quad |m(x)| \leq K,$$

from which it follows that these functions are integrable (L). It then remains to apply Lebesgue's theorem on passage to the limit under the integral sign.

We now paraphrase Corollary 2. To do this, we note that

$$(L)\int_a^b \varphi_i(x)\,dx = \sum_{k=0}^{n_i-1} \int_{x_k^{(i)}}^{x_{k+1}^{(i)}} \varphi_i(x)\,dx = \sum_{k=0}^{n_i-1} m_k^{(i)}[x_{k+1}^{(i)} - x_k^{(i)}] = s_i,$$

where s_i is the lower Darboux sum corresponding to the i-th method of subdivision. Corollary 2 thus means that for $i \to \infty$,

$$s_i \to (L)\int_a^b m(x)\,dx.$$

In like manner, one can establish that the upper Darboux sum S_i approaches the integral of the upper Baire function:

$$S_i \to (L)\int_a^b M(x)\,dx$$

as $i \to \infty$. These relations being true, we have

$$S_i - s_i \to (L)\int_a^b [M(x) - m(x)]\,dx.$$

On the other hand, it is established in elementary analysis courses that a necessary and sufficient condition that a bounded function $f(x)$ be integrable (R) is that

$$S_i - s_i \to 0.$$

Comparing this with the foregoing, we see that in order for the function $f(x)$ to be integrable (R), it is necessary and sufficient that

$$(L)\int_a^b [M(x) - m(x)]\,dx = 0. \tag{1}$$

The condition (1) is satisfied in every case where the difference $M(x) - m(x)$ is equivalent to zero, but, since this difference is non-negative, it follows from (1) that

$$m(x) \sim M(x). \tag{2}$$

Integrability (R) of a bounded function $f(x)$ is therefore equivalent to relation (2). Comparing this result with Theorem 1, we obtain the following theorem.

THEOREM 2. *A necessary and sufficient condition that the bounded function $f(x)$ be integrable (R) is that it be continuous almost everywhere.*

This remarkable theorem is the simplest and clearest test for integrability (R). In particular, it justifies the remark made in § 1 that only functions which are not very discontinuous can be integrable (R).

Now suppose that the function $f(x)$ is integrable (R). Then it is necessarily bounded and

$$m(x) = M(x).$$

almost everywhere. However,

$$m(x) \leqslant f(x) \leqslant M(x).$$

Therefore

$$f(x) = m(x)$$

almost everywhere, and $f(x)$, being equivalent to the measurable function $m(x)$, is itself measurable. Since every bounded measurable function is integrable (L), $f(x)$ is integrable (L); i.e., the integrability of a function in the Riemann sense implies its integrability in the Lebesgue sense.

Finally, it follows from the equivalence of the functions $f(x)$ and $m(x)$ that

$$(L)\int_a^b f(x)\,dx = (L)\int_a^b m(x)\,dx.$$

It is known from elementary analysis that under the hypotheses of the fundamental lemma, for an (R)-integrable function $f(x)$, we have

$$s_i \to (R)\int_a^b f(x)\,dx,$$

where s_i is the lower Darboux sum corresponding to the i-th subdivision. Since

$$s_i \to (L)\int_a^b m(x)\,dx,$$

as we have already shown, we see that

$$(R)\int_a^b f(x)\,dx = (L)\int_a^b f(x)\,dx.$$

We have accordingly proved the following result.

THEOREM 3. *Every function integrable (R) is necessarily integrable (L), and the (R) and (L) integrals are equal.*

In conclusion, we note that the Dirichlet function $\psi(x)$ (equal to zero at irrational and to one at rational points) is integrable (L) (since it is equivalent to zero), but as noted in § 1, it is not integrable (R), so that the converse of Theorem 3 fails.

§ 5. RECONSTRUCTION OF THE PRIMITIVE FUNCTION

Let there be defined on the closed interval $[a, b]$ a continuous function $f(x)$ whose derivative $f'(x)$ is defined at every point of $[a, b]$ (at the endpoints a and b we consider one-sided derivatives). We now ask how, knowing the derivative $f'(x)$, can we reconstruct the primitive function $f(x)$? In elementary analysis texts, it is shown that if the derivative $f'(x)$ is (R)-integrable, then

$$f(x) = f(a) + \int_a^x f'(t)\,dt,$$

but examples [2] are known of bounded derivatives which are not integrable (R). Hence we cannot use the Riemann integral to give a complete answer to our question. The Lebesgue integral, however, is a more powerful tool for this purpose.

THEOREM. *Let the function $f(x)$ have a finite derivative $f'(x)$ at every point of $[a, b]$. If $f'(x)$ is bounded, then it is integrable (L) and we also have*

$$f(x) = f(a) + \int_a^x f'(t)\,dt.$$

---

[2] See, for example, E. W. HOBSON, *The Theory of Functions of a Real Variable and the Theory of Fourier Series*, Cambridge Univ. Press, 2nd Ed. 1926, Vol. 2, pp. 412-421.

Proof. First of all, we note that the function $f(x)$ is necessarily continuous, since it has a finite derivative at every point of $[a, b]$. Extend the definition of the function to the larger interval $[a, b + 1]$ by setting

$$f(x) = f(b) + (x - b) f'(b)$$

for $b < x \leqslant b + 1$. The function $f(x)$ is now continuous and has a finite derivative on $[a, b + 1]$. For $x \in [a, b]$ and $n = 1, 2, 3, \ldots$, set

$$\varphi_n(x) = n \left[f\left(x + \frac{1}{n}\right) - f(x) \right].$$

At every point $x \in [a, b]$, we have

$$\lim_{n \to \infty} \varphi_n(x) = f'(x),$$

and since each of the functions $\varphi_n(x)$ is measurable, being continuous, $f'(x)$ is also measurable; since $f(x)$ is bounded by hypothesis, it is integrable (L). Further, by Lagrange's formula, we have

$$\varphi_n(x) = n \left[f\left(x + \frac{1}{n}\right) - f(x) \right] = f'\left(x + \frac{\theta}{n}\right) \qquad (0 < \theta < 1),$$

so that all the functions $\varphi_n(x)$ have a common upper bound. By Lebesgue's theorem on passage to the limit under the integral sign, we may write

$$\int_a^b f'(x) \, dx = \lim_{n \to \infty} \int_a^b \varphi_n(x) \, dx. \tag{1}$$

However, we also have

$$\int_a^b \varphi_n(x) \, dx = n \int_a^b f\left(x + \frac{1}{n}\right) dx - n \int_a^b f(x) \, dx = n \int_{a+\frac{1}{n}}^{b+\frac{1}{n}} f(x) \, dx - n \int_a^b f(x) \, dx.$$

(The change of variable in the above integral of $f(x + \frac{1}{n})$ can be carried out because this function is continuous, and the integral can be taken in the Riemann sense. For integrals (R), the theory of substitution may be presupposed.) Hence

$$\int_a^b \varphi_n(x) \, dx = n \int_b^{b+\frac{1}{n}} f(x) \, dx - n \int_a^{a+\frac{1}{n}} f(x) \, dx.$$

Applying the first law of the mean to each of the last two integrals, we obtain

$$\int_a^b \varphi_n(x) \, dx = f\left(b + \frac{\theta'_n}{n}\right) - f\left(a + \frac{\theta''_n}{n}\right) \qquad (0 < \theta'_n < 1, \ 0 < \theta''_n < 1),$$

From this relation, on the basis of the continuity of the function $f(x)$, we infer that

$$\lim_{n \to \infty} \int_a^b \varphi_n(x) \, dx = f(b) - f(a).$$

Combining this with (1), we find that

$$f(b) - f(a) = \int_a^b f'(x)\,dx.$$

The present theorem follows on replacing b by an arbitrary x from $[a, b]$.*

---

*The Lebesgue integral for functions defined on unbounded measurable sets is a very useful tool of analysis. It is convenient to postpone a discussion of this integral to the end of Chapter VI. (See § 4, Ch. VI.)

CHAPTER VI

SUMMABLE FUNCTIONS

§ 1. THE INTEGRAL OF A NON-NEGATIVE MEASURABLE FUNCTION

In this chapter, we generalize the definition of the Lebesgue integral to include unbounded functions; in the present section, we consider non-negative functions.

LEMMA 1. *Let the function $f(x)$ be measurable and non-negative on the measurable set E and let N be a natural number. If the function $[f(x)]_N$* [1] *is defined in the following manner:*

$$[f(x)]_N = \begin{cases} f(x), & \text{for} \quad f(x) \leqslant N, \\ N, & \text{for} \quad f(x) > N, \end{cases}$$

then $[f(x)]_N$ is measurable also.

Proof. It is easily verified that

$$E([f]_N > a) = \begin{cases} E(f > a), & \text{for} \quad a < N, \\ 0, & \text{for} \quad a \geqslant N, \end{cases}$$

and from this the lemma follows.

Under the conditions of the preceding lemma, the function $[f(x)]_N$ is obviously bounded and is hence integrable (L). Since

$$[f(x)]_1 \leqslant [f(x)]_2 \leqslant [f(x)]_3 \leqslant \cdots,$$

we have

$$\int_E [f]_1 \, dx \leqslant \int_E [f]_2 \, dx \leqslant \int_E [f]_3 \, dx \leqslant \cdots$$

and a definite (finite or infinite) limit exists:

$$\lim_{N \to +\infty} \int_E [f(x)]_N \, dx. \tag{*}$$

DEFINITION. The limit (*) is called the *Lebesgue integral of the function $f(x)$ on the set E* and is denoted by the symbol

$$\int_E f(x) \, dx.$$

If this integral is finite, the function is said to be integrable (L) or summable on the set E.

Thus, *every* measurable, non-negative function has an integral, but only functions having a finite integral are called summable. When we wish to emphasize that the integral is to be understood in the Lebesgue sense, we denote it by the symbol

$$(L) \int_E f(x) \, dx.$$

---

[1] Sometimes this function is called the " section of the function by the number N."

1. THE INTEGRAL OF A NON-NEGATIVE MEASURABLE FUNCTION

For the case $E = [a, b]$, we use the notation

$$\int_a^b f(x)\,dx.$$

It is not difficult to see that for a bounded, measurable, non-negative function $f(x)$, the new definition of the integral coincides with that given earlier, because for sufficiently large N, we have

$$[f(x)]_N \equiv f(x).$$

It follows that every bounded, measurable, non-negative function is summable.

THEOREM 1. *If a function $f(x)$ is summable on the set E, then it is finite almost everywhere on E.*

Proof. Set

$$A = E(f = +\infty).$$

On the set A, the function $[f(x)]_N$ equals N, so that

$$\int_E [f]_N\,dx \geq \int_A [f]_N\,dx = N \cdot mA,$$

and if mA were > 0, then the integral $\int_E [f]_N\,dx$ would go to infinity as $N \to \infty$, which contradicts the hypothesis that the function $f(x)$ is summable.

THEOREM 2. *If $mE = 0$, then every non-negative function $f(x)$ is summable on the set E and*

$$\int_E f(x)\,dx = 0.$$

This theorem is obvious.

THEOREM 3. *If the functions $f(x)$ and $g(x)$ are equivalent on the set E, then*

$$\int_E f(x)\,dx = \int_E g(x)\,dx.$$

In fact, at every point where the functions $f(x)$ and $g(x)$ are equal, the functions $[f(x)]_N$ and $[g(x)]_N$ will also be equal, so that these latter functions are likewise equivalent to one another. The remainder of the proof is obvious.

THEOREM 4. *If a function $f(x)$ is non-negative and measurable on the set E and E_0 is a measurable subset of E, then*

$$\int_{E_0} f(x)\,dx \leq \int_E f(x)\,dx.$$

The inequality to be verified is obvious if $f(x)$ is replaced by $[f(x)]_N$, after which it remains merely to take the limit as $N \to \infty$. In particular, it follows from the summability of the function $f(x)$ on the set E that it is summable on every measurable subset of E.

THEOREM 5. *Let the functions $f(x)$ and $F(x)$ be non-negative and measurable on the set E. If $f(x) \leq F(x)$, then*

$$\int_E f(x)\,dx \leq \int_E F(x)\,dx.$$

This property is established by integrating the obvious inequality
$$[f(x)]_N \leq [F(x)]_N$$
and taking the limit as $N \to \infty$.

Theorem 5 implies that if $F(x)$ is summable, then $f(x)$ is also summable.

THEOREM 6. *If (under the usual assumptions),*
$$\int_E f(x)\,dx = 0,$$
the function $f(x)$ is equivalent to zero.

Proof. Since
$$0 \leq \int_E [f(x)]_N\,dx \leq \int_E f(x)\,dx,$$
the function $[f(x)]_N$ (for arbitrary N) is equivalent to zero. Let
$$A = \sum_{N=1}^{\infty} E([f]_N \neq 0),$$
then $mA = 0$. It is easy to see that $f(x) = 0$ for $x \in E - A$ because
$$\lim_{N \to +\infty} [f(x)]_N = f(x)$$
for every $x \in E$.

THEOREM 7. *Let $f'(x)$ and $f''(x)$ be two non-negative measurable functions defined on the set E. If $f(x) = f'(x) + f''(x)$, then*
$$\int_E f(x)\,dx = \int_E f'(x)\,dx + \int_E f''(x)\,dx.$$

Proof. Since, for arbitrary N,
$$[f'(x)]_N + [f''(x)]_N \leq f(x),$$
it is clear that
$$\int_E [f']_N\,dx + \int_E [f'']_N\,dx \leq \int_E f\,dx.$$
Taking the limit as $N \to \infty$, we obtain
$$\int_E f'\,dx + \int_E f''\,dx \leq \int_E f\,dx. \tag{1}$$
In order to verify the inverse inequality, we show that
$$[f(x)]_N \leq [f'(x)]_N + [f''(x)]_N \tag{2}$$
for every N. Let $x_0 \in E$. If
$$f'(x_0) \leq N, \quad f''(x_0) \leq N,$$
then
$$[f(x_0)]_N \leq f(x_0) = f'(x_0) + f''(x_0) = [f'(x_0)]_N + [f''(x_0)]_N.$$
If at least one of the numbers $f'(x_0)$ and $f''(x_0)$ is greater than N, then
$$[f(x_0)]_N = N \leq [f'(x_0)]_N + [f''(x_0)]_N,$$

1. The Integral of a Non-negative Measurable Function

because then one of the terms of the right-hand member equals N and the other is non-negative. This establishes (2).

Integrating the inequality (2), we obtain

$$\int_E [f]_N \, dx \leq \int_E [f']_N \, dx + \int_E [f'']_N \, dx.$$

Consequently

$$\int_E [f]_N \, dx \leq \int_E f' \, dx + \int_E f'' \, dx$$

and, in the limit as $N \to \infty$,

$$\int_E f \, dx \leq \int_E f' \, dx + \int_E f'' \, dx. \tag{3}$$

Combining (1) and (3), we obtain the present theorem. In particular, if each of the functions $f'(x)$ and $f''(x)$ is summable, then their sum is also summable.

THEOREM 8. *If $f(x)$ is a measurable non-negative function defined on the set E and if k is a finite non-negative number, then*

$$\int_E kf(x) \, dx = k \int_E f(x) \, dx.$$

Proof. The theorem is trivial for $k = 0$. For every natural number k, it is a consequence of the preceding theorem. If $k = \frac{1}{m}$, where m is a natural number, then again, by Theorem 7,

$$\int_E f(x) \, dx = m \int_E \frac{1}{m} f(x) \, dx$$

and

$$\int_E \frac{1}{m} f(x) \, dx = \frac{1}{m} \int_E f(x) \, dx.$$

From this, the validity of the theorem for every positive rational value of k follows. Finally, let k be a positive irrational number. Take positive rational numbers r and R such that $r < k < R$. By Theorem 5,

$$r \int_E f(x) \, dx \leq \int_E kf(x) \, dx \leq R \int_E f(x) \, dx,$$

and we have only to take the limit as $r \to k$ and $R \to k$.

In particular, the summability of $kf(x)$ follows from the summability of the function $f(x)$.

The next theorem is very important. Before formulating it, we prove an almost obvious lemma.

LEMMA 2. *Let*

$$\lim_{n \to \infty} f_n(x_0) = F(x_0)$$

at the point x_0. Then for every integer N

$$\lim_{n \to \infty} [f_n(x_0)]_N = [F(x_0)]_N.$$

Proof. If $F(x_0) > N$, for every sufficiently large n, we will have $f_n(x_0) > N$ and (for these n)
$$[f_n(x_0)]_N = N = [F(x_0)]_N.$$

In exactly the same way, if $F(x_0) < N$, for sufficiently large n, we will have $f_n(x_0) < N$, and, hence
$$[f_n(x_0)]_N = f_n(x_0) \to F(x_0) = [F(x_0)]_N.$$

It remains to consider the case when $F(x_0) = N$. In this case, for arbitrary $\varepsilon > 0$, we can find an n_0 such that for $n > n_0$,
$$f_n(x_0) > N - \varepsilon$$
and, hence (for $n > n_0$),
$$N - \varepsilon < [f_n(x_0)]_N \leqslant N,$$
i.e.,
$$|[F(x_0)]_N - [f_n(x_0)]_N| < \varepsilon \qquad (n > n_0). $$

Thus, the lemma holds in all cases.

THEOREM 9 (P. FATOU). *If the sequence of measurable and non-negative functions $f_1(x), f_2(x), \ldots$ converges to the function $F(x)$ almost everywhere on the set E, then*
$$\int_E F(x)\,dx \leqslant \sup\left\{\int_E f_n(x)\,dx\right\}.\,{}^2 \qquad (*)$$

Proof: By the lemma, we have
$$[f_n(x)]_N \to [F(x)]_N$$
as $n \to \infty$, almost everywhere on the set E. Inasmuch as each of the functions $[f_n(x)]_N$ is bounded by the number N, we can apply Lebesgue's theorem on passage to the limit under the integral sign and Lemma 2, so that
$$\int_E [F]_N\,dx = \lim_{n \to \infty} \int_E [f_n]_N\,dx.$$
But for all n,
$$\int_E [f_n]_N\,dx \leqslant \int_E f_n\,dx \leqslant \sup\left\{\int_E f_n\,dx\right\},$$
so that upon taking the limit $n \to \infty$, we have
$$\int_E [F]_N\,dx \leqslant \sup\left\{\int_E f_n\,dx\right\}.$$

---

[2] It is not difficult to see that (*) is also true when the sequence $\{f_n(x)\}$ converges to $F(x)$ not almost everywhere but only in measure. In fact, in this case, a subsequence $\{f_{n_k}(x)\}$ converging to $F(x)$ almost everywhere can be selected from $\{f_n(x)\}$, and it is sufficient to note that
$$\sup\left\{\int_E f_{n_k}(x)\,dx\right\} \leqslant \sup\left\{\int_E f_n(x)\,dx\right\}.$$
We point out that this remark is not a generalization of the theorem because in the theorem the case when $F(x) \equiv +\infty$ is not excluded and then it is impossible to speak of convergence in measure.

1. The Integral of a Non-negative Measurable Function

Taking the limit as $N \to \infty$, we obtain the theorem.

In particular, if all $f_n(x)$ are summable and
$$\int_E f_n(x)\, dx \leqslant A < +\infty,$$
then the limit function $F(x)$ is also summable.

COROLLARY. *If, under the hypotheses of Theorem 9, the limit*
$$\lim_{n \to \infty} \int_E f_n(x)\, dx, \tag{4}$$
exists, then
$$\int_E F(x)\, dx \leqslant \lim_{n \to \infty} \int_E f_n(x)\, dx. \tag{5}$$

Proof. The inequality under consideration is trivial if the limit (4) equals $+\infty$. Suppose then that
$$\lim_{n \to \infty} \int_E f_n\, dx = l < +\infty.$$
Then, for an arbitrary $\varepsilon > 0$, there exists a natural number n_0 such that for $n \geqslant n_0$,
$$\int_E f_n\, dx < l + \varepsilon.$$
Applying Theorem 9 to the sequence of functions $f_{n_0}(x), f_{n_0+1}(x), \ldots$, we obtain
$$\int_E F\, dx \leqslant l + \varepsilon,$$
and this implies (5), since ε is arbitrary.

With the aid of this corollary, it is easy to obtain another theorem concerning passage to the limit under the integral sign.

THEOREM 10 (B. LEVI). *Let an increasing sequence of measurable non-negative functions*
$$f_1(x) \leqslant f_2(x) \leqslant f_3(x) \leqslant \ldots$$
be defined on the set E. If
$$F(x) = \lim_{n \to \infty} f_n(x),$$
then
$$\int_E F(x)\, dx = \lim_{n \to \infty} \int_E f_n(x)\, dx.$$

Proof. First of all, the limit
$$\lim \int_E f_n\, dx$$
exists and, by the preceding corollary,
$$\int_E F\, dx \leqslant \lim_{n \to \infty} \int_E f_n\, dx.$$

On the other hand, we have $f_n(x) \leq F(x)$ for every n, whence

$$\int_E f_n(x)\,dx \leq \int_E F(x)\,dx,$$

and this implies that

$$\lim \int_E f_n\,dx \leq \int_E F\,dx.$$

This completes the proof.

THEOREM 11. *Let a sequence of measurable non-negative functions $u_1(x), u_2(x), \ldots$ be defined on the set E. If*

$$\sum_{k=1}^\infty u_k(x) = F(x),$$

then

$$\int_E F(x)\,dx = \sum_{k=1}^\infty \int_E u_k(x)\,dx.$$

To prove this it suffices to set

$$f_n(x) = \sum_{k=1}^n u_k(x)$$

and apply the preceding theorem.

COROLLARY. *Under the hypotheses of the preceding theorem, suppose that*

$$\sum_{k=1}^\infty \int_E u_k(x)\,dx < +\infty,$$

Then

$$\lim_{k \to \infty} u_k(x) = 0. \tag{6}$$

almost everywhere on the set E.

In fact, in the case under consideration, $F(x)$ is summable and is therefore finite almost everywhere. In other words, the series $\Sigma u_k(x)$ converges almost everywhere and at points of convergence of this series, it is clear that (6) holds.

THEOREM 12 (COUNTABLE ADDITIVITY OF THE INTEGRAL). *Let a measurable set E be the sum of a finite or denumerable family of pairwise disjoint measurable sets E_k:*

$$E = \sum_k E_k \qquad (E_k E_{k'} = 0, \quad k \neq k').$$

For every non-negative measurable function $f(x)$ defined on the set E, we have

$$\int_E f(x)\,dx = \sum_k \int_{E_k} f(x)\,dx.$$

Proof. Introduce the functions $u_k(x)$ ($k = 1, 2, \ldots$), letting

$$u_k(x) = \begin{cases} f(x), & \text{if } x \in E_k \\ 0, & \text{if } x \in E - E_k. \end{cases}$$

It is easy to see that

$$f(x) = \sum_k u_k(x),$$

and hence (by Theorem 7, if the number of terms is finite, and by Theorem 11 otherwise),

$$\int_E f(x)\,dx = \sum_k \int_E u_k(x)\,dx. \tag{7}$$

We now evaluate the integral $\int_E u_k(x)\,dx$. To do this, we note that

$$[u_k(x)]_N = \begin{cases} [f(x)]_N, & \text{if } x \in E_k \\ 0, & \text{if } x \in E - E_k. \end{cases}$$

This implies that

$$\int_E [u_k]_N\,dx = \int_{E_k} [f]_N\,dx.$$

Taking the limit as $N \to \infty$, we find that

$$\int_E u_k\,dx = \int_{E_k} f\,dx,$$

which, together with (7), proves the theorem.

§ 2. SUMMABLE FUNCTIONS OF ARBITRARY SIGN

We generalize the definition of the Lebesgue integral to unbounded functions of arbitrary sign. As we will see, it turns out that this is not possible for all measurable functions.

Let $f(x)$ be a measurable function defined on a measurable set E. Introduce the functions $f_+(x)$ and $f_-(x)$, setting

$$f_+(x) = \begin{cases} f(x), & \text{if } f(x) \geq 0 \\ 0, & \text{if } f(x) < 0 \end{cases}; \quad f_-(x) = \begin{cases} 0, & \text{if } f(x) \geq 0 \\ -f(x), & \text{if } f(x) > 0. \end{cases}$$

These functions are measurable and non-negative, so that both of the integrals

$$\int_E f_+(x)\,dx, \quad \int_E f_-(x)\,dx$$

exist. It is easy to see that

$$f(x) = f_+(x) - f_-(x).$$

Therefore, it is natural to agree to call the difference

$$\int_E f_+(x)\,dx - \int_E f_-(x)\,dx$$

the integral of the function $f(x)$. However, the difference

$$+\infty - (+\infty)$$

is meaningless. The symbol

$$\int_E f_+(x)\,dx - \int_E f_-(x)\,dx$$

has meaning, therefore, if and only if at least one of the functions $f_+(x)$ and $f_-(x)$ is summable.

DEFINITION 1. If at least one of the functions $f_+(x)$ and $f_-(x)$ is summable on the set E, then the (finite or infinite) difference

$$\int_E f_+(x)\,dx - \int_E f_-(x)\,dx$$

is called the Lebesgue integral of the function $f(x)$ on the set E and is denoted by the symbol

$$\int_E f(x)\, dx. \qquad (1)$$

If the measurable function $f(x)$ is bounded, then both functions $f_+(x)$ and $f_-(x)$ are also bounded; hence the new definition of the integral of the function $f(x)$ amounts to that given earlier. Similarly, if the measurable function $f(x)$ is non-negative (although perhaps unbounded), it is clear that

$$f_+(x) = f(x), \quad f_-(x) = 0,$$

and we arrive again at the old definition of integral. It is obvious that the integral (1) exists and is finite if and only if both of the functions $f_+(x)$ and $f_-(x)$ are summable.

DEFINITION 2. *The function $f(x)$ is said to be integrable (L) or summable on the set E if the integral $\int_E f(x)\, dx$ exists and is finite.*

Every measurable bounded function is summable; for a non-negative function, the new definition of summability is equivalent to that given earlier.

The set of all summable functions is usually designated by the letter L; thus the assertion that $f(x)$ is summable can be written as $f(x) \in L$.

THEOREM 1. *A measurable function $f(x)$ is summable if and only if the function $|f(x)|$ is summable. In this case,*

$$\left| \int_E f(x)\, dx \right| \leq \int_E |f(x)|\, dx.$$

Proof. It is easy to see that

$$|f(x)| = f_+(x) + f_-(x)$$

and hence (Theorem 7, §1)

$$\int_E |f|\, dx = \int_E f_+\, dx + \int_E f_-\, dx.$$

The theorem follows at once.

We note some simple consequences of the preceding theorem.

 I. A summable function is finite almost everywhere.

 II. If $mE = 0$, then every function $f(x)$ is summable on E and $\int_E f(x)\, dx = 0$.

 III. A function which is summable on a set E is summable on every measurable subset of E.

 IV. Let the functions $f(x)$ and $F(x)$ be measurable on a set E and suppose that $|f(x)| \leq F(x)$. If the function $F(x)$ is summable, then $f(x)$ is also summable.

If functions $f(x)$ and $g(x)$ are equivalent on a set E, then, obviously, $f_+(x)$ and $g_+(x)$ are equivalent as well as $f_-(x)$ and $g_-(x)$. This implies the following fact.

THEOREM 2. *Let the functions $f(x)$ and $g(x)$ be equivalent. If one of the integrals $\int_E f(x)\, dx$, $\int_E g(x)\, dx$ exists, then so does the other, and the two integrals are equal.*

In particular, the functions $f(x)$ and $g(x)$ are either both summable or both not summable. In the sequel, in general, we shall not usually distinguish among equivalent functions. Such a convention is very useful. For example, it permits summable functions to be added without any exceptions. The fact is that in forming the sum $f'(x) + f''(x)$, we must exclude from consideration those points where the terms take on infinite

2. Summable Functions of Arbitrary Sign

values of different sign. In order not to be forced to make this exclusion, we can alter the value of one of the terms at these points because the set of such points has measure zero (the terms are summable !). In doing this, it is immaterial which term is changed and what new value we assign to it—the new sum is equivalent to the old.

THEOREM 3 (FINITE ADDITIVITY OF THE INTEGRAL). *Let the set E be the sum of a finite number of pairwise disjoint measurable sets*

$$E = \sum_{k=1}^{n} E_k \qquad (E_k E_{k'} = 0, \quad k \neq k').$$

If a function $f(x)$ is summable on each of the sets E_k, then it is also summable on their sum E, and

$$\int_E f(x)\,dx = \sum_{k=1}^{n} \int_{E_k} f(x)\,dx.$$

Proof. By Theorem 12, §1, the equalities

$$\int_E f_+\,dx = \sum_{k=1}^{n} \int_{E_k} f_+\,dx,$$

$$\int_E f_-\,dx = \sum_{k=1}^{n} \int_{E_k} f_-\,dx$$

are valid, where the right members (and hence the left members also) are finite. It remains only to subtract the second equality from the first.

In the case of a denumerable family of sets, the summability of a function $f(x)$ on each of the terms does not imply its summability on their sum.

EXAMPLE. Let the function $f(x)$ be defined on $(0, 1]$ in the following way:

$$f(x) = \begin{cases} n & \text{for } \dfrac{2n+1}{2n(n+1)} < x \leq \dfrac{1}{n}, \\ -n & \text{for } \dfrac{1}{n+1} < x \leq \dfrac{2n+1}{2n(n+1)} \end{cases} \qquad (n = 1, 2, 3, \ldots).$$

Then $f(x)$ is summable on each of the half-open intervals $\left(\dfrac{1}{n+1}, \dfrac{1}{n}\right]$, and

$$\int_{\frac{1}{n+1}}^{\frac{1}{n}} f(x)\,dx = 0.$$

Nevertheless the function $f(x)$ is not summable on the set $(0, 1]$, because

$$\int_0^1 |f(x)|\,dx = \sum_{n=1}^{\infty} \int_{\frac{1}{n+1}}^{\frac{1}{n}} |f(x)|\,dx = \sum_{n=1}^{\infty} \frac{1}{n+1} = +\infty.$$

However, the following theorems on complete additivity of the integral are valid.

THEOREM 4. *If a function $f(x)$ is summable on a set E and if E is representable as the sum of a denumerable family of pairwise disjoint measurable sets:*

$$E = \sum_{k=1}^{\infty} E_k \qquad (E_k E_{k'} = 0, \quad k \neq k')$$

then

$$\int_E f(x)\,dx = \sum_{k=1}^{\infty} \int_{E_k} f(x)\,dx. \qquad (2)$$

THEOREM 5. *Let the measurable set E be as in the preceding theorem. If $f(x)$ is summable on each of the sets E_k and if*

$$\sum_{k=1}^{\infty} \int_{E_k} |f(x)|\,dx < +\infty,$$

then $f(x)$ is summable on the set E and equality (2) holds.

Proof. To prove Theorem 4, we apply Theorem 12, §1, and find that

$$\int_E f_+\,dx = \sum_{k=1}^{\infty} \int_{E_k} f_+\,dx, \quad \int_E f_-\,dx = \sum_{k=1}^{\infty} \int_{E_k} f_-\,dx.$$

Since the left-hand members of these expressions are finite, so are the right-hand members, and the equalities indicated are valid. Theorem 4 follows upon subtracting the second equality from the first.

To prove Theorem 5, we again apply Theorem 12, §1 to establish the equality

$$\int_E |f|\,dx = \sum_{k=1}^{\infty} \int_{E_k} |f|\,dx.$$

This implies that $|f(x)|$ is summable on E. Therefore $f(x)$ is also summable on E, and we can now apply Theorem 4 above.

We note that it is impossible to replace the hypothesis of Theorem 5 by the requirement that the series

$$\sum_{k=1}^{\infty} \int_{E_k} f(x)\,dx$$

converge. The last example given makes this clear.

THEOREM 6. *If a function $f(x)$ is summable on a set E, and if k is a finite constant, then the function $kf(x)$ also is summable on E and*

$$\int_E kf(x)\,dx = k \int_E f(x)\,dx.$$

Proof. The theorem is trivial for $k = 0$. For $k > 0$, the obvious equalities

$$(kf)_+ = kf_+, \quad (kf)_- = kf_-,$$

reduce the theorem to Theorem 8, §1 (simply integrate the above equalities and subtract the second from the first).

2. Summable Functions of Arbitrary Sign

To prove the theorem for $k < 0$, we consider first the case $k = -1$. It is easy to see that
$$(-f)_+ = f_-, \quad (-f)_- = f_+.$$
Therefore
$$\int_E -f(x)\,dx = \int_E f_-(x)\,dx - \int_E f_+(x)\,dx = -\int_E f(x)\,dx.$$
Thus, the factor (-1) can be taken outside the integral sign. Finally, let k be an arbitrary negative number. Then
$$\int_E kf\,dx = -\int_E |k|f\,dx = -|k|\int_E f\,dx = k\int_E f\,dx,$$
and the theorem is proved.

COROLLARY. *If a function $f(x)$ is summable on a set E and if $\varphi(x)$ is measurable and bounded on E, then the product $\varphi(x)f(x)$ is summable on E.*

In fact, the absolute value $|\varphi(x)f(x)|$ (which is obviously a measurable function) does not exceed the summable function $K|f(x)|$, where $K = \sup\{|\varphi(x)|\}$.

THEOREM 7. *If two functions $f'(x)$ and $f''(x)$ are summable on a set E, then the function $f(x) = f'(x) + f''(x)$ is also summable on E, and*
$$\int_E f(x)\,dx = \int_E f'(x)\,dx + \int_E f''(x)\,dx. \tag{3}$$

Proof. The summability of the function $f(x)$ follows from the fact that
$$|f(x)| \leqslant |f'(x)| + |f''(x)|,$$
and Theorem 7, §1. Equality (3) remains to be proved. To do this, introduce the sets

$E_1 = E(f' \geqslant 0, f'' \geqslant 0);\qquad E_2 = E(f' < 0, f'' < 0);$
$E_3 = E(f' \geqslant 0, f'' < 0, f \geqslant 0);\quad E_4 = E(f' \geqslant 0, f'' < 0, f < 0);$
$E_5 = E(f' < 0, f'' \geqslant 0, f \geqslant 0);\quad E_6 = E(f' < 0, f'' \geqslant 0, f < 0).$

It is clear that
$$E = \sum_{k=1}^{6} E_k \qquad (E_k E_{k'} = 0,\ k \neq k'),$$
and it is sufficient to prove that
$$\int_{E_k} f\,dx = \int_{E_k} f'\,dx + \int_{E_k} f''\,dx \qquad (k = 1, 2, \ldots, 6).$$

This is done in the same way for all k. As an example, we carry out the reasoning for $k = 6$. Rewriting the equality
$$f(x) = f'(x) + f''(x)$$
as
$$-f'(x) = f''(x) + [-f(x)],$$

we obtain an equality in which both members of the right side are non-negative on the set E_6. Theorem 7, §1, implies that

$$\int_{E_6} (-f')|dx = \int_{E_6} f''dx + \int_{E_6} (-f)\,dx,$$

and consequently

$$\int_{E_6} f\,dx = \int_{E_6} f'\,dx + \int_{E_6} f''\,dx.$$

The theorem is proved.

The following theorem is very important.

THEOREM 8. *Let a function $f(x)$ be summable on a set E. For every $\varepsilon > 0$, there exists a $\delta > 0$ such that for every measurable set $e \subset E$ of measure $me < \delta$, the inequality*

$$\left|\int_e f(x)\,dx\right| < \varepsilon$$

obtains.

Proof. Since $f(x)$ is summable, by hypothesis, it follows that $|f(x)|$ is also summable. Let ε be any positive number. Then, by the definition of the Lebesgue integral of $|f(x)|$, there exists a positive number N_0 such that

$$\int_E |f(x)|\,dx - \int_E [|f(x)|]_{N_0} dx < \frac{\varepsilon}{2}.$$

Set

$$\delta = \frac{\varepsilon}{2N_0}.$$

This δ turns out to be the number required. We note first that the function

$$|f(x)| - [|f(x)|]_{N_0}$$

is non-negative on the set E. Hence, if e is a measurable subset of the set E, we necessarily have

$$\int_e \{|f(x)| - [|f(x)|]_{N_0}\}\,dx \leq \int_E \{|f(x)| - [|f(x)|]_{N_0}\}\,dx.$$

Therefore

$$\int_e |f(x)|\,dx - \int_e [|f(x)|]_{N_0}\,dx < \frac{\varepsilon}{2}$$

and we have

$$\int_e |f(x)|\,dx < \frac{\varepsilon}{2} + \int_e [|f(x)|]_{N_0}\,dx.$$

Since

$$[|f(x)|]_{N_0} \leq N_0,$$

it is clear that

$$\int_e [|f(x)|]_{N_0}\,dx \leq N_0 \cdot me,$$

and, consequently,
$$\int_e |f(x)|\,dx < \tfrac{\varepsilon}{2} + N_0 \cdot me.$$

From this it is clear that for $me < \delta$ we have
$$\int_e |f(x)|\,dx < \varepsilon,$$

and, *a fortiori*,
$$\left|\int_e f(x)\,dx\right| < \varepsilon,$$

as was to be proved.

The property of the integral just established is called *absolute continuity*.*

§ 3. PASSAGE TO THE LIMIT UNDER THE INTEGRAL SIGN

Lebesgue's theorem proved in §3, Chapt. V, admits the following generalization.

THEOREM 1 (H. LEBESGUE). *Let a sequence of measurable functions $f_1(x), f_2(x), f_3(x), \ldots$, converging in measure to a function $F(x)$, be defined on a set E. If there exists a function $\Phi(x)$ summable on E such that for all n and x,*
$$|f_n(x)| \leqslant \Phi(x), \tag{*}$$

---

*A different definition of summable functions of arbitrary sign can be given, which makes the calculations of the present § somewhat easier. Take the definition of non-negative summable functions just as in §1 of the present chapter. Then define the class of all summable functions on a measurable set E as being those functions f which can be written as the difference of two non-negative summable functions:
$$f = f_1 - f_2 \qquad f_1, f_2 \geqslant 0, \tag{4}$$
with f_1 and f_2 summable. It is obvious that every function summable in this new sense is summable in the sense of §2 of the present chapter. Define $\int_E f\,dx$ by
$$\int_E f\,dx = \int_E f_1\,dx - \int_E f_2\,dx. \tag{5}$$
It is not hard to see that the right-hand side of (5) is independent of what pair of functions f_1, f_2 is used to represent f by (4). Suppose that
$$f = f_1 - f_2 = g_1 - g_2,$$
where f_1, f_2, g_1 and g_2 are $\geqslant 0$ and are summable on E. We then have
$$f_1 + g_2 = g_1 + f_2.$$
Applying Theorem 7, §1, to this equality and transposing, we find that
$$\int_E f_1\,dx - \int_E f_2\,dx = \int_E g_1\,dx - \int_E g_2\,dx,$$
so that (5) does in fact define $\int_E f\,dx$ unambiguously. It is now easy to show that Theorems 7 and 6 of the present § hold for this integral, and, finally, that if f can be written in the form (4), then f_+ and f_- are both summable. Verification of these assertions is left as an exercise to the reader.—E.H.

then

$$\lim_{n\to\infty} \int_E f_n(x)\,dx = \int_E F(x)\,dx.$$

Proof. First of all, we note that the hypothesis (*) implies the summability of each of the functions $f_n(x)$. Further, it is easy to see that

$$|F(x)| \leqslant \Phi(x) \qquad (1)$$

for almost all x. To show this, it is sufficient to extract from $\{f_n(x)\}$ a subsequence $\{f_{n_k}(x)\}$ which converges to $F(x)$ almost everywhere, using Riesz's theorem (Theorem 4, § 3, Ch. IV), and then pass to the limit in the inequality

$$|f_{n_k}(x)| \leqslant \Phi(x).$$

By changing the values of the function $F(x)$ on a set of measure zero if necessary, we can make inequality (1) valid at every point of the set E. In particular, (1) implies that the function $F(x)$ is summable on E.

Now choose an arbitrary $\sigma > 0$ and set

$$A_n(\sigma) = E(|f_n - F| \geqslant \sigma), \quad B_n(\sigma) = E(|f_n - F| < \sigma).$$

Then

$$E = A_n(\sigma) + B_n(\sigma), \quad A_n(\sigma) \cdot B_n(\sigma) = 0,$$

and

$$mA_n(\sigma) \to 0$$

as $n \to \infty$. Noting all this, we make the following evaluations:

$$\left| \int_E f_n\,dx - \int_E F\,dx \right| \leqslant \int_E |f_n - F|\,dx = \int_{A_n(\sigma)} |f_n - F|\,dx + \int_{B_n(\sigma)} |f_n - F|\,dx.$$

On the set $B_n(\sigma)$ we have $|f_n - F| < \sigma$, therefore

$$\int_{B_n(\sigma)} |f_n - F|\,dx \leqslant \sigma m B_n(\sigma) \leqslant \sigma \cdot mE.$$

Furthermore,

$$|f_n - F| \leqslant 2\Phi(x),$$

so that

$$\int_{A_n(\sigma)} |f_n - F|\,dx \leqslant 2 \int_{A_n(\sigma)} \Phi(x)\,dx.$$

Combining these inequalities, we obtain

$$\left| \int_E f_n\,dx - \int_E F\,dx \right| \leqslant 2 \int_{A_n(\sigma)} \Phi(x)\,dx + \sigma mE. \qquad (2)$$

Finally, let $\varepsilon > 0$. Choose $\delta > 0$ so small that

$$\sigma \cdot mE < \frac{\varepsilon}{2}. \qquad (3)$$

Then, making use of the absolute continuity of the integral of the function $\Phi(x)$

3. Passage to the Limit under the Integral Sign

(Theorem 8, § 2) we find a $\delta > 0$ such that for every measurable set $e \subset E$ of measure $me < \delta$, we have
$$\int_e \Phi(x)\,dx < \frac{\varepsilon}{4}.$$
There exists a natural number n_0 such that if $n > n_0$, then (for σ fixed in advance)
$$mA_n(\sigma) < \delta,$$
and
$$2\int_{A_n(\sigma)} \Phi(x)\,dx < \frac{\varepsilon}{2} \tag{4}$$
for $n > n_0$. Combining (2), (3) and (4), we see that for $n > n_0$,
$$\left|\int_E f_n(x)\,dx - \int_E F(x)\,dx\right| < \varepsilon,$$
which proves the theorem.

COROLLARY. *Under the hypotheses of the preceding theorem, we have*
$$\lim_{n\to\infty} \int_E \varphi(x) f_n(x)\,dx = \int_E \varphi(x) F(x)\,dx,$$
$\varphi(x)$ *being an arbitrary bounded measurable function.*

Proof. If $|\varphi(x)| \leqslant K$, then
$$|\varphi(x) f_n(x)| \leqslant K\Phi(x),$$
and condition (*) is fulfilled. It remains to show that
$$\varphi(x) f_n(x) \Rightarrow \varphi(x) F(x).$$
This follows from the fact that
$$E(|\varphi f_n - \varphi F| \geqslant \sigma) \subset E\left(|f_n - F| \geqslant \frac{\sigma}{K}\right).$$

Theorem 1 admits a further generalization. To state this generalization, we introduce the following important concept. Let $M = \{f(x)\}$ be a family of summable functions defined on a measurable set E. If we fix one of these functions, say $f_0(x)$, then for every $\varepsilon > 0$, there exists a $\delta > 0$ such that the relations
$$e \subset E, \quad me < \delta$$
imply that
$$\left|\int_e f_0(x)\,dx\right| < \varepsilon.$$
This δ depends on the function $f_0(x)$ and in general there is no one common δ for all functions of the family M. This situation gives rise to the following notion.

DEFINITION. *Let $M = \{f(x)\}$ be a family of summable functions defined on a set E. If for every $\varepsilon > 0$ there exists a $\delta > 0$ such that the relations*
$$e \subset E, \quad me < \delta$$
imply the relation
$$\left|\int_e f(x)\,dx\right| < \varepsilon$$

for all functions of the family M, then the functions of the family M are said to have equi-absolutely continuous integrals.

THEOREM 2 (D. VITALI). *Let a sequence of summable functions $f_1(x), f_2(x), f_3(x), \ldots$ converging in measure to the function $F(x)$, be defined on a measurable set E. If the functions of the sequence $\{f_n(x)\}$ have equi-absolutely continuous integrals, then $F(x)$ is summable and*

$$\lim_{n \to \infty} \int_E f_n(x)\, dx = \int_E F(x)\, dx.$$

Proof. It is first necessary to show that the limit function $F(x)$ is summable on the set E. To do this, we choose an arbitrary $\varepsilon > 0$ and choose $\delta > 0$ such that for $me < \delta$, we have

$$\left| \int_e f_n(x)\, dx \right| < \frac{\varepsilon}{2} \qquad (n = 1, 2, 3, \ldots).$$

Let e be any measurable subset of E of measure $me < \delta$. Then, setting

$$e_+ = e\,(f_n \geqslant 0), \quad e_- = e\,(f_n < 0),$$

we obviously have

$$me_+ < \delta, \quad me_- < \delta$$

and accordingly

$$\int_{e_+} |f_n|\, dx = \left| \int_{e_+} f_n\, dx \right| < \frac{\varepsilon}{2}, \quad \int_{e_-} |f_n|\, dx = \left| \int_{e_-} f_n\, dx \right| < \frac{\varepsilon}{2}$$

for all n. Hence

$$\int_e |f_n(x)|\, dx < \varepsilon. \tag{5}$$

In other words, the functions $|f_n(x)|$ also have [3] equi-absolutely continuous integrals.

If we construct (according to Riesz's theorem, Theorem 4, § 3, Ch. IV) a subsequence $\{f_{n_k}(x)\}$ converging to $F(x)$ almost everywhere and write down the inequalities (5) for $f_{n_k}(x)$, then Fatou's theorem (Theorem 9, § 1) allows us to state that

$$\int_e |F(x)|\, dx \leqslant \varepsilon, \tag{6}$$

so that $F(x)$ is summable on the set e. Here e is an arbitrary subset of E of measure $< \delta$. From this it is clear that $F(x)$ is summable on the initial set E also, for E is the sum of a finite number of parts of measure $< \delta$.

We turn now to the proof of the main assertion of the theorem. Choosing $\sigma > 0$ and writing

$$A_n(\sigma) = E\,(|f_n - F| \geqslant \sigma), \quad B_n(\sigma) = E\,(|f_n - F| < \sigma),$$

---

[3] We have thus shown that if the functions $f_n(x)$ have equi-absolutely continuous integrals, then the absolute values $|f_n(x)|$ also have equi-absolutely continuous integrals.

3. Passage to the Limit under the Integral Sign

we again obtain the evaluation

$$\left| \int_E f_n \, dx - \int_E F \, dx \right| \leqslant \int_{A_n(\sigma)} |f_n - F| \, dx + \sigma m E$$

just as in the proof of Theorem 1. Hence

$$\left| \int_E f_n \, dx - \int_E F \, dx \right| \leqslant \int_{A_n(\sigma)} |f_n| \, dx + \int_{A_n(\sigma)} |F| \, dx + \sigma m E. \tag{7}$$

Let $\varepsilon > 0$ and let σ be determined so that

$$\sigma m E < \frac{\varepsilon}{3}.$$

As noted at the beginning of the proof, there corresponds to every $\varepsilon > 0$, a $\delta > 0$ such that

$$e \subset E \text{ and } me < \delta$$

imply that

$$\int_e |f_n| \, dx < \frac{\varepsilon}{3}, \quad \int_e |F| \, dx < \frac{\varepsilon}{3}$$

(see (5) and (6)). There exists a natural number n_0 such that

$$m A_n(\sigma) < \delta,$$

for $n > n_0$. For all such n, (7) implies that

$$\left| \int_E f_n \, dx - \int_E F \, dx \right| < \varepsilon,$$

and the present theorem is proved.

COROLLARY. *Under the hypotheses of the preceding theorem, the relation*

$$\lim_{n \to \infty} \int_E \varphi(x) f_n(x) \, dx = \int_E \varphi(x) F(x) \, dx$$

holds, $\varphi(x)$ being an arbitrary bounded measurable function.

In fact, if $|\varphi(x)| \leqslant K$, then

$$\left| \int_e \varphi(x) f_n(x) \, dx \right| \leqslant K \int_e |f_n(x)| \, dx,$$

so that the functions $\varphi(x) f_n(x)$ also have equi-absolutely continuous integrals.

It happens that the converse of the above corollary is true. To show this, we need the following theorem, which is important in itself.

THEOREM 3 (H. LEBESGUE). *Let a sequence of summable functions $\{f_n(x)\}$ be defined on a measurable set E. If, for every measurable subset of the set E, we have*

$$\lim_{n \to \infty} \int_e f_n(x) \, dx = 0, \tag{8}$$

then the functions $f_n(x)$ have equi-absolutely continuous integrals.

Proof. Assume that the theorem is not true. This means that there exists a number $\varepsilon_0 > 0$ having the following property: for every $\delta > 0$, there can be found a measurable set $e \subset E$ of measure $me < \delta$ and a natural number n such that

$$\left|\int_e f_n(x)\,dx\right| \geq \varepsilon_0. \tag{9}$$

Choose a number $\delta > 0$ and consider the first N functions of the sequence: $f_1(x)$, $f_2(x), \ldots, f_N(x)$. For each of these functions $f_k(x)$, a $\delta_k > 0$ can be found such that $me < \delta_k$ ($e \subset E$) implies

$$\left|\int_e f_k(x)\,dx\right| < \varepsilon_0. \tag{10}$$

Let δ^* be the minimum of the numbers $\delta, \delta_1, \delta_2, \ldots, \delta_n$. According to the above, there exist a set $e \subset E$ of measure $me < \delta^*$ and an index n such that inequality (9) holds. On the other hand, $me < \delta_k$ ($k = 1, 2, \ldots, N$) so that (10) holds for $k = 1, 2, \ldots, N$, and therefore we must have $n > N$.

The number ε_0 thus possesses the following property. For all $\delta > 0$ and $N = 1, 2, 3, \ldots$, there exists a measurable set $e \subset E$ and a natural number n such that $n > N$, $me < \delta$, $|\int_e f_n(x)dx| \geq \varepsilon_0$. We next fix on a set $e \subset E$ and a natural number n_1 for which

$$\left|\int_{e_1} f_{n_1}(x)\,dx\right| \geq \varepsilon_0.$$

Using the absolute continuity of the integral of the function $f_{n_1}(x)$, we choose a $\delta_1 > 0$ such that

$$\left|\int_e f_{n_1}(x)\,dx\right| < \frac{\varepsilon_0}{4}$$

for all sets $e \subset E$ of measure $me < \delta_1$. We next choose a set $e_2 \subset E$ and a natural number n_2 such that

$$n_2 > n_1, \quad me_2 < \frac{\delta_1}{2}, \quad \left|\int_{e_2} f_{n_2}(x)\,dx\right| \geq \varepsilon_0$$

after which we choose a $\delta_2 > 0$ such that $me < \delta_2$ ($e \subset E$) implies

$$\left|\int_e f_{n_2}(x)\,dx\right| < \frac{\varepsilon_0}{4}.$$

It is easy to see that $\delta_2 < \frac{\delta_1}{2}$.

Having done this, we choose a set $e_3 \subset E$ and a natural number n_3 for which

$$n_3 > n_2, \quad me_3 < \frac{\delta_2}{2}, \quad \left|\int_{e_3} f_{n_3}(x)\,dx\right| \geq \varepsilon_0,$$

after which we choose a $\delta_3 > 0$ such that $me < \delta_3$ ($e \subset E$) implies

$$\left|\int_e f_{n_3}(x)\,dx\right| < \frac{\varepsilon_0}{4}.$$

It is obvious that $\delta_3 < \frac{\delta_2}{2}$.

Continuing this process, we construct three sequences: a sequence of measurable

3. Passage to the Limit under the Integral Sign

sets $e_k \subset E$; a sequence of strictly increasing natural numbers n_k; and a sequence of numbers $\delta_k > 0$ such that

1) $$\left| \int_{e_k} f_{n_k}(x)\,dx \right| \geq \varepsilon_0;$$

2) $$me_{k+1} < \frac{\delta_k}{2};$$

3) if $e \subset E$ and $me < \delta_k$, then
$$\left| \int_e f_{n_k}(x)\,dx \right| < \frac{\varepsilon_0}{4}.$$

It follows from these properties that $\delta_{k+1} < \frac{\delta_k}{2}$, and hence

$$m(e_{k+1} + e_{k+2} + e_{k+3} + \ldots) < \frac{\delta_k}{2} + \frac{\delta_{k+1}}{2} + \frac{\delta_{k+2}}{2} + \ldots < \delta_k.$$

Therefore
$$\left| \int_{e_k(e_{k+1}+e_{k+2}+\ldots)} f_{n_k}(x)\,dx \right| < \frac{\varepsilon_0}{4}.$$

We now introduce new sets
$$A_k = e_k - (e_{k+1} + e_{k+2} + \ldots).$$

It is clear that
$$\left| \int_{A_k} f_{n_k}(x)\,dx \right| \geq \frac{3}{4}\varepsilon_0. \tag{11}$$

Moreover (which is the purpose of introducing the sets A_k and which distinguishes them from the sets e_k), the sets A_k are pairwise disjoint. Note further that $A_k \subset e_k$ and hence

$$m(A_{k+1} + A_{k+2} + A_{k+3} + \ldots) < \delta_k. \tag{12}$$

It is now an easy matter to complete the proof. Set $k_1 = 1$ and define k_2 as any one of the natural numbers $m > 1$ for which

$$\left| \int_{A_{k_1}} f_{n_m}(x)\,dx \right| < \frac{\varepsilon_0}{4}.$$

The existence of such m follows from (8). Having done this, designate by k_3 any one of the indices $m > k_2$ for which

$$\left| \int_{A_{k_1} + A_{k_2}} f_{n_m}(x)\,dx \right| < \frac{\varepsilon_0}{4}.$$

Continuing this process, we obtain a strictly increasing sequence of natural numbers $k_1 < k_2 < k_3 < \ldots$, of such a nature that

$$\left| \int_{A_{k_1} + \ldots A_{k_{i-1}}} f_{n_{k_i}}(x)\,dx \right| < \frac{\varepsilon_0}{4}. \tag{13}$$

The inequality

$$\left|\int_{A_{k_i}} f_{n k_i}(x)\,dx\right| \geq \frac{3}{4}\varepsilon_0, \tag{14}$$

which is a special case of (11), is to be compared with inequality (13). Finally, in view of (12),

$$m(A_{k_i+1} + A_{k_i+2} + A_{k_i+3} + \ldots) \leq m(A_{k_i+1} + A_{k_i+2} + \ldots) < \delta_{k_i}$$

Therefore

$$\left|\int_{A_{k_i+1}+A_{k_i+2}+\ldots} f_{n k_i}(x)\,dx\right| < \frac{\varepsilon_0}{4}. \tag{15}$$

Set $Q = A_{k_1} + A_{k_2} + A_{k_3} + \ldots$. Then

$$\int_Q f_{n k_i}(x)\,dx = \int_{A_{k_1}+\ldots A_{k_{i-1}}} f_{n k_i}(x)\,dx + \int_{A_{k_i}} f_{n k_i}(x)\,dx + \int_{A_{k_i+1}+A_{k_i+2}+\ldots} f_{n k_i}(x)\,dx,$$

whence, in connection with (13), (14) and (15) it follows that

$$\left|\int_Q f_{n k_i}(x)\,dx\right| \geq \frac{\varepsilon_0}{4}, \qquad (i = 1, 2, 3\ldots)$$

and this contradicts the hypothesis (8). The theorem is proved.

This somewhat complicated method [4] of proof is encountered very often; the reader should carefully study the reasoning used.

COROLLARY 1. *Let a sequence of summable functions $\{f_n(x)\}$ and a summable function $F(x)$ be defined on a measurable set E. If for every arbitrary measurable set $e \subset E$, we have*

$$\lim_{n\to\infty} \int_e f_n(x)\,dx = \int_e F(x)\,dx, \tag{16}$$

then the functions $f_n(x)$ have equi-absolutely continuous integrals.

According to Theorem 3, the differences $f_n(x) - F(x)$ have equi-absolutely continuous integrals, and our assertion follows from the inequality

$$\left|\int_e f_n(x)\,dx\right| \leq \left|\int_e \{f_n(x) - F(x)\}\,dx\right| + \left|\int_e F(x)\,dx\right|.$$

COROLLARY 2. *Let a sequence of summable functions $f_n(x)$ and a summable function $F(x)$ be defined on a measurable set E. If, for all bounded measurable functions $\varphi(x)$, we have*

$$\lim_{n\to\infty} \int_E \varphi(x) f_n(x)\,dx = \int_E \varphi(x) F(x)\,dx,$$

then the functions $f_n(x)$ have equi-absolutely continuous integrals.

---

[4] It seems appropriate to call it the "sliding hump method."

3. Passage to the Limit under the Integral Sign

For the function $\varphi(x)$ we can, in particular, choose the characteristic function of an arbitrary measurable subset of E, and this brings us back to Corollary 1. This justifies our remark about the converse of the corollary to Theorem 2.

We obtain the following result from the foregoing.

Theorem 4 (D. Vitali). *Let a sequence of summable functions $\{f_n(x)\}$, converging in measure to the summable function $F(x)$, be defined on a measurable set E. The equality* (16) *holds for every measurable set $e \subset E$ if and only if the functions $f_n(x)$ have equi-absolutely continuous integrals.*

We note that the condition of equi-absolute continuity of the integrals of the functions $f_n(x)$ is needed only to show that passage to the limit can be accomplished under the integral extended over an *arbitrary* measurable subset of the set E. We cannot infer the equi-absolute continuity of the integrals mentioned from the single relation

$$\lim_{n \to \infty} \int_E f_n(x)\,dx = \int_E F(x)\,dx.$$

For example, let the functions $f_n(x)$ be defined on [0, 1] in the following way: $f_n(0) = 0$ and

$$f_n(x) = \begin{cases} n & \text{for } 0 < x \leqslant \dfrac{1}{2n} \\ -n & \text{for } \dfrac{1}{2n} < x \leqslant \dfrac{1}{n} \\ 0 & \text{for } \dfrac{1}{n} < x \leqslant 1. \end{cases}$$

It is clear that this sequence of functions converges to zero and that

$$\lim_{n \to \infty} \int_0^1 f_n(x)\,dx = 0.$$

At the same time,

$$\int_0^{\frac{1}{2n}} f_n(x)\,dx = \frac{1}{2},$$

thus, the functions f_n are not equi-absolutely continuous.

It is interesting that the situation is simpler for functions of fixed sign.

Theorem 5. *Let $\{f_n(x)\}$ be a sequence of non-negative summable functions converging in measure to the function $F(x)$, defined on a measurable set E. If* (16) *holds for $e = E$, then* (16) *holds for all measurable subsets e of E.*[5]

This theorem is proved easily from Fatou's[6] theorem (Theorem 9, § 1). Assume the present theorem is false. Then a measurable set $A \subset E$ can be found for which the equality

$$\lim_{n \to \infty} \int_A f_n(x)\,dx = \int_A F(x)\,dx$$

---

[5] The preceding example shows that the theorem fails without the hypothesis $f_n(x) \geqslant 0$.
[6] More exactly, from the footnote to this theorem.

fails. Then there exists a number $\sigma > 0$ such that an infinite number of the integrals $\int_A f_n(x)\, dx$ lie outside the interval

$$\left(\int_A F\, dx - 2\sigma,\ \int_A F\, dx + 2\sigma \right).$$

If an infinite number of the integrals $\int_A f_n(x)\, dx$ are less than $\int_A F(x)\, dx - 2\sigma$, then, by taking a subsequence, we infer that the inequality

$$\int_A f_n(x)\, dx \leqslant \int_A F(x)\, dx - 2\sigma$$

holds for all n. This leads at once to a contradiction of Fatou's theorem. It must therefore be the case that

$$\int_A f_n(x)\, dx \geqslant \int_A F(x)\, dx + 2\sigma \qquad (17)$$

for an infinite number of values of n. Again, upon passing to a subsequence of $\{f_n\}$, we may suppose that (17) holds for all n. By hypothesis, however, we have

$$\lim_{n \to \infty} \int_E f_n(x)\, dx = \int_E F(x)\, dx.$$

Hence the inequality

$$\left| \int_E f_n(x)\, dx - \int_E F(x)\, dx \right| < \sigma$$

obtains for all sufficiently large n. Again choosing a subsequence, we may suppose that

$$\int_E f_n(x)\, dx < \int_E F(x)\, dx + \sigma$$

for all n. Using the inequality (17) and writing $E - A = B$, we find that

$$\int_B f_n(x)\, dx < \int_B F(x)\, dx - \sigma$$

for all n; this leads again to an immediate contradiction to Fatou's theorem.

Theorem 5 leads to the following result.

THEOREM 6 (G. M. FICHTENHOLZ). *Let a sequence of summable functions $\{f_n(x)\}$, converging in measure to a summable function $F(x)$, be defined on a measurable set E. Equality (16) holds for all measurable sets $e \subset E$ if and only if the relation*

$$\lim_{n \to \infty} \int_E |f_n(x)|\, dx = \int_E |F(x)|\, dx \qquad (18)$$

obtains.

In fact, if (18) holds, then by Theorem 5, the set E can be replaced in (18) by an arbitrary measurable set $e \subset E$. Theorem 3 implies that the functions $|f_n(x)|$, and hence the functions $f_n(x)$, have equi-absolutely continuous integrals. It follows that (16) holds for every measurable $e \subset E$.

Conversely, if the equality (16) holds for every measurable set $e \subset E$, the functions $f_n(x)$ have equi-absolutely continuous integrals. Then absolute values $|f_n(x)|$ also have equi-absolutely continuous integrals, and it remains only to apply Theorem 2 to the sequence $\{|f_n(x)|\}$.

In conclusion, we describe a test for establishing equi-absolute continuity of integrals.

THEOREM 7 (DE LA VALLÉE-POUSSIN): *Let a family of measurable functions $M = \{f(x)\}$ be defined on the measurable set E. If there exists a positive increasing function $\Phi(u)$ defined for $u \geq 0$ and tending to $+\infty$ with u, and for which*

$$\int_E |f(x)| \cdot \Phi(|f(x)|) \, dx < A,$$

where $f(x)$ is an arbitrary function from M and A is a finite constant not depending on the choice of $f(x)$, then the functions $f(x)$ are integrable on E and their integrals are equi-absolutely continuous.

In explanation of the conditions of the theorem, we note that the composite function $\Phi(|((x)|))$, where $\Phi(u)$ is monotonic and $|f(x)|$ is measurable, is again a measurable function.[7] Turning to the proof itself, we take $\varepsilon > 0$. To this ε there corresponds a K such that

$$\frac{A}{\Phi(K)} < \frac{\varepsilon}{2}.$$

Fixing this K, we consider an arbitrary measurable set $e \subset E$. Let $f(x)$ be an arbitrary function from M. Set $e_1 = e(|f(x)| > K)$, $e_2 = e(|f(x)| \leq K)$. Then

$$\int_e |f(x)| \, dx = \int_{e_1} |f(x)| \, dx + \int_{e_2} |f(x)| \, dx \leq \frac{1}{\Phi(K)} \int_{e_1} |f(x)| \cdot \Phi(|f(x)|) \, dx + \int_{e_2} |f(x)| \, dx.$$

Hence

$$\int_e |f(x)| \, dx \leq \frac{A}{\Phi(K)} + K \cdot me_2 < \frac{\varepsilon}{2} + Kme.$$

This proves the summability of $f(x)$. In addition to this, setting $\delta = \frac{\varepsilon}{2K}$, we see that for $me < \delta$,

$$\int_e |f(x)| \, dx < \varepsilon.$$

Theorem 7 implies, for example, that if for all functions $f(x)$ of the family M, we have

$$\int_E f^2(x) \, dx < A,$$

then these functions have equi-absolutely continuous integrals.

§ 4. EDITOR'S APPENDIX TO CHAPTER VI.

We shall discuss here the definition of the Lebesgue integral for functions defined on unbounded sets, which may have infinite measure. Let $f(x)$, then, be a measurable function defined on an unbounded measurable set E. (The case $E = (-\infty, \infty)$ is particularly important). We could start with bounded non-negative measurable functions, define Lebesgue sums as in § 1, Chapter V, of the form s and S, and define the Lebesgue integral $\int f(x) \, dx$ as the supremum of all *lower* sums s, since every upper

---

[7] If $a > 0$, $\underset{u}{E}(\Phi(u) > a)$ is an interval of the form $(b, +\infty)$. Therefore, $E(\Phi(f) > a) = E(f > b)$.

sum S may be infinite, if E has infinite measure. (Consider the function $f(x) = \frac{1}{1+x^2}$ on the entire line, for example). We should then have to establish all of the needed properties, such as additivity and homogeneity, for this integral, and then proceed by stages to general summable functions.

It is quicker and simpler to use the properties of summable functions already established in Chapters V and VI. For this purpose, we make the following definition. Let $f(x)$ be a measurable function defined on the measurable set E, which may be bounded or unbounded. If E is bounded, then the definition of summability given in § 2 is perfectly adequate. If E is unbounded, then no set $E[-n, n]$ is equal to E. Define the function $f_{[n]}(x)$ as being equal to $f(x)$ on the interval $[-n, n]$ and being undefined elsewhere.

Now suppose that $f(x)$ is non-negative and measurable and defined on E: $f(x)$ may be bounded or unbounded. Then we make the following definition.

DEFINITION 1. $\int_E f(x)\, dx = \lim_{n\to\infty} \int_{E[-n, n]} f_{[n]}(x)\, dx.$

Each of the integrals on the right-hand side in the above definition is defined, although it may be equal to $+\infty$ (see Definition, § 1). It is obvious that the integrals on the right-hand side form an increasing sequence of non-negative numbers, and so the limit must exist, as a finite number or as $+\infty$. (We make the assumption that an increasing sequence which from some point on is equal to $+\infty$ has limit equal to $+\infty$).

Next, suppose that $f(x)$ is of arbitrary sign on the set E. Then, introducing the functions $f_+(x)$ and $f_-(x)$ as in § 2, we make a second definition.

DEFINITION 2. $\int_E f(x)\, dx = \int_E f_+(x)\, dx - \int_E f_-(x)\, dx,$

provided that at least one of the integrals appearing on the right-hand side is finite. Otherwise, the integral $\int_E f(x)\, dx$ is undefined. If $\int_E f(x)\, dx$ exists and is finite, then $f(x)$ is said to be summable on the set E.

All of the theorems, corollaries, and remarks of §§ 1 and 2 are true for integrals taken over unbounded sets. Verification of most of these involves only a simple passage to the limit on both sides of an equality or inequality and may be left to the reader. Some of the arguments, however, are a little subtle. For example, let us give a proof of Fatou's theorem (Theorem 9, § 1).

Proof of Fatou's theorem for unbounded E.

Let n be an arbitrary positive integer. Then, clearly, $\lim_{m\to\infty} f_{m[n]}(x) = F_{[n]}(x)$ for almost all $x \in E[-n, n]$. By Fatou's theorem for bounded sets, therefore, we have

$$\int_{E[-n, n]} F_{[n]}(x)\, dx \leqslant \sup_m \{ \int_{E[-n, n]} f_{m[n]}(x)\, dx \}. \tag{1}$$

Taking limits of both sides of (1) as $n \to \infty$, we obtain

$$\int_E F(x)\, dx \leqslant \lim_{n\to\infty} [\sup_m \{ \int_{E[-n, n]} f_{m[n]}(x)\, dx \}]. \tag{2}$$

Since each integral $\int_{E[-n, n]} f_{m[n]}(x)\, dx$ is an increasing function of n, it is clear that

$$\sup_m \{ \int_{E[-n, n]} f_{m[n]}(x)\, dx \}$$

increases as n increases. Hence the limit in (2) exists. It is also obvious that

$$\int_{E[-n, n]} f_{m[n]}(x)\, dx \leqslant \int_E f_m(x)\, dx$$

for all m and n. Therefore
$$\sup_m \{ \int_{E[-n,n]} f_{m[n]}(x)\,dx \} \leq \sup_m \{ \int_E f_m(x)\,dx \},$$
for all n, and hence
$$\lim_{n\to\infty} [\sup_m \{ \int_{E[-n,n]} f_{m[n]}(x)\,dx \}] \leq \sup_m \{ \int_E f_m(x)\,dx \}. \tag{3}$$
Combining (2) and (3), we obtain Fatou's theorem.

Having Fatou's theorem for functions defined on unbounded sets E, it is easy to prove the theorem of B. Levi (Theorem 10, § 1) by repeating verbatim the argument given in the text. The proof of Theorem 12, § 1, can be carried over to the case of unbounded E and not necessarily bounded E_k by a slight paraphrasing.

The fundamental theorem on passage to the limit under the integral sign, and the one most widely used in applications of Lebesgue integration, differs slightly from Theorem 1, § 3. We therefore state and prove it in detail. The reader should note that the set E referred to may have either finite or infinite measure.

THEOREM 1 (LEBESGUE'S THEOREM ON DOMINATED CONVERGENCE). *Let E be a measurable set, and let $f_n(x)$ be a sequence of summable functions defined on E having the property that $|f_n(x)| \leq s(x)$ for all n and almost all x, where $s(x)$ is a summable non-negative function on E. Then if $\lim_{n\to\infty} f_n(x) = F(x)$ exists almost everywhere on E, the relation*
$$\lim_{n\to\infty} \int_E f_n(x)\,dx = \int_E F(x)\,dx \tag{4}$$
holds.

Proof. First of all, it is obvious that $F(x)$ is summable and that $|F(x)| \leq s(x)$ almost everywhere. Now define the sequence of functions $p_n(x)$ by the following rule:
$$p_n(x) = \lim_{k\to\infty} \{\min[f_n(x), f_{n+1}(x), \ldots, f_{n+k-1}(x)]\} \qquad n = 1, 2, 3, \ldots.$$
It is easy to see that the functions $p_n(x)$ are a non-decreasing sequence of measurable functions, that $|p_n(x)| \leq s(x)$ almost everywhere, and that
$$\lim_{n\to\infty} p_n(x) = \varliminf_{n\to\infty} f_n(x)$$
$$= F(x)$$
for almost all $x \in E$. The sequence of functions $s(x) + p_n(x)$ is therefore an increasing sequence of non-negative summable functions. Applying the theorem of B. Levi (Theorem 10, § 1), we have
$$\lim_{n\to\infty} \int_E [s(x) + p_n(x)]\,dx = \int_E \lim_{n\to\infty} [s(x) + p_n(x)]\,dx. \tag{5}$$
The right-hand side of (5) is obviously equal to
$$\int_E s(x)\,dx + \int_E F(x)\,dx. \tag{6}$$
The left-hand side of (5) is equal to
$$\int_E s(x)\,dx + \lim_{n\to\infty} \int_E p_n(x)\,dx. \tag{7}$$
Since $f_n(x) \geq p_n(x)$ for almost all $x \in E$, we have
$$\varliminf_{n\to\infty} \int_E f_n(x)\,dx \geq \lim_{n\to\infty} \int_E p_n(x)\,dx. \tag{8}$$

Assembling (5), (6), (7), and (8), we obtain

$$\lim_{n\to\infty} \int_E f_n(x)\,dx \geq \int_E F(x)\,dx. \tag{9}$$

We next define a sequence of functions $q_n(x)$ by the following rule:

$$q_n(x) = \lim_{k\to\infty} \{ \max [f_n(x), f_{n+1}(x), \ldots, f_{n+k-1}(x)] \}, \qquad n = 1, 2, 3, \ldots.$$

The functions $q_n(x)$ obviously form a non-increasing sequence of functions such that $|q_n(x)| \leq s(x)$ and such that

$$\lim_{n\to\infty} q_n(x) = \overline{\lim_{n\to\infty}} f_n(x)$$
$$= F(x)$$

for almost all $x \in E$. We consider then the increasing sequence of non-negative functions $s(x) - p_n(x)$. Again applying the theorem of B. Levi and repeating the argument used above, we infer that

$$\overline{\lim_{n\to\infty}} \int_E f_n(x)\,dx \leq \int_E F(x)\,dx. \tag{10}$$

Comparing (9) and (10), we see that

$$\lim_{n\to\infty} \int_E f_n(x)\,dx$$

exists and that (4) holds. This of course completes the proof.

COROLLARY. *Let $f(x)$ be a summable function defined on E. Extend the definition of $[f]_N$ given in § 1 by the rule*

$$[f(x)]_N = \begin{cases} -N & \text{if } f(x) < -N, \\ f(x) & \text{if } -N \leq f(x) \leq N, \\ N & \text{if } f(x) > N. \end{cases}$$

Then

$$\int_E f(x)\,dx = \lim_{N\to\infty} \int_E [f(x)]_N\,dx.$$

One can now prove Theorem 8, § 2, by applying the preceding corollary and repeating the proof of Theorem 8, § 2.

The theorems of § 3, barring Theorem 1, are of a somewhat specialized nature and are needed comparatively rarely in applications. We therefore omit a discussion of their extensions to sequences of functions defined on sets of infinite measure. We note merely that all of them are true for arbitrary sets of finite measure, whether bounded or unbounded.

Exercises for Chapters V and VI.

1. If $f_n(x) \geq 0$ and $\int_E f_n(x)\,dx \to 0$, then $f_n(x) \Rightarrow 0$, but it is not necessarily the case that $f_n(x)$ tends to 0 almost everywhere.

2. The relation

$$\int_E \frac{|f_n|}{1+|f_n|}\,dx \to 0$$

is equivalent to $f_n(x) \Rightarrow 0$.

EXERCISES

3. If $a_n \to 0$, there exists a sequence of non-negative measurable functions $u_n(x)$ such that $\sum_{n=1}^{\infty} a_n \int_E u_n(x)\, dx < +\infty$, but the functions $u_n(x)$ do not tend to 0 at any of the points of E.

4. If the integral
$$\int_E \varphi(x) f(x)\, dx$$
exists for every summable function $f(x)$, then the function $\varphi(x)$ is bounded almost everywhere (Lebesgue).

5. Let a measurable finite function $f(x)$ be defined on the set E. Consider a doubly infinite sequence of numbers
$$\ldots y_{-3}, y_{-2}, y_{-1}, y_0, y_1, y_2, y_3, \ldots \quad (y_k \to +\infty,\ y_{-k} \to -\infty,\ y_{k+1} - y_k < \lambda)$$
and set $e_k = E(y_k \leqslant f < y_{k+1})$. A necessary and sufficient condition that the function $f(x)$ be summable is that the series $\sum_{k=-\infty}^{\infty} y_k m e_k$ be absolutely convergent.

6. If, under the conditions of Exercise 5, the series $\Sigma y_k m e_k$ converges absolutely, then for $\lambda \to 0$, its sum tends to $\int_E f(x)\, dx$.

7. The limit of a uniformly convergent sequence of functions integrable (R) is a function integrable (R).

8. The characteristic function of the Cantor perfect set P_0 is integrable (R). What is its integral over $[0, 1]$?

9. Let $f(x)$ and $g(x)$ be two non-negative measurable functions defined on the set E. If $E_y = E(g > y)$, then
$$\int_E f(x) g(x)\, dx = \int_0^{+\infty} \Phi(y)\, dy,$$
where $\Phi(y) = \int_{E_y} f(x)\, dx$. (D. K. Faddeyev)

10. Let $E_1, E_2, \ldots, E_n$ be measurable subsets of $[0, 1]$. If each of the points of $[0, 1]$ belongs to at least q of these sets, then at least one of them has measure $\geqslant \frac{q}{n}$. (L. V. Kantorovič)

11. Let a summable function $f(x)$ be defined on $[a, b]$. Further, let a be a constant such that $0 < a < b - a$. If for every set e of measure a, $\int_e f(x)\, dx = 0$, then $f(x) \sim 0$. (M. K. Gavurin)

12. Let $f(x)$ be summable on $[a, b]$ and equal to 0 outside of $[a, b]$. If
$$\varphi(x) = \frac{1}{2h} \int_{x-h}^{x+h} f(t)\, dt,$$
then
$$\int_a^b |\varphi(x)|\, dx \leqslant \int_a^b |f(x)|\, dx \qquad \text{(A. N. Kolmogorov).}$$

13. Let a summable function $f(x)$ be defined on $[a, b]$. If $\int_a^c f(x)\, dx = 0$ for all $c\ (a \leqslant c \leqslant b)$, then $f(x) \sim 0$.

14. Let a strictly positive summable function $f(x)$ be defined on $[a, b]$. Let $0 < q \leqslant b - a$ and S be the family of all measurable subsets $e \subset [a, b]$ for which $me \geqslant q$. Prove that
$$\inf_{e \in S} \left\{ \int_e f(x)\, dx \right\} > 0.$$

15 Let $M = \{f(x)\}$ be a family of functions summable on $[a, b]$. If the functions of the family have equi-absolutely continuous integrals then there exists an increasing positive function $\Phi(u)$ defined for $0 \leqslant u < +\infty$ which tends to $+\infty$ with c and such that

$$\int_a^b |f(x)| \cdot \Phi(|f(x)|)\, dx \leqslant A < +\infty,$$

for all $f(x)$ of M, where the constant A does not depend on the choice of $f(x)$. (de la Vallée-Poussin)

CHAPTER VII

SQUARE-SUMMABLE FUNCTIONS

§ 1. FUNDAMENTAL DEFINITIONS. INEQUALITIES. NORM

In the present chapter, we consider a highly important class of functions: functions with summable squares. For simplicity, we will assume that all functions under consideration are defined on a certain closed interval $E = [a, b]$.

DEFINITION. A measurable function $f(x)$ is said to be a *function with summable square* or a *square-summable function* if

$$\int_a^b f^2(x)\, dx < +\infty.$$

The set of all functions with summable square is usually designated by the symbol L_2.

THEOREM 1. *Every function with summable square is summable, i.e., $L_2 \subset L$.*

This theorem follows from the obvious inequality

$$|f(x)| \leqslant \frac{1 + f^2(x)}{2}.$$

In exactly the same way, the inequality

$$|f(x)g(x)| \leqslant \frac{f^2(x) + g^2(x)}{2}$$

implies the following result.

THEOREM 2. *The product of two functions, each with summable square, is a summable function.*

Theorem 2 and the identity

$$(f \pm g)^2 = f^2 \pm 2fg + g^2$$

yield another simple fact.

THEOREM 3. *The sum and the difference of two functions in L_2 are in L_2.*

Finally, it is completely obvious that if $f(x)$ is in L_2 then all functions of the form $kf(x)$, where k is a finite constant, are also in L_2.

THEOREM 4 (THE INEQUALITY OF CAUCHY-BUNYAKOVSKI-SCHWARZ, OR CBS INEQUALITY). *If $f(x) \in L_2$ and $g(x) \in L_2$, then*

$$\left[\int_a^b f(x)g(x)\, dx\right]^2 \leqslant \left[\int_a^b f^2(x)\, dx\right] \cdot \left[\int_a^b g^2(x)\, dx\right]. \tag{1}$$

Proof. Consider the quadratic function

$$\psi(u) = Au^2 + 2Bu + C,$$

in which the coefficients A, B, C, are real and $A > 0$. If this function is non-negative for all real values of u, then

$$B^2 \leqslant AC. \tag{2}$$

If this were not so, we would have

$$\psi\left(-\frac{B}{A}\right) = \frac{1}{A}(AC - B^2) < 0.$$

Having made this observation, we set

$$\psi(u) = \int_a^b [uf(x) + g(x)]^2 \, dx = u^2 \int_a^b f^2 \, dx + 2u \int_a^b fg \, dx + \int_a^b g^2 \, dx.$$

This quadratic is non-negative and hence satisfies condition (2), which is equivalent to the present theorem.[1]

COROLLARY. *If $f(x) \in L_2$, then*

$$\int_a^b |f(x)| \, dx \leqslant \sqrt{b-a} \cdot \sqrt{\int_a^b f^2(x) \, dx}. \tag{3}$$

Upon setting $g = 1$ in (1) and replacing $f(x)$ by $|f(x)|$, we obtain (3).

THEOREM 5 (MINKOWSKI'S INEQUALITY). *If $f(x) \in L_2$ and $g(x) \in L_2$, then*

$$\sqrt{\int_a^b [f(x) + g(x)]^2 \, dx} \leqslant \sqrt{\int_a^b f^2(x) \, dx} + \sqrt{\int_a^b g^2(x) \, dx}.$$

Proof. Taking the square root of both sides of the CBS-inequality, we find

$$\int_a^b fg \, dx \leqslant \sqrt{\int_a^b f^2 \, dx} \cdot \sqrt{\int_a^b g^2 \, dx}.$$

Multiplying this inequality by 2 and adding

$$\int_a^b f^2 \, dx + \int_a^b g^2 \, dx,$$

to both sides, we obtain

$$\int_a^b (f+g)^2 \, dx \leqslant \left(\sqrt{\int_a^b f^2 \, dx} + \sqrt{\int_a^b g^2 \, dx}\right)^2,$$

which implies Minkowski's inequality.

Minkowski's inequality permits us to consider the space of functions L_2 from a new point of view. Namely, if we associate with every function $f(x) \in L_2$ the number

$$\|f\| = \sqrt{\int_a^b f^2(x) \, dx},$$

---

[1] We assume $\int_a^b f^2 \, dx > 0$. If $\int_a^b f^2 \, dx = 0$, $f(x)$ is equivalent to zero and inequality (1) becomes the trivial identity $0 = 0$.

then the following assertions are valid:
  I. $\|f\| \geqslant 0$, and $\|f\| = 0$ if and only if $f(x) \sim 0$;
  II. $\|kf\| = |k| \cdot \|f\|$ and, in particular, $\|-f\| = \|f\|$;
  III. $\|f + g\| \leqslant \|f\| + \|g\|$.

The number $\|f\|$ is called the *norm* of the function $f(x)$. The analogy between $\|f\|$ and the absolute value $|x|$ of a real (or complex) number x is very obvious. This analogy is the source of a number of important and beautiful constructions. Roughly speaking, the essential significance of absolute values in analysis is that with their aid, we can carry out the measurement of distances on the line:

$$\rho(x, y) = |x - y|.$$

The norm introduced in L_2 permits us to regard the set L_2 as a space in which it is also possible to measure distances, if we take the number

$$\rho(f, g) = \|f - g\|$$

as the distance between the elements f and g of L_2.

If we agree to consider functions which are equivalent as identical, then the distance $\rho(f, g)$ just defined possesses the usual properties:
  1) $\rho(f, g) \geqslant 0$, and $\rho(f, g) = 0$ if and only if $f = g$;
  2) $\rho(f, g) = \rho(g, f)$;
  3) $\rho(f, g) \leqslant \rho(f, h) + \rho(h, g)$.

If such a function $\rho(x, y)$ is defined for every pair of elements of a set A of elements of an arbitrary nature, then the set A is called a *metric space*.

Thus L_2 is a metric space. D. Hilbert first developed this point of view regarding L_2, so that L_2 is frequently called a *Hilbert space*.*

§ 2. MEAN CONVERGENCE

The concept of norm permits us to introduce the idea of limit in Hilbert space with the aid of almost the same expressions as used in the ordinary case of the line.

DEFINITION 1. The element f of the space L_2 is called a *limit of the sequence* $f_1, f_2, f_3, \ldots$ of elements of the same space, if for every $\varepsilon > 0$, there exists a natural number N such that

$$\|f_n - f\| < \varepsilon$$

for all $n > N$.

We express this situation by saying that the sequence $\{f_n\}$ *converges* to the element f: we write in the usual way

$$\lim_{n \to \infty} f_n = f \text{ and } f_n \to f.$$

---

* The space L_2 described here is only one example of a large number of spaces which are known as Hilbert spaces. In fact, for distinct pairs of real numbers $a < b$ and $a' < b'$, it is plain that one must distinguish between $L_2(a, b)$ and $L_2(a', b')$. There are a great many Hilbert spaces distinct from these spaces $L_2(a, b)$, as well. For an axiomatic treatment of not quite the most general Hilbert space, see M. H. Stone, *Linear Transformations in Hilbert Space and Their Applications to Analysis*, New York, Amer. Math. Soc. Coll. Publ., 1932.—E. H.

VII. SQUARE-SUMMABLE FUNCTIONS

We draw the reader's attention to the profound difference between the relations
$$f_n(x) \to f(x) \text{ and } f_n \to f.$$
The first means that for fixed x, the *numerical* sequence $\{f_n(x)\}$ converges to the limit $f(x)$ in the usual sense. The second expression, however, means that the sequence of elements of L_2 converges to the element $f \in L_2$ in the sense of Definition 1. In the usual symbols of the theory of functions, the relation $f_n \to f$ means that
$$\lim_{n \to \infty} \int_a^b [f_n(x) - f(x)]^2 \, dx = 0.$$

This new form of the convergence of a sequence of functions is called *mean convergence* or *convergence in the mean*.

THEOREM 1. *If a sequence $\{f_n(x)\}$ converges in the mean to the function $f(x)$, then it converges in measure to $f(x)$.*

Proof. Let σ be a fixed positive number and let
$$A_n(\sigma) = E(|f_n - f| \geq \sigma).$$
Then
$$\int_a^b (f_n - f)^2 \, dx \geq \int_{A_n(\sigma)} (f_n - f)^2 \, dx \geq \sigma^2 m A_n(\sigma),$$
and, since σ is fixed,
$$m A_n(\sigma) \to 0.$$

This implies that $f_n \Rightarrow f$.

COROLLARY. *If the sequence $\{f_n(x)\}$ converges in the mean to $f(x)$, then there exists a subsequence $\{f_{n_k}(x)\}$ of $\{f_n(x)\}$ which converges to $f(x)$ almost everywhere.*

This corollary is established by referring to Theorem 1 and Riesz's theorem (Theorem 4, §3, Ch. IV). However, it can be proved without referring to convergence in measure at all. Namely, if
$$\lim_{n \to \infty} \int_a^b (f_n - f)^2 \, dx = 0,$$
then it is possible to find $n_1 < n_2 < n_3 < \ldots$, such that
$$\int_a^b (f_{n_k} - f)^2 \, dx < \frac{1}{2^k}.$$
Then the series
$$\sum_{k=1}^{\infty} \int_a^b (f_{n_k} - f)^2 \, dx$$
converges and by the corollary of Theorem 11, §1, Chapter VI
$$f_{n_k}(x) \to f(x)$$
almost everywhere on $[a, b]$.

We note that convergence in the mean of the sequence $\{f_n(x)\}$ to the function $f(x)$ does not imply its convergence to $f(x)$ almost everywhere. This is illustrated, say, by

2. Mean Convergence

the example constructed in §3, Chapter IV. Conversely, it is possible that $f_n(x) \to f(x)$ at every point of $[a, b]$ while f_n does not converge to f in the mean.

EXAMPLE. Let a sequence $\{f_n(x)\}$ be defined on $[0, 1]$ by the requirements

$$f_n(x) = n \quad \text{for} \quad 0 < x < \frac{1}{n}$$

and $f_n(x) = 0$ at all other points of $[0, 1]$. Then it is clear that for arbitrary $x \in [0, 1]$

$$\lim f_n(x) = 0,$$

but at the same time

$$\int_0^1 f_n^2(x)\, dx = \int_0^{1/n} n^2\, dx = n \to +\infty.$$

THEOREM 2 (UNIQUENESS OF THE LIMIT). *A sequence $f_1, f_2, f_3, \ldots$ of elements of L_2 can have at most one limit.*

Proof. Assume that

$$f_n \to f \text{ and } f_n \to g;$$

then

$$\|f - g\| \leqslant \|f - f_n\| + \|f_n - g\|,$$

and since the right member of this inequality has limit zero and the left member is constant and non-negative, it follows that

$$\|f - g\| = 0,$$

Therefore $f - g = 0$ and $f = g$, as was to be proved.

It is possible to give another proof of the theorem. If $f_n \to f$ and $f_n \to g$, then the sequence $\{f_n(x)\}$ converges in measure to $f(x)$ and to $g(x)$, so that $f(x) \sim g(x)$ but we have agreed to consider equivalent functions to be one element of the space.

THEOREM 3 (CONTINUITY OF THE NORM). *If $f_n \to f$, then $\|f_n\| \to \|f\|$.*

Proof. The obvious inequalities

$$\|f_n\| \leqslant \|f\| + \|f_n - f\|,$$
$$\|f\| \leqslant \|f_n\| + \|f_n - f\|$$

imply that

$$|\, \|f_n\| - \|f\|\, | \leqslant \|f_n - f\|,$$

this inequality implies the present theorem.

COROLLARY. *The norms of the elements of a convergent sequence in L_2 are bounded.*

DEFINITION 2. A sequence $\{f_n\}$ of points of the space L_2 is said to be a *Cauchy sequence* or a *fundamental sequence*, if to every $\varepsilon > 0$, there corresponds a natural number N such that

$$\|f_n - f_m\| < \varepsilon$$

for all $m, n > N$.

THEOREM 4. *If the sequence $\{f_n\}$ has a limit, then it is a Cauchy sequence.*

Proof. Let

$$\lim_{n \to \infty} f_n = f.$$

VII. SQUARE-SUMMABLE FUNCTIONS

Having taken an arbitrary $\varepsilon > 0$, we find an N such that for $n > N$,
$$\|f_n - f\| < \frac{\varepsilon}{2}.$$

Now if $n > N$ and $m > N$, then
$$\|f_n - f_m\| \leqslant \|f_n - f\| + \|f - f_m\| < \varepsilon,$$
which proves the theorem.

The converse theorem is much deeper.

THEOREM 5. *If $\{f_n\}$ is a Cauchy sequence, then it has a limit.*

Proof. Consider the convergent series $\sum_{k=1}^{\infty} \frac{1}{2^k}$. For every k, choose an n_k such that
$$\|f_n - f_m\| < \frac{1}{2^k}.$$
for $n \geqslant n_k$ and $m \geqslant n_k$. We may suppose without loss of generality that
$$n_1 < n_2 < n_3 < \cdots$$
so that
$$\|f_{n_{k+1}} - f_{n_k}\| < \frac{1}{2^k},$$
and hence
$$\sum_{k=1}^{\infty} \|f_{n_{k+1}} - f_{n_k}\| < +\infty.$$

By inequality (3) of §1,
$$\int_a^b |f_{n_{k+1}} - f_{n_k}| \, dx \leqslant \sqrt{b-a} \, \|f_{n_{k+1}} - f_{n_k}\|,$$
so that the series
$$\sum_{k=1}^{\infty} \int_a^b |f_{n_{k+1}} - f_{n_k}| \, dx$$
converges. By Theorem 11, §1, Chapter VI, it follows that the series
$$|f_{n_1}(x)| + \sum_{k=1}^{\infty} |f_{n_{k+1}}(x) - f_{n_k}(x)|,$$
converges almost everywhere. Therefore the series
$$f_{n_1}(x) + \sum_{k=1}^{\infty} \{f_{n_{k+1}}(x) - f_{n_k}(x)\}$$
converges almost everywhere. The convergence of this last series almost everywhere is obviously equivalent to the existence of the finite limit
$$\lim_{n \to \infty} f_{n_k}(x)$$
almost everywhere. We introduce the function $f(x)$, equal to this limit everywhere where

2. Mean Convergence

it exists and is finite and equal to zero at those points where this limit does not exist or is infinite. The function $f(x)$ is plainly measurable and by definition,

$$f_{n_k}(x) \to f(x).$$

almost everywhere on $[a, b]$.

Our problem is to establish that this function is an element of Hilbert space and that it is the L_2-limit of the sequence $\{f_n\}$. For this purpose, having selected an arbitrary $\varepsilon > 0$, we find a natural number N such that

$$\|f_n - f_m\| < \varepsilon.$$

for all $m, n > N$.

If k_0 is a number such that $n_{k_0} > N$, then

$$\int_a^b (f_n - f_{n_k})^2 \, dx < \varepsilon^2.$$

for all $n > N$ and arbitrary $k > k_0$. From this, upon applying Fatou's theorem (Theorem 9, §1, Chapter VI) to the sequence of functions $\{(f_n - f_{n_k})^2\}$ $(k > k_0)$, we find that

$$\int_a^b (f_n - f)^2 \, dx \leq \varepsilon^2,$$

i.e.,

$$\|f_n - f\| \leq \varepsilon$$

for all $n > N$. Since $f_n - f \in L_2$, it follows that $f \in L_2$. Also $\lim_{n \to \infty} f = f_n$. This completes the proof.

The property of Hilbert space L_2, established in the preceding theorem, is called *completeness*. The reader has, of course, noticed that Theorems 4 and 5 are analogues of the Bolzano-Cauchy test for convergence. The Bolzano-Cauchy test is one of the numerous forms of the property of continuity of the real line Z. This property can be expressed by any one of the following assertions:

A. If the points of the real line Z are divided into two classes X and Y such that every point of the class X lies to the left of every point of the class Y, then either there is a largest element in the class X or a smallest element in the class Y.

B. A set which is bounded above admits a least upper bound.

C. A bounded monotonic increasing sequence has a finite limit.

D. If $\{d_n\}$ is a sequence of nested closed intervals whose lengths tend to zero, then there exists a point belonging to all the segments d_n.

E. The Bolzano-Cauchy property: every Cauchy sequence $\{x_n\}$ has a finite limit.

It is sufficient to remove one point from the straight line Z to make all the theorems indicated above untrue.

Of the theorems A, B, C, D, E, only the last one E is formulated without the aid of the concept of *order* of the points on the straight line. It uses only the notion of distance. Therefore, it is natural that E be carried over as the definition of the property of completeness to more complicated spaces than the real line.

DEFINITION 3. *A set A contained in L_2 is said to be* everywhere dense *in L_2 if every point of L_2 is the limit of a sequence of points belonging to A.*

In function-theoretical language, Definition 3 reads as follows: a class of functions

$A \subset L_2$ is everywhere dense in L_2 if every function in L_2 is the limit (in the sense of mean convergence) of a sequence of functions selected from A.

It is easy to see that a necessary and sufficient condition that the set $A = \{g\}$ be everywhere dense in L_2 is that for an arbitrary point $f \in L_2$ and arbitrary $\varepsilon > 0$, it is possible to find a point $g \in A$ such that

$$\|f - g\| < \varepsilon.$$

THEOREM 6. *Each of the following classes of functions*:
M, the class of bounded measurable functions;
C, the class of continuous functions;
P, the class of polynomials;
S, the class of step functions;
is everywhere dense in L_2. If the closed interval [a,b] is $[-\pi, \pi]$, then T, the class of trigonometric polynomials is also everywhere dense.

Proof. 1) Let $f(x) \in L_2$. Choosing an arbitrary $\varepsilon > 0$, we infer from the absolute continuity of the integral that there exists a $\delta > 0$ such that the relations

$$e \subset [a, b], \quad me < \delta$$

imply the inequality

$$\int_e f^2(x)\, dx < \varepsilon^2.$$

By theorem 1, §4, Chapter IV, there exists a measurable bounded function $g(x)$ such that

$$mE(f \neq g) < \delta,$$

we may define $g(x)$ as 0 at all points of the set $E(f \neq g)$. Then

$$\|f - g\|^2 = \int_a^b (f - g)^2\, dx = \int_{E(f \neq g)} (f - g)^2\, dx = \int_{E(f \neq g)} f^2\, dx < \varepsilon^2,$$

i.e.,

$$\|f - g\| < \varepsilon.$$

This proves the theorem for the class M.

2) Let $f(x) \in L_2$ and $\varepsilon > 0$. Choose a function $g(x) \in M$ such that

$$\|f - g\| < \frac{\varepsilon}{2}.$$

Let $|g(x)| \leq K$. By Luzin's theorem (Theorem 4, §5 Chapter IV), there exists a continuous function $\varphi(x)$ such that

$$mE(g \neq \varphi) < \frac{\varepsilon^2}{16K^2}, \quad |\varphi(x)| \leq K.$$

For this function φ, we have

$$\|g - \varphi\|^2 = \int_a^b (g - \varphi)^2\, dx = \int_{E(g \neq \varphi)} (g - \varphi)^2\, dx \leq 4K^2 \cdot mE(g \neq \varphi) < \frac{\varepsilon^2}{4},$$

Obviously

$$\|g - \varphi\| < \frac{\varepsilon}{2},$$

2. MEAN CONVERGENCE

and hence
$$\|f - \varphi\| < \varepsilon.$$

This proves the present theorem for the class C.

3) Let $f(x) \in L_2$ and $\varepsilon > 0$. Let $\varphi(x) \in C$ have the property that
$$\|f - \varphi\| < \frac{\varepsilon}{2}.$$

In accordance with Weierstrass's theorem (Theorem 2, §5, Chapter IV), there is a polynomial $P(x)$ such that
$$|\varphi(x) - P(x)| < \frac{\varepsilon}{2\sqrt{b-a}}$$

for all $x \in [a, b]$. It follows that
$$\|\varphi - P\|^2 = \int_a^b (\varphi - P)^2 \, dx \leq \frac{\varepsilon^2}{4(b-a)} \cdot (b - a) = \frac{\varepsilon^2}{4},$$

from which the inequalities
$$\|\varphi - P\| < \frac{\varepsilon}{2}$$

and
$$\|f - P\| < \varepsilon$$

are obvious. The theorem is thus proved for the class P.

4) Let $f(x) \in L_2$ and $\varepsilon > 0$. Let $\varphi(x) \in C$ have the property that
$$\|f - \varphi\| < \frac{\varepsilon}{2}.$$

Since $\varphi(x)$ is continuous on $[a, b]$, we can subdivide $[a, b]$ by means of points
$$c_0 = a < c_1 < c_2 < \ldots < c_n = b$$

into parts such that the oscillation of $\varphi(x)$ is less than $\frac{\varepsilon}{2\sqrt{b-a}}$, in each of the subintervals $[c_k, c_{k+1}]$. Now define a step function $s(x)$ by setting

$$\begin{aligned} s(x) &= \varphi(c_k) & (c_k \leq x < c_{k+1},\ k = 0, 1, \ldots, n-2) \\ s(x) &= \varphi(c_{n-1}) & (c_{n-1} \leq x \leq b). \end{aligned}$$

We then have $|s(x) - \varphi(x)| < \frac{\varepsilon}{2\sqrt{b-a}}$, everywhere on $[a, b]$, and therefore
$$\|s - \varphi\| < \frac{\varepsilon}{2}.$$

It follows that
$$\|f - s\| < \varepsilon,$$

and the theorem is proved for the class S.

5) Finally, let $[a, b] = [-\pi, \pi]$ and $f(x) \in L_2$. Selecting an arbitrary $\varepsilon > 0$, we find, as above, a function continuous on $[-\pi, \pi]$ such that
$$\|f - \varphi\| < \frac{\varepsilon}{2}.$$

Let
$$|\varphi(x)| \leq K.$$

For $0 < \delta < 2\pi$, we define the continuous function $\psi(x)$ on the segment $[-\pi, \pi]$ by setting
$$\psi(x) = \varphi(x) \text{ for } x \in [-\pi + \delta, \pi],$$
$$\psi(-\pi) = \varphi(\pi),$$
and defining $\psi(x)$ as a linear function on the closed interval $[-\pi, -\pi + \delta]$; we choose $\delta > 0$ so that
$$\delta < \frac{\varepsilon^2}{64K^2}.$$

The function $\psi(x)$ is obviously continuous on $[-\pi, \pi]$, but what is now the main thing, it has the property that
$$\psi(-\pi) = \psi(\pi).$$
Furthermore, $|\psi(x)| \leq K$, so that
$$||\varphi - \psi||^2 = \int_{-\pi}^{\pi} (\varphi - \psi)^2 dx = \int_{-\pi}^{-\pi+\delta} (\varphi < \psi)^2 dx \leq 4K^2\delta < \frac{\varepsilon^2}{16}.$$

Accordingly,
$$||f - \psi|| < \frac{3\varepsilon}{4}.$$

By Weierstrass's Theorem (Theorem 2, §5, Ch. IV), there exists a trigonometric polynomial $T(x)$ such that for all $x \in [-\pi, \pi]$,
$$|\psi(x) - T(x)| < \frac{\varepsilon}{4\sqrt{2\pi}}.$$

Then
$$||\psi - T||^2 = \int_{-\pi}^{\pi} (\psi - T)^2 dx < \frac{\varepsilon^2}{16},$$

and consequently
$$||f - T|| < \varepsilon.$$

This completes the proof.

The concept of *weak convergence* of a sequence of functions plays an important role in many questions.

DEFINITION 4. *A sequence of functions $f_1(x), f_2(x), \ldots$, in L_2, is said to converge weakly to a function $f(x) \in L_2$ if the equality*
$$\lim_{n \to \infty} \int_a^b g(x) f_n(x) \, dx = \int_a^b g(x) f(x) \, dx$$

holds for every function $g(x) \in L_2$.

We will not study this new form of convergence in detail and will limit ourselves to one theorem.

THEOREM 7. *If a sequence of functions $\{f_n(x)\}$ converges in the mean to the function $f(x)$, then it also converges weakly to this function.*

3. ORTHOGONAL SYSTEMS

Proof. Let $g(x) \in L_2$. Then the CBS inequality yields

$$\left\{\int_a^b g(x)[f_n(x) - f(x)]\, dx\right\}^2 \leq \left[\int_a^b g^2(x)\, dx\right] \cdot \left[\int_a^b [f_n(x) - f(x)]^2\, dx\right].$$

Therefore

$$\left|\int_a^b gf_n\, dx - \int_a^b gf\, dx\right| \leq \|g\| \cdot \|f_n - f\| \to 0,$$

as was to be proved.

§ 3. ORTHOGONAL SYSTEMS

DEFINITION 1. Two measurable functions $f(x)$ and $g(x)$ defined on the closed interval $[a, b]$ are said to be *orthogonal* if

$$\int_a^b f(x)\, g(x)\, dx = 0.$$

DEFINITION 2. A measurable function $f(x)$ defined on $[a, b]$ is said to be *normalized* if

$$\int_a^b f^2(x)\, dx = 1.$$

DEFINITION 3. A system of functions $\omega_1(x), \omega_2(x), \omega_3(x), \ldots$, defined on the segment $[a, b]$, is said to be an *orthonormal* system if every function of the system is normalized and every pair of distinct functions of the system are orthogonal.

In other words, a system of functions $\{\omega_k(x)\}$ is orthonormal if

$$\int_a^b \omega_i(x)\, \omega_k(x)\, dx = \begin{cases} 1 & (i = k) \\ 0 & (i \neq k) \end{cases}.$$

It is clear that every orthonormal system is contained in L_2. A classical example of an orthonormal system is the trigonometric system

$$\frac{1}{\sqrt{2\pi}}, \quad \frac{\cos x}{\sqrt{\pi}}, \quad \frac{\sin x}{\sqrt{\pi}}, \quad \frac{\cos 2x}{\sqrt{\pi}}, \quad \frac{\sin 2x}{\sqrt{\pi}}, \quad \ldots, \tag{1}$$

defined on the segment $[-\pi, +\pi]$.

Suppose that $f(x) \in L_2$ and that $f(x)$ is a linear combination of functions of a given orthonormal system:

$$f(x) = c_1 \omega_1(x) + \ldots + c_n \omega_n(x).$$

Multiplying this equality by $\omega_k(x)$ ($k = 1, \ldots, n$) and integrating, we find

$$c_k = \int_a^b f(x)\, \omega_k(x)\, dx,$$

i.e., the coefficients $c_1, \ldots, c_n$ are completely determined. In particular, if
$$T(x) = A + \sum_{k=1}^{n}(a_k \cos kx + b_k \sin kx),$$
then
$$A = \frac{1}{2\pi}\int_{-\pi}^{\pi} T(x)\,dx, \quad a_k = \frac{1}{\pi}\int_{-\pi}^{\pi} T(x)\cos kx\,dx, \quad b_k = \frac{1}{\pi}\int_{-\pi}^{\pi} T(x)\sin kx\,dx.$$

These formulas were discovered by Fourier for the trigonometric system. It is natural to give the following general definition.

DEFINITION 4. Let $\{\omega_k(x)\}$ be an orthonormal system and let $f(x)$ be a function in L_2. The numbers
$$c_k = \int_a^b f(x)\,\omega_k(x)\,dx$$
are called the *Fourier coefficients* of the function $f(x)$ in the system $\{\omega_k(x)\}$.

The series
$$\sum_{k=1}^{\infty} c_k \omega_k(x)$$
is called the *Fourier series* of the function $f(x)$ in the system $\{\omega_k(x)\}$.

We now consider how near the partial sum of Fourier series of the function $f(x)$,
$$S_n(x) = \sum_{k=1}^{n} c_k \omega_k(x)$$
is in L_2, to the function $f(x)$, *i.e.*, we will compute
$$\|f - S_n\|.$$
To do this, we first evaluate the integrals
$$\int_a^b f(x) S_n(x)\,dx \quad \text{и} \quad \int_a^b S_n^2(x)\,dx.$$
For the first integral we have
$$\int_a^b f(x) S_n(x)\,dx = \sum_{k=1}^{n} c_k \int_a^b f(x)\,\omega_k(x)\,dx = \sum_{k=1}^{n} c_k^2.$$
In exactly the same way,
$$\int_a^b S_n^2(x)\,dx = \sum_{i,k} c_i c_k \int_a^b \omega_i(x)\,\omega_k(x)\,dx = \sum_{k=1}^{n} c_k^2. \tag{2}$$
It follows that
$$\|f - S_n\|^2 = \int_a^b (f^2 - 2fS_n + S_n^2)\,dx = \int_a^b f^2\,dx - \sum_{k=1}^{n} c_k^2,$$

3. ORTHOGONAL SYSTEMS

that is,
$$\|f-S_n\|^2 = \|f\|^2 - \sum_{k=1}^{n} c_k^2. \tag{3}$$

Equation (3) is called *Bessel's identity*. Since its left member is non-negative, we obtain *Bessel's inequality* from it:
$$\sum_{k=1}^{n} c_k^2 \leqslant \|f\|^2.$$

Since the number n in Bessel's inequality is an arbitrary positive integer, Bessel's inequality can be written in the strengthened form
$$\sum_{k=1}^{\infty} c_k^2 \leqslant \|f\|^2. \tag{4}$$

In particular, we may have
$$\sum_{k=1}^{\infty} c_k^2 = \|f\|^2, \tag{5}$$

this equation is called Parseval's identity. It has a very simple meaning. Namely, Bessel's identity (3) and (5) permit us to write
$$\lim_{n \to \infty} \|f - S_n\| = 0.$$

In other words, Parseval's identity implies that the partial sums $S_n(x)$ of the Fourier series of the function $f(x)$ converge (in the sense of the metric in L_2, that is, in the mean) to $f(x)$.

DEFINITION 5. *The orthogonal system* $\{\omega_k(x)\}$ *is said to be closed if Parseval's identity holds for all functions in* L_2.

THEOREM 1. *If the system* $\{\omega_k(x)\}$ *is closed, and if* $f(x)$ *and* $g(x)$ *are functions in* L_2, *we have*
$$\int_a^b f(x)g(x)\,dx = \sum_{k=1}^{\infty} a_k b_k,$$

where
$$a_k = \int_a^b f(x)\omega_k(x)\,dx, \quad b_k = \int_a^b g(x)\omega_k(x)\,dx.$$

Proof. The numbers $a_k + b_k$ are the Fourier coefficients for the sum $f(x) + g(x)$. Therefore
$$\|f + g\|^2 = \sum_{k=1}^{\infty} (a_k + b_k)^2,$$

and direct computation gives
$$\int_a^b f^2\,dx + 2\int_a^b fg\,dx + \int_a^b g^2\,dx = \sum_{k=1}^{\infty} a_k^2 + 2\sum_{k=1}^{\infty} a_k b_k + \sum_{k=1}^{\infty} b_k^2.$$

This is equivalent to the theorem.

The formula just established is a generalized form of Parseval's identity.

COROLLARY. *If the orthonormal system $\{\omega_k(x)\}$ is closed and $f(x) \in L_2$, then the Fourier series of $f(x)$ in the system $\{\omega_k(x)\}$ can be integrated term by term over an arbitrary measurable set $E \subset [a, b]$, i.e.,*

$$\int_E f(x)\,dx = \sum_{k=1}^{\infty} c_k \int_E \omega_k(x)\,dx.$$

In fact, if we take the characteristic function of the set E for $g(x)$, then it is clear that $g(x)$ is square-summable. We thus have a special case of the generalized Parseval's identity.

It is interesting to note that the Fourier series itself, $\sum c_k \omega_k(x)$, need not converge pointwise to the function $f(x)$.

THEOREM 2 (V. A. STEKLOV). *Let A be a class of functions everywhere dense in L_2. If Parseval's identity holds for all functions of the class A, then the system $\{\omega_k(x)\}$ is closed.*

Proof. Let $f(x)$ be a function of the class L_2. We form the partial sums of its Fourier series

$$\sum_{k=1}^{n} c_k \omega_k(x)$$

and emphasize their dependence on the function $f(x)$ by writing them as $S_n(f)$. It is easy to verify that

1) $S_n(kf) = kS_n(f)$,
2) $S_n(f_1 + f_2) = S_n(f_1) + S_n(f_2)$,
3) $\|S_n(f)\| \leqslant \|f\|$.

The first two properties are trivial. The third follows from Bessel's inequality

$$\|S_n\|^2 = \sum_{k=1}^{n} c_k^2 \leqslant \|f\|^2.$$

Next, we select a function $f(x) \in L_2$ and an $\varepsilon > 0$. Since the class A is everywhere dense in L_2, a function $g(x) \in A$ can be found such that

$$\|f - g\| < \frac{\varepsilon}{3}.$$

Then

$$\|f - S_n(f)\| \leqslant \|f - g\| + \|g - S_n(g)\| + \|S_n(g) - S_n(f)\|.$$

Futhermore,

$$\|S_n(g) - S_n(f)\| = \|S_n(g - f)\| \leqslant \|g - f\| < \frac{\varepsilon}{3},$$

and hence

$$\|f - S_n(f)\| < \frac{2}{3}\varepsilon + \|g - S_n(g)\|.$$

Since Parseval's identity holds for $g(x)$, we have

$$\|g - S_n(g)\| < \frac{\varepsilon}{3},$$

for $n > n_0$, and, consequently

$$\|f - S_n(f)\| < \varepsilon$$

for $n > n_0$. This proves the theorem.

COROLLARY 1. *If Parseval's identity holds for all of the functions $1, x, x^2, x^3, \ldots$, then the system $\{\omega_k(x)\}$ is closed.*

Consider, in fact, the polynomial

$$P(x) = A_0 + A_1 x + \ldots + A_m x^m.$$

Then

$$S_n(P) = A_0 S_n(1) + A_1 S_n(x) + \ldots A_m S_n(x^m)$$

and accordingly

$$\|P - S_n(P)\| \leq \sum_{k=0}^{m} |A_k| \cdot \|x^k - S_n(x^k)\|.$$

The right-hand member of this equation has limit 0 as $n \to \infty$. Therefore Parseval's identity holds for every polynomial, and the class of polynomials P is everywhere dense in L_2.

Do closed systems exist? Another corollary of Steklov's theorem supplies the answer.

COROLLARY 2. *The trigonometric system (1) is closed.*

It suffices to verify Parseval's identity for an arbitrary trigonometric polynomial

$$T(x) = A + \sum_{k=1}^{n} (a_k \cos kx + b_k \sin kx),$$

but this is obvious, [2] because $T(x)$ is a linear combination of functions of the system (1).

THEOREM 3 (F. RIESZ-E. FISCHER). *Let an orthonormal system $\{\omega_k(x)\}$ be defined on the closed interval $[a, b]$. If the numbers $c_1, c_2, c_3, \ldots$ are such that*

$$\sum_{k=1}^{\infty} c_k^2 < +\infty,$$

there is a function $f(x) \in L_2$ such that:
1) *the numbers c_k are the Fourier coefficients of $f(x)$;*
2) *Parseval's identity holds for $f(x)$.*

---

[2] Let $f(x) = \sum_{k=1}^{n} c_k \omega_k(x)$. Then, multiplying this equation by $f(x)$ and integrating, we obtain at once the completeness formula

$$\int_a^b f^2(x)\, dx = \sum_{k=1}^{n} c_k^2.$$

Proof. Let
$$S_n(x) = \sum_{k=1}^{n} c_k \omega_k(x).$$

We shall show that the sequence $S_1, S_2, \ldots$ is a Cauchy sequence. For this purpose, let us evaluate $\|S_m - S_n\|$ for $m > n$:

$$\|S_m - S_n\|^2 = \int_a^b [\sum_{k=n+1}^{m} c_k \omega_k(x)]^2 dx = \sum_{i,k} c_i c_k \int_a^b \omega_i(x) \omega_k(x) dx = \sum_{k=n+1}^{m} c_k^2.$$

For every $\varepsilon > 0$, there exists a natural number N such that for $m > n > N$,

$$\sum_{k=n+1}^{m} c_k^2 < \varepsilon^2$$

or, equivalently,

$$\|S_m - S_n\| < \varepsilon.$$

This implies that $\{S_n\}$ is a Cauchy sequence. Since L_2 is a complete metric space, there exists a function $f(x) \in L_2$ such that
$$\|S_n - f\| \to 0.$$

This function f satisfies our requirements. In fact, by Theorem 7, §2, the sequence $\{S_n(x)\}$ converges weakly to $f(x)$, i.e.,

$$\lim_{n \to \infty} \int_a^b S_n(x) g(x) dx = \int_a^b f(x) g(x) dx$$

for every $g(x) \in L_2$. In particular upon setting
$$g(x) = \omega_i(x),$$

we obtain

$$\int_a^b f(x) \omega_i(x) dx = \lim_{n \to \infty} \int_a^b S_n(x) \omega_i(x) dx.$$

For $n > i$, we have

$$\int_a^b S_n(x) \omega_i(x) dx = \int_a^b [\sum_{k=1}^{n} c_k \omega_k(x)] \omega_i(x) dx = c_i.$$

Accordingly,
$$\int_a^b f(x) \omega_i(x) dx = c_i$$

and the function $f(x)$ satisfies requirement (1). It also follows that $S_n(x)$ is the partial sum of the Fourier series of the functions $f(x)$, and the relation
$$\|S_n - f\| \to 0,$$

shows that Parseval's identity holds for $f(x)$.

3. ORTHOGONAL SYSTEMS

REMARK. There exists only one function satisfying both conditions of the Riesz-Fischer theorem. Assume that there are two such functions, $f(x)$ and $g(x)$. By the first condition, they have a common Fourier series. The second condition implies that

$$S_n \to f, \quad S_n \to g,$$

which implies that $f = g$.

It is interesting to see if the preceding remark is still valid when we omit the second condition of the theorem. We need the following definition to answer this question.

DEFINITION 6. A system of functions $\{\varphi_k(x)\}$, defined on the closed interval $[a, b]$, and belonging to L_2, is said to be *complete* if there exists no function different from zero[3] in L_2 which is orthogonal to all the functions $\varphi_k(x)$.

We note that in the foregoing definition, no requirement is made that the system $\{\varphi_k(x)\}$ be orthonormal.

It is easy to see that a function satisfying the first condition of the Riesz-Fischer theorem is unique if and only if the orthonormal system $\{\omega_k(x)\}$ is complete. In fact, if this system is complete and two functions $f(x)$ and $g(x)$ have the same Fourier coefficients in it,

$$\int_a^b f(x)\omega_k(x)\,dx = \int_a^b g(x)\omega_k(x)\,dx \qquad (k = 1, 2, 3 \ldots),$$

then their difference, being orthogonal to all functions of the system, must be identically zero. Conversely, if the system is not complete and $h(x)$ is a function different from zero and orthogonal to all functions of the system, then it is sufficient to add to it any function $f(x)$ satisfying the first condition in order to obtain a function distinct from $f(x)$ also satisfying the first condition.

In the case of orthonormal systems, the concepts of closure and completeness coincide.

THEOREM 4. *An orthonormal system $\{\omega_k(x)\}$ is complete if and only if it is closed.*

Proof. Suppose that the system $\{\omega_k(x)\}$ is closed. If a function $f(x) \in L_2$ is orthogonal to all functions $\omega_k(x)$, then all of its Fourier coefficients are zero. The Parseval identity gives

$$\|f\|^2 = \sum_{k=1}^{\infty} c_k^2 = 0,$$

and the function $f(x)$ is identically zero, *i.e.*, the system is complete. Conversely, suppose that the system $\{\omega_k(x)\}$ is complete. Assume that Parseval's identity fails for some function $g(x) \in L_2$. Then, necessarily

$$\sum_{k=1}^{\infty} c_k^2 < \|g\|^2,$$

where the numbers

$$c_k = \int_a^b g(x)\omega_k(x)\,dx$$

are the Fourier coefficients of the function $g(x)$. On the basis of the Riesz-Fischer

---

[3] Recall that a function equivalent to zero is taken to be identically zero.

theorem, a function $f(x)$ can be found such that

$$\int_a^b f(x)\,\omega_k(x)\,dx = c_k, \quad \|f\|^2 = \sum_{k=1}^{\infty} c_k^2.$$

In this case, the difference $f(x) - g(x)$ is orthogonal to all functions of the system and

$$f(x) = g(x),$$

since the system is complete. This contradicts the condition

$$\|f\| < \|g\|.$$

COROLLARY. *The trigonometric system* (1) *is complete.*

In conclusion, we take up one more problem, connected with the theory of orthonormal systems.

Let $\{\omega_k(x)\}$ be an orthonormal system on the closed interval $[a, b]$ and let the series

$$\sum_{k=1}^{\infty} c_k^2 \qquad (6)$$

converge. By the Riesz-Fischer theorem, the series

$$\sum_{k=1}^{\infty} c_k \omega_k(x) \qquad (7)$$

is the Fourier series of some function $f(x) \in L_2$, and its partial sums

$$S_n(x) = \sum_{k=1}^{n} c_k \omega_k(x)$$

converge in the mean to $f(x)$. Therefore, a partial sequence $\{S_{n_i}(x)\}$ of them can be formed which converges [to the function $f(x)$] almost everywhere on $[a, b]$. It turns out that the choice of the indices n_i can be carried out without specifying the system $\{\omega_k(x)\}$, simply by using the series (6). A large number of investigations have been devoted to this problem. We introduce only the simplest results which are pertinent here:

THEOREM 5 (S. KACZMARZ). *Let*

$$r_n = \sum_{k=n}^{\infty} c_k^2.$$

If the natural numbers $n_1 < n_2 < \ldots$ have the property that

$$\sum_{i=1}^{\infty} r_{n_i} < +\infty, \qquad (K)$$

then the sequence $\{S_{n_i}(x)\}$ converges almost everywhere.

Proof. Bessel's identity shows that

$$\|S_{n-1} - f\|^2 = \|f\|^2 - \sum_{k=1}^{n-1} c_k^2 = r_n$$

(We assume that $f(x)$ is the function which satisfies both conditions of the Riesz-Fischer

3. ORTHOGONAL SYSTEMS

theorem.) Therefore, it follows from condition (K) that

$$\sum_{i=1}^{\infty} \int_a^b (S_{n_i-1} - f)^2 \, dx < +\infty$$

and, by the corollary to Theorem 11, §1, Chapter VI, we have

$$S_{n_i-1}(x) \to f(x).$$

almost everywhere on $[a, b]$. On the other hand,

$$\sum_{k=1}^{\infty} \int_a^b [c_k \omega_k(x)]^2 \, dx = \sum_{k=1}^{\infty} c_k^2 < +\infty$$

and, by virtue of Theorem 11, §11, Chapter VI, we have

$$c_{n_i} \omega_{n_i}(x) \to 0$$

almost everywhere on $[a, b]$. It follows that

$$S_{n_i}(x) \to f(x),$$

as we wished to prove.

THEOREM 6 (H. RADEMACHER). *Let $\psi(k)$ be a positive, increasing function defined for $k = 1, 2, 3, \ldots$, such that $\lim_{k \to \infty} \psi(k) = \infty$, and such that*

$$\sum_{k=1}^{\infty} \psi(k) \, c_k^2 < +\infty. \tag{8}$$

If the natural numbers $1 = n_1 < n_2 < n_3 < \ldots$ have the property that

$$\psi(n_i) \geqslant i, \tag{R}$$

then the sequence $\{S_{n_i}(x)\}$ converges almost everywhere.

Proof. We shall show that the numbers n_i satisfying condition (R) also satisfy condition (K), and the theorem follows. For this purpose, we note that condition (8) can be written as

$$\sum_{i=1}^{\infty} \sum_{k=n_i}^{n_{i+1}-1} \psi(k) \, c_k^2 < +\infty. \tag{9}$$

Condition (R) implies that

$$\sum_{i=1}^{\infty} i \sum_{k=n_i}^{n_{i+1}-1} c_k^2 < +\infty. \tag{10}$$

Consequently the double series

$$\left.\begin{array}{l}\sum_{k=n_1}^{n_2-1} c_k^2 + \sum_{k=n_2}^{n_3-1} c_k^2 + \sum_{k=n_3}^{n_4-1} c_k^2 + \cdots \\ \qquad + \sum_{k=n_2}^{n_3-1} c_k^2 + \sum_{k=n_3}^{n_4-1} c_k^2 + \cdots \\ \qquad \qquad \qquad + \sum_{k=n_3}^{n_4-1} c_k^2 + \cdots \\ \qquad \qquad \qquad \qquad + \cdots\end{array}\right\} \qquad (11)$$

converges when summed by columns. Therefore it converges and is summable by rows. This is equivalent to condition (K) because the sum of the i-th row is r_{n_i}.

REMARK. Conditions (K) and (R) are equivalent. In fact, we have already seen that numbers n_i satisfying condition (R) also satisfy condition (K). Conversely, let the numbers n_i satisfy condition (K). Then, summing series (11) by rows, we obtain a finite sum. Summing it by columns, we see that (10) is fulfilled. If we set

$$\psi(k) = i \qquad (n_i \leqslant k < n_{i+1}; \quad i = 1, 2, \ldots),$$

then (10) can be written in the form (9) or (8). Therefore $\psi(k)$ satisfies the conditions of Rademacher's theorem and the numbers n_i satisfy condition (R).

§ 4. THE SPACE l_2

Points in two-dimensional Euclidean space R_2 are of course ordered pairs (a_1, a_2) of real numbers. With each point $M(a_1, a_2) \in R_2$, we may consider also the vector with initial point $(0, 0)$ and terminal point M. The co-ordinates a_1 and a_2 of the point M are the projections of the vector x on the two co-ordinate axes. Therefore, a pair of numbers (a_1, a_2) can be considered not only as a point M but also as a vector x. This point of view is very useful. Namely, given two vectors $x = (a_1, a_2)$ and $y = (b_1, b_2)$, one can form their sum

$$x + y = (a_1 + b_1, \quad a_2 + b_2),$$

and one can multiply a vector $x = (a_1, a_2)$ by a real number k:

$$kx = (ka_1, \quad ka_2).$$

Such operations cannot be carried out for points, if they are considered as purely geometric entities.

The length of the vector $x = (a_1, a_2)$ is the number

$$\|x\| = \sqrt{a_1^2 + a_2^2}.$$

(This is merely the Pythagorean theorem). The inner product (x, y) of two vectors $x = (a_1, a_2)$ and $y = (b_1, b_2)$ is the product of their lengths by the cosine of the angle between them:

$$(x, y) = \|x\| \cdot \|y\| \cdot \cos \theta,$$

4. The Space l_2

In terms of the projections of the vectors, we have the well-known formula

$$(x, y) = a_1 b_1 + a_2 b_2.$$

Knowing this product and the lengths of both vectors, it is easy to find the angle between them from the relation

$$\cos \theta = \frac{(x, y)}{\|x\| \cdot \|y\|} \qquad (0 \leqslant \theta \leqslant \pi).$$

In particular the vectors x and y are orthogonal if and only if

$$(x, y) = a_1 b_1 + a_2 b_2 = 0.$$

The above discussion can be repeated verbatim for three-dimensional space R_3.

1) A triple of numbers $x = (a_1, a_2, a_3)$, taken in a definite order, can be considered either as a *point* of the space R_3 or as a *vector* lying in R_3.

2) In the second interpretation, we can multiply a vector by a number and add two vectors. The length $\|x\|$ of the vector $x = (a_1, a_2, a_3)$, as the Pythagorean theorem shows, is

$$\|x\| = \sqrt{a_1^2 + a_2^2 + a_3^2}.$$

3) Given vectors $x = (a_1, a_2, a_3)$ and $y = (b_1, b_2, b_3)$, we can form their scalar product

$$(x, y) = \|x\| \cdot \|y\| \cdot \cos \theta = a_1 b_1 + a_2 b_2 + a_3 b_3.$$

4) Knowing the scalar product (x, y) and the lengths of the vectors, we can find the angle θ between them

$$\cos \theta = \frac{(x, y)}{\|x\| \cdot \|y\|} \qquad (0 \leqslant \theta \leqslant \pi).$$

5) Finally, the condition of orthogonality of vectors is

$$(x, y) = a_1 b_1 + a_2 b_2 + a_3 b_3 = 0.$$

Generalizing these relations, we can introduce the concept of n-dimensional Euclidean space R_n, whose points and vectors are ordered n-tuples of real numbers

$$x = (a_1, a_2, \ldots, a_n).$$

The length of the vector $x = (a_1, \ldots, a_n)$ is defined to be the number

$$\|x\| = \sqrt{a_1^2 + \ldots + a_n^2}.$$

The inner product of vectors x and y cannot be defined by means of the angle between $\vec{x}$ and $\vec{y}$, and we must take the equation

$$(x, y) = \sum_{k=1}^{n} a_k b_k$$

for its definition. The angle θ can be defined, on the other hand, from the scalar product by the relation

$$\cos\theta = \frac{(x, y)}{\|x\| \cdot \|y\|} \qquad (0 \leqslant \theta \leqslant \pi),$$

For this definition to be reasonable, we must prove that

$$|(x, y)| \leqslant \|x\| \cdot \|y\|.$$

When this has been proved (this is done below), it is natural to consider the vectors $x = (a_1, \ldots, a_n)$ and $y = (b_1, \ldots, b_n)$ as orthogonal if

$$(x, y) = \sum_{k=1}^{n} a_k b_k = 0.$$

Continuing this process of generalization, we naturally come to the concept of the infinite-dimensional space R_∞, which is also designated by l_2. In this, we restrict ourselves to the vector treatment of the space in question.

DEFINITION. An infinite sequence of real numbers

$$x = (a_1, a_2, a_3, \ldots)$$

is said to be an element of the *space l_2* if

$$\|x\| = \sqrt{\sum_{k=1}^{\infty} a_k^2} < +\infty.$$

The number $\|x\|$ is called the *length* or the *norm* of the vector x.

It is easy to see that if $x \in l_2$, then, for all real numbers k, the vector

$$kx = (ka_1, ka_2, ka_3, \ldots)$$

is also in l_2. Furthermore,

$$\|kx\| = |k| \cdot \|x\|$$

and in particular $\|-x\| = \|x\|$.

If the vector

$$y = (b_1, b_2, b_3, \ldots)$$

is in l_2 as well as x, then the vector sum

$$x + y = (a_1 + b_1, a_2 + b_2, a_3 + b_3, \ldots)$$

also lies in l_2, because

$$(a_k + b_k)^2 \leqslant 2(a_k^2 + b_k^2).$$

4. The Space l_2

It follows from the inequality

$$|a_k b_k| \leq a_k^2 + b_k^2$$

that the series

$$(x, y) = \sum_{k=1}^{\infty} a_k b_k$$

converges absolutely. Its sum is called the inner product of the vectors x and y.

A close connection exists between the spaces l_2 and L_2. Let $\{\omega_k(x)\}$ be any complete orthogonal system in L_2, and let $f(x)$ be a function in L_2. Computing the Fourier coefficients

$$c_k = \int_a^b f(x) \omega_k(x)\, dx,\, ^4$$

we obtain an infinite sequence

$$x = (c_1, c_2, c_3, \ldots).$$

It follows from Parseval's identity,

$$\|x\| = \sqrt{\sum_{k=1}^{\infty} c_k^2} = \|f\| < \infty,$$

that x is an element of l_2. In this way, we obtain a single-valued mapping of L_2 into l_2. It is easy to see from the foregoing discussion of L_2, however, that this mapping is one-to-one and that the image of L_2 is all of l_2. First, it is clear that distinct elements of L_2 have distinct corresponding elements of l_2, by virtue of the completeness of the system $\{\omega_k(x)\}$. By the Riesz-Fischer theorem, every element of l_2 is the sequence of Fourier coefficients of a function in L_2.

The above correspondence possesses other properties than that of being one-to-one onto. If

$$x \sim f, y \sim g,$$

then

$$x + y \sim f + g,$$
$$kx \sim kf.$$

In other words, no linear relation

$$k_1 f_1 + \ldots + k_n f_n = 0$$

among the elements of L_2 is altered if we replace the elements $f_1, \ldots, f_n$ by elements of l_2 which correspond to them [the vector $(0, 0, 0, \ldots)$ in l_2 is written as 0]. Upon combining this fact and the fact that

$$\|x\| = \|f\|,$$

the complete geometric identity of the spaces L_2 and l_2 becomes clear. For this reason, l_2 is also called Hilbert space.

---

[4] We hope that the reader will not be confused by the fact that we use the letter x to denote the argument of the functions $f(x)$, $\omega_k(x)$, ... and also to denote vectors of the space l_2.

Let
$$x = (a_1, a_2, a_3, \ldots), \quad y = (b_1, b_2, b_3, \ldots)$$
be two vectors in l_2, and let f and g be the elements of L_2 corresponding to them. Parseval's identity yields
$$\int_a^b f(x) g(x) \, dx = \sum_{k=1}^{\infty} a_k b_k = (x, y).$$
It is thus natural to call the integral
$$\int_a^b f(x) g(x) \, dx$$
the scalar product of the elements f and g and to write it as
$$(f, g).$$
Thus
$$(f, g) = (x, y).$$
In this notation, the CBS inequality is written
$$|(f, g)| \leqslant \|f\| \cdot \|g\|.$$
This implies the possibility of defining an angle θ between any two non-zero elements f and g of the space L_2:
$$\cos \theta = \frac{(f, g)}{\|f\| \cdot \|g\|} \qquad (0 \leqslant \theta \leqslant \pi).$$
In particular, the definition given above of orthogonality of functions $f(x)$ and $g(x)$,
$$(f, g) = 0$$
is equivalent to the condition that the angle between them be $\frac{\pi}{2}$.

Further, if $\omega(x)$ is a normalized function,
$$\|\omega\| = 1,$$
then it can be considered as a unit vector of the space L_2 (or l_2). In this case, we can define the projection of the vector f in the direction of ω in the usual way:
$$Pr_\omega f = \|f\| \cdot \cos \theta,$$
where θ is the angle between the vectors f and ω. In other words,
$$Pr_\omega f = \int_a^b f(x) \omega(x) \, dx.$$
Thus the Fourier coefficients of the function $f(x)$ in an orthonormal system [5] $\{\omega_k(\lambda)\}$ are projections of the vector f in the directions characterized by the functions of the system.

---

[5] This is not necessarily the system used to establish the relation between L_2 and l_2.

4. The Space l_2

In n-dimensional Euclidean space, the length of a vector $x = (a_1, a_2, \ldots, a_n)$ is the number

$$\|x\| = \sqrt{\sum_{k=1}^{n} a_k^2}.$$

This is a generalization of the Pythagorean theorem because the numbers a_k are the projections of the vector x on the co-ordinate axes. Let us consider m ($m \leqslant n$) of these projections. In order to find out whether all n projections on the axes have been taken into consideration, we can simply compare the numbers m and n, and also we can observe whether for every vector x, the equality

$$\|x\|^2 = \sum_{k=1}^{m} a_k^2$$

is true (because if $m < n$, then necessarily vectors x exist for which $\sum_{k=1}^{m} a_k^2 < \|x\|^2$). Finally, we can see whether there exist directions orthogonal to all m axes taken into consideration.

For the infinite-dimensional space L_2, every orthonormal system $\{\omega_k(x)\}$ is a system of orthogonal co-ordinates of the axes. In checking to see whether all possible directions have been taken into consideration in setting up this system, one cannot simply count, as in the finite-dimensional case. Generalizing the two other methods indicated for n-dimensional space, we naturally arrive at the definitions of closed and complete orthonormal systems. The basic reason for the equivalence of these two definitions now becomes clear.

Up to this point, the connection between the spaces l_2 and L_2 has been used to establish certain new points of view regarding L_2 (which, of course, is very important). We will show that this connection is also useful for obtaining new facts.

First of all, the inequality

$$|(x, y)| \leqslant \|x\| \cdot \|y\|,$$

is equivalent to the CBS inequality

$$|(f, g)| \leqslant \|f\| \cdot \|g\|.$$

Therefore

$$\left(\sum_{k=1}^{\infty} a_k b_k\right)^2 \leqslant \left(\sum_{k=1}^{\infty} a_k^2\right)\left(\sum_{k=1}^{\infty} a_k^2\right), \tag{1}$$

and the valid inequality

$$\|x+y\| \leqslant \|x\| + \|y\|$$

can be written in the form

$$\sqrt{\sum_{k=1}^{\infty} (a_k + b_k)^2} \leqslant \sqrt{\sum_{k=1}^{\infty} a_k^2} + \sqrt{\sum_{k=1}^{\infty} b_k^2}. \tag{2}$$

Inequalities (1) and (2) are of a purely algebraic character. Further, the completeness of the space l_2 follows automatically from the completeness of the space L_2 (i.e., if $x_1, x_2, \ldots$ is a sequence for which $\|x_n - x_m\| \to 0$ as $m, n \to \infty$, then $x_1, x_2, \ldots$, has a limit in l_2.

Since every n-dimensional space R_n is a closed subset of the space l_2, everything stated above (inequalities (1) and (2) and completeness) is applicable to R_n as well.*

In conclusion, we treat another problem in which the connection between L_2 and l_2 turns out to be very useful. Let g be a fixed element of L_2. Consider the function

$$\Phi(f) = (f, g) \tag{3}$$

defined for all $f \in L_2$. The function Φ possesses the obvious properties

1) $\Phi(f_1 + f_2) = \Phi(f_1) + \Phi(f_2)$,
2) $|\Phi(f)| \leq K \cdot \|f\|$ $(K = \|g\|)$.

Every function $\Phi(f)$ defined for an element of L_2, whose values are real numbers and which possesses properties 1) and 2) is said to be a bounded *linear functional* on the space L_2. Every bounded linear functional on the space L_2 has the form (3).

THEOREM (F. RIESZ). *If $\Phi(f)$ is a linear functional in the space L_2, then there exists one and only one element $g \in L_2$ such that for every $f \in L_2$,*

$$\Phi(f) = (f, g).$$

Proof. Suppose that we have set up, using some complete orthonormal system in L_2, a one-to-one mapping of L_2 onto l_2 which preserves norms and carries sums and scalar multiples into the corresponding sums and scalar multiples. It is clear that the linear functional Φ on L_2 can be considered as a functional on l_2, where $\Phi(x)$ is defined as being equal to $\Phi(f)$, whenever x is the element of l_2 corresponding to $f \in L_2$. It is also clear that

$$\Phi(x_1 + x_2) = \Phi(x_1) + \Phi(x_2), \quad |\Phi(x)| \leq K \cdot \|x\|,$$

i.e., that Φ is a bounded linear functional on l_2. We shall show that there exists an element $y \in l_2$ such that

$$\Phi(x) = (x, y) \tag{4}$$

for all $x \in l_2$. We prove first that the functional Φ is homogeneous, *i.e.*, that

$$\Phi(ax) = a\Phi(x) \tag{5}$$

for all real a. Relation (5) obviously holds if a is a natural number. From this, it is easy to verify that it is also fulfilled in the case when a has the form $\frac{1}{m}$, where m is a natural number, and hence (5) holds for every positive rational number a. Further, designating the vector $(0, 0, \ldots)$ by 0, we have

$$\Phi(0) = \Phi(0 + 0) = 2\Phi(0),$$

which implies that $\Phi(0) = 0$. Thus (5) holds for $a = 0$. Finally, it follows from the equalities

$$0 = \Phi(0) = \Phi[x + (-x)] = \Phi(x) + \Phi(-x)$$

that $\Phi(-x) = -\Phi(x)$ and hence (5) is true for every rational number a. It remains

---

* The author seems here to be flogging a dead horse. The inequality $(|x, y)| \leq \|x\| : \|y\|$, true for all $x, y \in l_2$, can be proved very simply by a direct method. The proof of Theorem 4, § 1, can be imitated directly to obtain this inequality for elements of R_n, and a passage to the limit gives the inequality for l_2. Completeness of R_n is obvious, and completeness of l_2 can be proved by a simple argument, very much like the proof of Theorem 5, § 2.—E. H.

4. THE SPACE l_2

to consider the case when a is irrational. In this case, let r be any rational number. Then the equality

$$\Phi(rx) = r\Phi(x) \qquad (6)$$

goes over into (5) in the limit as $r \to a$. This is obvious for the right member of (6). On the other hand,

$$|\Phi(rx) - \Phi(ax)| = |\Phi[(r-a)x]| \leqslant K \cdot |r-a| \cdot \|x\|,$$

so that the left member of (6) tends to the left member of (5). This proves (5) in all cases.

We next introduce the vectors

$$e_k = (0, \ldots 0, 1, 0, \ldots)$$

(unit in the k-th place) and set

$$\Phi(e_k) = A_k \qquad (k = 1, 2, \ldots).$$

The vector

$$y = (A_1, A_2, A_3, \ldots)$$

is in l_2. In fact, if

$$y_n = (A_1, A_2, \ldots, A_n, 0, 0, \ldots),$$

then $y_n = \sum_{k=1}^{n} A_k e_k$, and

$$\Phi(y_n) = \sum_{k=1}^{n} A_k \Phi(e_k) = \sum_{k=1}^{n} A_k^2.$$

The inequality $|\Phi(x)| \leqslant K \cdot \|x\|$ applied to y_n gives

$$\sqrt{\sum_{k=1}^{n} A_k^2} \leqslant K,$$

and this shows that

$$\|y\| \leqslant K.$$

since n is arbitrary.

The element y is the one sought, i.e., (4) holds for all $x \in l_2$. Let

$$x = (a_1, a_2, a_3, \ldots)$$

be an element of l_2. We set

$$x_n = (a_1, a_2, \ldots, a_n, 0, 0, \ldots).$$

Then $x_n = \sum_{k=1}^{n} a_k e_k$ and

$$\Phi(x_n) = \sum_{k=1}^{n} a_k \Phi(e_k) = \sum_{k=1}^{n} A_k a_k. \qquad (7)$$

As $n \to \infty$, the right member of this equation tends to (x, y). On the other hand,

$$|\Phi(x) - \Phi(x_n)| = |\Phi(x - x_n)| \leqslant K \cdot \|x - x_n\| = K \cdot \sqrt{\sum_{k=n+1}^{\infty} a_k^2},$$

so that $\Phi(x_n)$ converges to $\Phi(x)$ as $n \to \infty$, and (7) goes over into (4) in the limit.

Let g be the element of L_2 which corresponds to the element y in the correspondence described above. Let f be an arbitrary element of L_2 and let x be the element of l_2 corresponding to it. Then

$$\Phi(f) = \Phi(x) = (x, y) = (f, g).$$

It remains to verify that there is only one element g in L_2 satisfying the relation

$$\Phi(f) = (f, g)$$

for arbitrary f. If there were two such element g_1 and g_2, then we would have

$$(f, g_1 - g_2) = (f, g_1) - (f, g_2) = \Phi(f) - \Phi(f) = 0$$

for all $f \in L_2$. Setting $f = g_1 - g_2$, we have

$$(g_1 - g_2, g_1 - g_2) = \| g_1 - g_2 \|^2 = 0,$$

and therefore $g_1 = g_2$. This completes the present proof.

§ 5. LINEARLY INDEPENDENT SYSTEMS

DEFINITION 1. A system of functions $\varphi_1(x), \varphi_2(x), \ldots, \varphi_n(x)$, defined on $[a, b]$, is said to be *linearly dependent* if it is possible to find a set of constants $A_1, A_2, \ldots, A_n$, at least one of which is different from zero, such that

$$A_1 \varphi_1(x) + A_2 \varphi_2(x) + \ldots + A_n \varphi_n(x) \sim 0. \tag{1}$$

If, however, there is no such set of constants, *i.e.*, if (1) implies that

$$A_1 = A_2 = \ldots = A_n = 0,$$

then the system $\{\varphi_k(x)\}$ is said to be *linearly independent*.

It is easy to see that if at least one function of the system $\{\varphi_k(x)\}$ is equivalent to zero, the system is linearly dependent and that every system which is a subsystem of a linearly independent system is linearly independent.

THEOREM 1. *Every orthonormal system is linearly independent.*

In fact, if $\{\omega_k(x)\}$ $(k = 1, 2, \ldots, n)$ is an orthonormal system on $[a, b]$ and if

$$\sum_{k=1}^{n} A_k \omega_k(x) \sim 0,$$

then, multiplying this equality by $\omega_i(x)$ and integrating, we find that

$$A_i = 0 \qquad (i = 1, 2, \ldots, n).$$

This implies that $\{\omega_k(x)\}$ is a linearly independent system.

THEOREM 2. *The system of functions $x^{n_1}, x^{n_2}, \ldots, x^{n_i}$, where the exponents $n_1, n_2, \ldots, n_i$ are distinct integers, is linearly independent on every closed interval.*

This theorem follows from the fact that an integral polynomial can have only a finite number of roots.

DEFINITION 2. A denumerable system of functions $\varphi_1(x), \varphi_2(x), \ldots$ is said to be *linearly independent* if every finite subset of it is linearly independent.

5. Linearly Independent Systems

For example, every denumerable orthonormal system and the system $1, x, x^2, \ldots$ are linearly independent.

Let $\varphi_1(x), \ldots, \varphi_n(x)$ be a system of functions in L_2, defined on the closed interval $[a, b]$. As above, we set

$$(f, g) = \int_a^b f(x) g(x) dx$$

for two arbitrary functions $f(x)$ and $g(x)$ in L_2. Let us form the determinant

$$\Delta_n = \begin{vmatrix} (\varphi_1, \varphi_1) & (\varphi_1, \varphi_2) & \cdots & (\varphi_1, \varphi_n) \\ (\varphi_2, \varphi_1) & (\varphi_2, \varphi_2) & \cdots & (\varphi_2, \varphi_n) \\ \vdots & & & \vdots \\ (\varphi_n, \varphi_1) & (\varphi_n, \varphi_2) & \cdots & (\varphi_n, \varphi_n) \end{vmatrix}.$$

DEFINITION 3. *The determinant Δ_n is called the Gram determinant of the system of functions $\{\varphi_k(x)\}$.*

THEOREM 3. *The system of functions*

$$\varphi_1(x), \quad \varphi_2(x), \quad \ldots, \quad \varphi_n(x) \tag{2}$$

is linearly dependent if and only if its Gram determinant is zero.

Proof. Suppose that the system (2) is linearly independent. Then there exists a set of constants $A_1, \ldots, A_n$, at least one of which is different from zero, such that

$$A_1 \varphi_1(x) + A_2 \varphi_2(x) + \ldots + A_n \varphi_n(x) \sim 0. \tag{3}$$

Multiplying this equation by $\varphi_1(x), \varphi_2(x), \ldots, \varphi_n(x)$ in turn and integrating, we obtain

$$\left.\begin{array}{l} A_1(\varphi_1, \varphi_1) + A_2(\varphi_1, \varphi_2) + \ldots + A_n(\varphi_1, \varphi_n) = 0 \\ A_1(\varphi_2, \varphi_1) + A_2(\varphi_2, \varphi_2) + \ldots + A_n(\varphi_2, \varphi_n) = 0 \\ \cdots\cdots\cdots\cdots\cdots\cdots\cdots\cdots\cdots\cdots\cdots\cdots \\ A_1(\varphi_n, \varphi_1) + A_2(\varphi_n, \varphi_2) + \ldots + A_n(\varphi_n, \varphi_n) = 0 \end{array}\right\}. \tag{4}$$

If we regard the numbers A_k as unknowns in the equations (4), we see that these equations form a linear system of equations with determinant Δ_n. The homogeneous system (4) is satisfied for the numbers A_k and, since not all A_k are zero, it follows that

$$\Delta_n = 0, \tag{5}$$

which proves the necessity of this condition.

Suppose next that $\Delta_n = 0$. Then the homogeneous linear system of equations (4) has a solution $A_1, \ldots, A_n$ different from zero. Let $A_1, A_2, \ldots, A_n$ be this solution, so that equations (4) are identities. We rewrite these identities in the form

$$\int_a^b \varphi_1(x) [A_1 \varphi_1(x) + \ldots + A_n \varphi_n(x)] dx = 0$$
$$\cdots\cdots\cdots\cdots\cdots\cdots\cdots\cdots\cdots\cdots\cdots$$
$$\int_a^b \varphi_n(x) [A_1 \varphi_1(x) + \ldots + A_n \varphi_n(x)] dx = 0.$$

Multiplying these equations by $A_1, \ldots, A_n$, respectively, and adding, we find that

$$\int_a^b [A_1 \varphi_1(x) + \ldots + A_n \varphi_n(x)]^2 \, dx = 0,$$

(3) follows at once, and the system $\{\varphi_k(x)\}$ is linearly dependent.

COROLLARY. *If*

$$\Delta_n \neq 0,$$

then none of the determinants $\Delta_1, \Delta_2, \Delta_3, \ldots, \Delta_{n-1}$ *is zero.*[6]

In fact, if $\Delta_n \neq 0$, the system $\{\varphi_k(x)\}$ is linearly independent; consequently, every subset $\varphi_1, \ldots, \varphi_m$ $(m < n)$ of it is linearly independent also, and then $\Delta_n \neq 0$.

LEMMA. *Let a system of n functions* $\varphi_1(x), \ldots, \varphi_n(x)$ *be defined on* $[a, b]$. *Define* $\psi_n(x)$ *by the relation*

$$\psi_n(x) = \begin{vmatrix} (\varphi_1, \varphi_1) & (\varphi_1, \varphi_2) & \cdots & (\varphi_1, \varphi_{n-1}) & \varphi_1(x) \\ (\varphi_2, \varphi_1) & (\varphi_2, \varphi_2) & \cdots & (\varphi_2, \varphi_{n-1}) & \varphi_2(x) \\ \cdots & \cdots & \cdots & \cdots & \cdots \\ (\varphi_n, \varphi_1) & (\varphi_n, \varphi_2) & \cdots & (\varphi_n, \varphi_{n-1}) & \varphi_n(x) \end{vmatrix}. \qquad (6)$$

Then

$$(\psi_n, \varphi_k) = \begin{cases} 0 & (k < n) \\ \Delta_n & (k = n). \end{cases}$$

Proof. In order to multiply $\psi_n(x)$ by $\varphi_k(x)$, it is sufficient to multiply the last column of the determinant in (6) by $\varphi_k(x)$ and in order to find the integral of the product so obtained, it is necessary only to integrate its last column. The remainder of the proof is clear.

If we expand the determinant $\psi_n(x)$ in terms of the elements of the last column, it is clear that we obtain

$$\psi_n(x) = A_1 \varphi_1(x) + \ldots + A_{n-1} \varphi_{n-1}(x) + \Delta_{n-1} \varphi_n(x). \qquad (7)$$

Hence, if the system $\{\varphi_k(x)\}$ is linearly independent, then $\psi_n(x)$ is not equivalent to zero (since $\Delta_{n-1} \neq 0$). Multiplying equation (7) by $\psi_n(x)$, integrating and referring to the lemma, we find

$$\int_a^b \psi_n^2 \, dx = \Delta_{n-1} \Delta_n, \qquad (8)$$

so that the determinants Δ_n and Δ_{n-1} (not equal to zero) have the same sign. For the same reason, the determinants Δ_{n-1} and Δ_{n-2}, etc., have the same sign. Therefore Δ_n has the same sign as $\Delta_1 = (\varphi_1, \varphi_1) > 0$. We have thus established the following theorem.

THEOREM 4. *The Gram determinant of a linearly independent system is positive.*

The reasoning developed above leads to a simple proof of the following fact.

THEOREM 5 (E. SCHMIDT). *Let a finite or denumerable linearly independent system*

---

[6] Δ_1, is taken to be (φ_1, φ_1).

5. Linearly Independent Systems

$\varphi_1(x), \varphi_2(x), \ldots$ be given on the closed interval $[a, b]$. Then it is possible to construct an orthonormal system $\omega_1(x), \omega_2(x), \ldots$ such that 1) each function $\omega_n(x)$ is a linear combination of the first n functions of the system $\{\varphi_k(x)\}$ and, conversely, 2) every function $\varphi_n(x)$ is a linear combination of the first n functions of the system $\{\omega_k(x)\}$.

Proof. Let

$$\omega_1(x) = \frac{\varphi_1(x)}{\sqrt{\Delta_1}}, \qquad \omega_n(x) = \frac{\psi_n(x)}{\sqrt{\Delta_{n-1}\Delta_n}} \qquad (n \geqslant 2),$$

where $\psi_n(x)$ is defined by inequality (6). This defines the required system. In fact, it is clear from (7) that

$$\omega_n(x) = \sum_{k=1}^{n} a_k \varphi_k(x),$$

so that the system $\{\omega_k(x)\}$ satisfies the first requirement of the theorem. It follows from the lemma that $\psi_n(x)$, and hence, $\omega_n(x)$ also, is orthogonal to all the functions $\varphi_1(x), \ldots, \varphi_{n-1}(x)$. But then $\omega_n(x)$ is orthogonal to all linear combinations of the functions $\varphi_1(x), \ldots, \varphi_{n-1}(x)$, in particular to the functions $\omega_1(x), \ldots, \omega_{n-1}(x)$. Therefore the system $\{\omega_k(x)\}$ consists of pairwise orthogonal functions. By (8), all the functions $\omega_n(x)$ are normalized and so $\{\omega_k(x)\}$ is an orthonormal system.

It remains to show that

$$\varphi_n(x) = \sum_{k=1}^{n} b_k \omega_k(x). \tag{9}$$

This is trivial for $n = 1$. Suppose that it has been proved for all $n < m$. Then, by (7),

$$\varphi_m(x) = \frac{1}{\Delta_{m-1}} \psi_m(x) - \sum_{k=1}^{m-1} \frac{A_k}{\Delta_{m-1}} \varphi_k(x),$$

whence, replacing $\psi_m(x)$ in the right member by $\sqrt{\Delta_{m-1}\Delta_m}\,\omega_m(x)$, and each $\varphi_k(x)$ ($k = 1, \ldots, m-1$) by a linear combination of the functions $\omega_1(x), \ldots, \omega_k(x)$, we find that (9) is also true for $n = m$, as was to be proved.

REMARK. The systems $\{\varphi_k(x)\}$ and $\{\omega_k(x)\}$ are both complete or both incomplete. This follows from the fact that every function $h(x)$ orthogonal to all the functions of one system is orthogonal to all the functions of the other system also.

EXAMPLE. The system of functions

$$1, \ x, \ x^2, \ x^3, \ \ldots$$

is linearly independent on the closed interval $[-1, 1]$. Applying the process of orthogonalization indicated in the theorem to it, we obtain a system of polynomials

$$L_0(x), \ L_1(x), \ L_2(x), \ \ldots \tag{10}$$

orthonormal on $[-1, +1]$, where $L_n(x)$ is a polynomial of degree n.[7] The polynomials (10) are called *Legendre polynomials*.

---

[7] By Theorem 5, the degree of $L_n(x)$ is not greater than n. But it is also not less than n because

$$x^n = \sum_{k=0}^{n} a_k L_k(x).$$

THEOREM 6. *The system of Legendre polynomials is closed.*
Proof. According to Schmidt's theorem,

$$x^n = \sum_{k=0}^{n} a_k L_k(x). \tag{11}$$

At the beginning of §3, we showed that the coefficients a_k are the Fourier coefficients of the function x^n in the orthonormal system $\{L_k(x)\}$. Multiplying (11) by x^n and integrating between the limits -1 and $+1$, we find that

$$\| x^n \|^2 = \sum_{k=0}^{n} a_k^2,$$

i.e., Parseval's identity is valid for each of the functions x^n ($n = 0, 1, 2, \ldots$). By Corollary 1 of Theorem 2, §3 our theorem is proved.

COROLLARY. *The system of functions* $1, x, x^2, \ldots$ *is complete in* $L_2([-1, 1])$.

§ 6. THE SPACES L_p AND l_p

In this paragraph we will take up a certain generalization of the space L_2.

DEFINITION 1. A measurable function (as above, we restrict our discussion to functions defined on some closed interval $[a, b]$) is said to be *p-th power summable*, where $p \geqslant 1$, if

$$\int_a^b |f(x)|^p \, dx < +\infty.$$

It is customary to denote the set of such functions by the symbol L_p. Obviously, $L_1 = L$.

THEOREM 1. *If $f(x)$ is p-th power summable ($p > 1$), then $f(x)$ is summable, i.e., $L_p \subset L$.*

If we set $E = [a, b]$, $A = E(|f| < 1)$, $B = E - A$, the summability of the function $f(x)$ on the set A is obvious, and its summability on the set B follows from the fact that $|f(x)| \leqslant |f(x)|^p$ on this set.

The following theorem is proved similarly.

THEOREM 2. *The sum of two functions in L_p is again a function in L_p.*

Let $f(x) g$ and (x) be in L_p. Setting

$$E = [a, b], \quad A = E(|f| \leqslant |g|), \quad B = E - A,$$

we will have for $x \in A$:

$$|f(x) + g(x)|^p \leqslant \{|f(x)| + |g(x)|\}^p \leqslant 2^p |g(x)|^p,$$

and consequently

$$\int_A |f(x) + g(x)|^p \, dx \leqslant 2^p \int_A |g(x)|^p \, dx < +\infty.$$

We show similarly that the integral $\int_B |f + g|^p \, dx$ is finite. We note the further obvious fact that all functions $kf(x)$, where k is a finite constant, are in L_p if $f(x)$ is in L_p.

6. THE SPACES L_p AND l_p

Suppose that $p > 1$. The number

$$q = \frac{p}{p-1}$$

is called the *index conjugate to p*. Since

$$\frac{1}{p} + \frac{1}{q} = 1$$

the index conjugate to q, is p. That is, p and q are mutually conjugate indices. In particular, if $p = 2$, then $q = 2$ also. Thus, the index 2 is self-conjugate. (Certain important properties of the space L_2, which are not valid for other L_p, have their root in this fact.)

THEOREM 3 (HÖLDER'S INEQUALITY). *If $f(x) \in L_p$ and $g(x) \in L_p$, where p and q are mutually conjugate and $p > 1$, then the product $f(x) g(x)$ is summable and the inequality*

$$\left| \int_a^b f(x) g(x) \, dx \right| \leq \sqrt[p]{\int_a^b |f(x)|^p \, dx} \cdot \sqrt[q]{\int_a^b |g(x)|^q \, dx}. \tag{1}$$

is valid.

Proof. Let $0 < \alpha < 1$. Consider the function

$$\psi(x) = x^\alpha - \alpha x \qquad (0 < x < +\infty)$$

Its derivative, $\psi'(x) = \alpha[x^{\alpha-1} - 1]$, is positive for $0 < x < 1$ and negative for $x > 1$. Therefore the maximum value of the function $\psi(x)$ is attained for $x = 1$. Thus

$$\psi(x) \leq \psi(1) = 1 - \alpha,$$

whence it follows that

$$x^\alpha \leq \alpha x + (1 - \alpha). \tag{2}$$

for all $x > 0$. Let A and B be two positive numbers. If we substitute $x = \frac{A}{B}$ in (2) and multiply the inequality thus obtained by B, we obtain

$$A^\alpha B^{1-\alpha} \leq \alpha A + (1 - \alpha) B.$$

Let p and q be the two mutually conjugate indices mentioned above. If we set $\alpha = \frac{1}{p}$, $1 - \alpha = \frac{1}{q}$, we see that

$$A^{\frac{1}{p}} B^{\frac{1}{q}} \leq \frac{A}{p} + \frac{B}{q}. \tag{3}$$

The last inequality has been proved for $A > 0$ and $B > 0$, but it obviously holds also when A or B or both A and B are zero.

Having established (3), we consider the functions $f(x)$ and $g(x)$ of the present theorem. If at least one of them is equivalent to zero, all assertions of the theorem are obvious. Excluding this trivial case, we can assert that both the integrals

$$\int_a^b |f(x)|^p \, dx, \qquad \int_a^b |g(x)|^q \, dx$$

are strictly positive and therefore we may consider the functions

$$\varphi(x) = \frac{f(x)}{\sqrt[p]{\int_a^b |f(x)|^p \, dx}}, \quad \gamma(x) = \frac{g(x)}{\sqrt[q]{\int_a^b |g(x)|^q \, dx}}.$$

If we set

$$A = |\varphi(x)|^p, \quad B = |\gamma(x)|^q,$$

in (3), we obtain

$$|\varphi(x)\gamma(x)| \leqslant \frac{|\varphi(x)|^p}{p} + \frac{|\gamma(x)|^q}{q}, \tag{4}$$

whence the summability of the product $\varphi(x)\gamma(x)$ follows and with it that of the product $f(x)g(x)$. Besides, noting that

$$\int_a^b |\varphi(x)|^p \, dx = \int_a^b |\gamma(x)|^q \, dx = 1$$

and integrating (4), we find

$$\int_a^b |\varphi(x)\gamma(x)| \, dx \leqslant \frac{1}{p} + \frac{1}{q} = 1.$$

The inequality

$$\int_a^b |f(x)g(x)| \, dx \leqslant \sqrt[p]{\int_a^b |f(x)|^p \, dx} \cdot \sqrt[q]{\int_a^b |g(x)|^q \, dx},$$

follows at once; this is a stronger inequality than (1).

Hölder's inequality is a generalization of the CBS inequality, which is merely Hölder's inequality with $p = 2$.

THEOREM 4. (MINKOWSKI'S INEQUALITY). If $f(x) \in L_p$ and $g(x) \in L_p$ ($p \geqslant 1$), then

$$\sqrt[p]{\int_a^b |f(x)+g(x)|^p \, dx} \leqslant \sqrt[p]{\int_a^b |f(x)|^p \, dx} + \sqrt[p]{\int_a^b |g(x)|^p \, dx}. \tag{5}$$

Proof. The theorem is very simple for $p = 1$. Consider then the case $p > 1$. Let q be the index conjugate to p. By Theorem 2, the sum $f(x) + g(x)$ is in L_p, and, therefore, $|f(x) + g(x)|^{p/q}$ is in L_q. Now substitute $|f(x)|$ for $f(x)$ and $|f(x) + g(x)|^{p/q}$ for $g(x)$ in Hölder's inequality. This yields

$$\int_a^b |f(x)| \cdot |f(x)+g(x)|^{p/q} \, dx \leqslant \sqrt[p]{\int_a^b |f(x)|^p \, dx} \cdot \sqrt[q]{\int_a^b |f(x)+g(x)|^p \, dx}. \tag{6}$$

Similarly

$$\int_a^b |g(x)| \cdot |f(x)+g(x)|^{p/q} \, dx \leqslant \sqrt[p]{\int_a^b |g(x)|^p \, dx} \cdot \sqrt[q]{\int_a^b |f(x)+g(x)|^p \, dx}. \tag{7}$$

However, $p = 1 + \frac{p}{q}$. This implies that

$$|f+g|^p = |f+g| \cdot |f+g|^{p/q} \leq |f| \cdot |f+g|^{p/q} + |g| \cdot |f+g|^{p/q}.$$

From these relations and (6) and (7), it follows that

$$\int_a^b |f+g|^p \, dx \leq \left\{ \sqrt[p]{\int_a^b |f|^p \, dx} + \sqrt[p]{\int_a^b |g|^p \, dx} \right\} \cdot \sqrt[q]{\int_a^b |f+g|^p \, dx},$$

and the theorem follows [cancelling $\left(\int_a^b |f+g|^p \, dx \right)^{1/q}$ is permissible if this integral is different from zero, and otherwise the theorem is obvious].

Minkowski's inequality as given here is a generalization of the inequality of Theorem 5, §2, to which it reduces for $p = 2$.

The inequalities of Hölder and Minkowski for finite sums are established in a very similar way:

$$\left| \sum_{k=1}^n a_k b_k \right| \leq \sqrt[p]{\sum_{k=1}^n |a_k|^p} \cdot \sqrt[q]{\sum_{k=1}^n |b_k|^q} \quad \left(p > 1, \; q = \frac{p}{p-1} \right) \quad (8)$$

$$\sqrt[p]{\sum_{k=1}^n |a_k + b_k|^p} \leq \sqrt[p]{\sum_{k=1}^n |a_k|^p} + \sqrt[p]{\sum_{k=1}^n |b_k|^p}. \quad (p \geq 1) \quad (9)$$

DEFINITION 2. Let $f(x) \in L_p$. The number

$$\|f\| = \sqrt[p]{\int_a^b |f(x)|^p \, dx}$$

is called the *norm* of the function $f(x)$ (considered as an element of L_p).

The following properties of the norm are obvious:
  I. $\|f\| \geq 0$, and $\|f\| = 0$ if and only if $f(x) \sim 0$.
  II. $\|kf\| = |k| \cdot \|f\|$ and in particular $\|-f\| = \|f\|$.
  III. $\|f + g\| \leq \|f\| + \|g\|$.

The introduction of a norm allows us to establish the same geometric terminology for L_p as introduced above for L_2. The convergence of a sequence of elements $\{f_n(x)\}$ in L_p to a limit $f(x) \in L_p$ in the norm means that

$$\lim_{n \to \infty} \int_a^b |f_n(x) - f(x)|^p \, dx = 0.$$

This form of convergence is called *convergence in the mean of order p*.

As for L_2, we can show that the convergence of a sequence in the mean of order p implies its convergence to the same limit in measure. As for L_2, we can establish the continuity of the norm and the uniqueness of limits, if no distinction is made between equivalent functions. Exactly as in L_2, the concept of a Cauchy sequence is introduced. It is proved that a sequence of elements of L_p has a limit if and only if it is a Cauchy sequence (the space L_p is complete).

Since there is nothing essentially new in passing from L_2 to L_p, we will not stop to give proofs of all of these statements. We note also without proof that each of the classes M, C, P, S and T (the last for $b - a = 2\pi$), considered in Theorem 6, §2, is everywhere dense in L_p.

The concept of *weak convergence* for $p > 1$ is introduced as follows. A sequence $\{f_n(x)\} \subset L_p$ *converges weakly* to $f(x) \in L_p$ if the equality

$$\lim_{n \to \infty} \int_a^b f_n(x) g(x) dx = \int_a^b f(x) g(x) dx \tag{10}$$

holds for all functions $g(x)$ in L_q, where q is the index conjugate to p. With the aid of Hölder's inequality, it is easy to show that a sequence converging in the mean converges weakly (to the same limit).

If $p = 1$, the conjugate index does not exist. In this case, we say the sequence $\{f_n(x)\} \subset L$ converges weakly to $f(x) \in L$, if equality (10) holds for every measurable and *bounded* function $g(x)$. It is clear that here also convergence in the mean (of first order) implies weak convergence to the same limit.

We mention briefly one further class of spaces of importance in analysis. These are the spaces l_p, where $p \geqslant 1$. The space l_p is the set of all sequences

$$x = (x_1, x_2, x_3, \ldots)$$

of real numbers x_k for which

$$\|x\| = \sqrt[p]{\sum_{k=1}^{\infty} |x_k|^p} < +\infty.$$

The number $\|x\|$ is called the *norm* of the element $x \in l_p$. As in the case of l_2, we define the sum $x + y$ of two elements of l_p and the product kx of an element $x \in l_p$ by a number k. The norm possesses the usual properties.

    I. $\|x\| \geqslant 0$, where $\|x\| = 0$, if and only if $x = 0$.[8]
    II. $\|kx\| = |k| \cdot \|x\|$ and, in particular, $\|-x\| = \|x\|$.
    III. $\|x + y\| \leqslant \|x\| + \|y\|$.

The first two of these properties are clear; the third follows from (9).[9]

With the aid of the concept of norm, the ideas of the limit of elements of l_p, Cauchy sequences, sets everywhere dense in l_p, etc., can be introduced. One can prove that the limit of a sequence of elements in l_p is unique, that the norm is continuous, and that the space l_p possesses the property of completeness. We shall not discuss this in detail.

§7. EDITOR'S APPENDIX TO CHAPTER VII

The notion of square-summable function need not be confined to finite intervals $[a, b]$. In fact, one may consider any measurable set E on the line and study the class of all measurable functions $f(x)$ defined on E for which

$$\int_E f^2(x) \, dx$$

is finite. All of the definitions, notation, and results of §1, and of §2 up to and including

---

[8] That is, $x = (0, 0, 0, \ldots)$.
[9] Inequality (9) deals with finite sums. By taking limits the inequality (9) can be immediately generalized to sums of infinite series.

7. Editor's Appendix

Definition 3, can be carried over to this more general class of Hilbert spaces with absolutely no change. Details are left to the reader.

In discussing Theorem 6, §2, and its extension to spaces of square-summable functions on an arbitrary measurable set, it is convenient to restrict ourselves to the case $E = (-\infty, \infty)$. We have then the following result.

THEOREM 1. *The following classes of functions are dense subsets of $L_2(-\infty, \infty)$:*

the class of bounded measurable functions which are square-summable over $(-\infty, \infty)$;

the class of continuous functions $f(x)$ on $(-\infty, \infty)$ such that $f(x) = 0$ for $|x| \geq A$ (the constant A depending upon the function $f(x)$);

the class of step functions vanishing outside of finite intervals (the interval depending upon the particular function);

the set of all functions of the form $p(x) e^{-x^2}$, where $p(x)$ is a polynomial.

Proof. The proofs that the first three function classes mentioned are dense in $L_2(-\infty, \infty)$ are very similar and are quite simple. Let us show, for example, that the continuous functions vanishing outside of finite intervals form a dense subspace of $L_2(-\infty, \infty)$. If $f(x)$ is an arbitrary function in $L_2(-\infty, \infty)$, and if ε is an arbitrary positive number, then by Lebesgue's theorem on dominated convergence, (Theorem 1, §4, Chapter VI), we infer that there exists a positive number A such that

$$\int_{-\infty}^{-A} f^2(x)\, dx < \frac{\varepsilon^2}{25}, \quad \int_{A}^{\infty} f^2(x)\, dx < \frac{\varepsilon^2}{25}. \tag{1}$$

Now, by Theorem 6, §2, there exists a continuous function $\theta(x)$ defined on the closed interval $[-A, A]$ such that

$$\int_{-A}^{A} |f(x) - \theta(x)|^2\, dx < \frac{\varepsilon^2}{25}.$$

(Note that if $f \in L_2(-\infty, \infty)$, then $f \in L_2(a, b)$ for all numbers $a < b$). The function $\theta(x)$ can be extended to be a continuous function on $(-\infty, \infty)$ which vanishes identically outside of the interval $(-A - \delta, A + \delta)$ $(\delta > 0)$ by making $\theta(x)$ linear in the intervals $[-A - \delta, -A]$ and $[A, A + \delta]$. It is easy to see that δ can be chosen so small that

$$\int_{-\infty}^{-A} \theta^2(x)\, dx < \frac{\varepsilon^2}{25}, \quad \int_{A}^{\infty} \theta^2(x)\, dx < \frac{\varepsilon^2}{25}.$$

We now have, making free use of Theorem 5, §1, that

$$\|f - \theta\| \leq \{\int_{-\infty}^{-A} |f(x) - \theta(x)|^2\, dx\}^{\frac{1}{2}} + \{\int_{-A}^{A} |f(x) - \theta(x)|^2\, dx\}^{\frac{1}{2}} +$$

$$\{\int_{A}^{\infty} |f(x) - \theta(x)|^2\, dx\}^{\frac{1}{2}}$$

$$\leq \{\int_{-\infty}^{-A} f^2(x)\, dx\}^{\frac{1}{2}} + \{\int_{-\infty}^{-A} \theta^2(x)\, dx\}^{\frac{1}{2}} + \{\int_{-A}^{A} |f(x) - \theta(x)|^2\, dx\}^{\frac{1}{2}} +$$

$$\{\int_{A}^{\infty} f^2(x)\, dx\}^{\frac{1}{2}} + \{\int_{A}^{\infty} \theta^2(x)\, dx\}^{\frac{1}{2}}$$

$$< 5 \cdot \frac{\varepsilon}{5} = \varepsilon.$$

The proof that the class of all functions $p(x) e^{-x^2}$, where $p(x)$ is a polynomial, is a dense subclass of $L_2(-\infty, \infty)$ is somewhat more complicated. For a proof, we refer the reader to E. C. Titchmarsh, *Introduction to the Theory of Fourier Integrals*, Oxford University Press, 1937, page 79, Theorem 55.

We return to a discussion of the theorems and definitions of the text, as they are applied to $L_2(E)$ for unbounded E and in particular to $L_2(-\infty, \infty)$. The definition of weak convergence (Definition 4, §2), remains the same for general $L_2(E)$, and Theorem 7, §2, obviously remains true.

Orthogonality is defined in general $L_2(E)$ just as for $L_2(a, b)$. In fact, all of the definitions of §3 may be taken over without change to our general $L_2(E)$. All theorems of §3, except for the corollaries to Theorem 2, and the corollary to Theorem 4, are true for general $L_2(E)$, and the proofs can be repeated word for word.

In §4, a one-to-one correspondence is set up between $L_2(a, b)$ and the sequence space l_2, preserving sums, scalar multiples, norms, and inner products. It is clear that all that is needed for this purpose is a complete orthonormal sequence in L_2. Thus, if we can find a complete orthonormal system

$$\omega_1(x), \omega_2(x), \ldots, \omega_n(x), \ldots$$

in $L_2(E)$, we can assert that $L_2(E)$ is geometrically identical with l_2. There exist complete orthonormal systems in $L_2(E)$ for every measurable set E. For a proof of this fact, we refer the reader to M. H. Stone, *Linear Transformations in Hilbert Space and Their Applications to Analysis*, Amer. Math. Soc. Coll. Publ. Vol. XV, 1932, Chapter I.

All definitions and theorems of §5 can be carried over with no change from the case of $L_2(a, b)$ to $L_2(E)$ for arbitrary measurable E, barring Theorem 6 and its corollary, which refer specifically to $L_2(-1, 1)$.

The spaces L_p defined in §6 can be defined over every measurable set E, and all of the theorems of 6 dealing with $L_p(a, b)$ are also true for $L_p(E)$. One needs merely to replace integrals of the form

$$\int_a^b$$

by integrals of the form

$$\int_E$$

Exercises for Chapter VII

1. Let $\{f_n(x)\}$ be a system of functions in L_2 converging in measure to $F(x)$. If $\|f_n\| \leqslant K$, then the sequence $\{f_n(x)\}$ converges weakly to $F(x)$. (F. Riesz)

2. If the sequence $\{f_n(x)\}$ converges weakly in L_2 to $F(x)$, then $\|f_n\| \leqslant K$. (H. Lebesgue)

3. Weak convergence in L_2 of the sequence $\{f_n(x)\}$ to $F(x)$ does not imply convergence in measure of $\{f_n(x)\}$ to $F(x)$.

4. If the sequence $\{f_n(x)\}$ converges weakly in L_2 to $F(x)$, and, in addition, $\|f_n\| \to \|F\|$, then the sequence $\{f_n(x)\}$ converges to $F(x)$ in the mean. (F. Riesz)

5. If the integral $\int_a^b f(x)g(x)\,dx$ exists for every $f(x) \in L_2$, then $g(x) \in L_2$. (H. Lebesgue)

6. Every orthonormal system is at most denumerable.

7. If $\{\omega_k(x)\}$ is a closed orthonormal system on the segment $[a, b]$, then $\sum_{k=1}^{\infty} \omega_k^2(x) = +\infty$ almost everywhere on $[a, b]$. (W. Orlicz)

8. Under the same conditions, $\sum_{k=1}^{\infty} \int_e \omega_k^2\,dx = +\infty$ for an arbitrary measurable set e of measure $me > 0$. (W. Orlicz)

9. No finite system of functions is complete in L_2.

EXERCISES

10. Let $\{\omega_k(x)\}$ ($k = 1, 2, \ldots, n$) be an orthonormal system and let $f(x) \in L_2$. Of all linear combinations $\sum_{k=1}^{n} A_k \omega_k(x)$, the norm of the difference $\|f - \sum_{k=1}^{n} A_k \omega_k\|$ has the least value when $A_k = (f, \omega_k)$ ($k = 1, 2, \ldots, n$).

11. Let $\{\omega_k(x)\}$ be a complete orthonormal system of functions. If $\{\varphi_k(x)\}$ is a system of functions in L_2 such that $\sum_{k=1}^{\infty} \int_a^b [\omega_k(x) - \varphi_k(x)]^2 \, dx < 1$, then the system $\{\varphi_k(x)\}$ is also complete. (N. K. Bari)

12. Let the function $f(x) \in L_2$ be defined on $[-\pi, \pi)$; suppose that $f(x + 2\pi) = f(x)$ and let

$$g_n(x) = \int_{\frac{1}{n}}^{\pi} \frac{f(x+t) - f(x-t)}{t} \, dt.$$

The functions $g_n(x)$ converge in the mean on $[-\pi, \pi]$ to a function $q(x) \in L_2$, where

$$\|q\| \leq \|f\| \cdot \int_{-\pi}^{\pi} \frac{\sin t}{t} \, dt$$

and it is impossible to reduce the value of the multiplier of $\|f\|$. (I. P. Natanson)

13. Let $f(x) \in L_2$ on $[a, b]$ and let $f(x) = 0$ outside $[a, b]$. If

$$\varphi(x) = \frac{1}{2h} \int_{x-h}^{x+h} f(t) \, dt,$$

then $\|\varphi\| \leq \|f\|$. (A. N. Kolmogorov)

14. Using the same notation, the functions $\varphi(x)$ converge in the mean in L_2 to $f(x)$ as $h \to 0$. (A. N. Kolmogorov)

15. Extend the results of exercises 1, 2, 3, 4, 5, 13 and 14 from L_2 to L_p for $p > 1$.

16. Prove the completeness of space L_p for $p \geq 1$.

17. Prove the completeness of space l_p for $p \geq 1$.

18. Prove the completeness of the space m of all *bounded* sequences $x = \{x_k\}$, where $\|x\| = \sup\{|x_k|\} < +\infty$.

19. If in the set C of all functions continuous on $[a, b]$, we introduce the norm $\|f\| = \max |f(x)|$, and sums and scalar multiples are as usual, then the space obtained is complete.

20. The system of functions $\varphi_k(x)$ is said to be *complete* in the class of functions A if, in the latter, there are no functions distinct from zero which are orthogonal to all the $\varphi_k(x)$. It does not follow from the completeness of an orthogonal system in the class (R) of all Riemann-integrable functions that this system is closed. (G. M. Fichtenholz)

21. If $1 \leq r < p$, then $L_p \subset L_r$.

22. If $1 \leq r < p$, then $l_p \supset l_r$.

23. Let $\{f_n(x)\} \subset L_p$ ($p > 1$) be a sequence of functions, convergent in the mean of order p to the function $F(x) \in L_p$. Show that $\{f_n(x)\}$ converges to $F(x)$ in the mean of order r, where $1 \leq r < p$, also.

24. Let $1 \leq r < p$ and $x_n = (x_1^{(n)}, x_2^{(n)}, x_3^{(n)}, \ldots) \in l_r$. If x_n tends to the element x in the space l_r, then this holds also in the space l_p.

25. If the sequence $\{a_k\}$ is such that the series $\sum_{k=1}^{\infty} a_k x_k$ converges for every sequence $\{x_k\} \in l_p$ ($p > 1$), then $\{a_k\} \in l_q$, where $\frac{1}{p} + \frac{1}{q} = 1$.

26. If the sequence $\{a_k\}$ is such that the series $\sum_{k=1}^{\infty} a_k x_k$ converges for every sequence $\{x_k\} \in l = l_l$, then $\{a_k\} \in m$, i.e., $\sup\{|a_k|\} < +\infty$.

27. If $p > 1$ and the equality sign holds in Minkowski's inequality (5), then $g(x) = Kf(x)$, where $K \geq 0$.

CHAPTER VIII

FUNCTIONS OF FINITE VARIATION. THE STIELTJES INTEGRAL

§ 1. MONOTONIC FUNCTIONS

As is known, a function $f(x)$, defined on the closed interval $[a, b]$ is said to be *increasing* if
$$x < y, \tag{1}$$
for
$$f(x) \leqslant f(y).$$

If $f(x) < f(y)$ for $x < y$ then $f(x)$ is said to be a *strictly increasing* function. Analogously, a function $f(x)$ is said to be *decreasing (strictly decreasing)* if $f(x) \geqslant f(y)$ $[f(x) > f(y)]$ for $x < y$. Increasing and decreasing functions are called *monotonic* or *monotone (strictly monotonic)*. If the function $f(x)$ decreases, then $-f(x)$ increases. This simple remark permits us to consider only increasing functions in many problems involving monotonic functions. Monotonic functions will always be considered *finite*.

Let $f(x)$ be an increasing function defined on $[a, b]$ and let
$$a \leqslant x_0 < b.$$
It is an elementary and well-known fact that for any sequence of points $x_1, x_2, x_3, \ldots$, tending to x_0 and lying to the right of x_0,
$$x_n \to x_0, \quad x_n > x_0$$
the limit
$$\lim_{n \to \infty} f(x_n)$$
exists and is finite.

This limit is nothing but
$$\inf \{f(x)\} \qquad (x_0 < x \leqslant b),$$
and hence does not depend on the choice of the sequence $\{x_n\}$. It is denoted by
$$f(x_0 + 0).$$
The symbol
$$f(x_0 - 0) \qquad (a < x_0 \leqslant b)$$
is defined similarly. It is easy to see that
$$f(x_0 - 0) \leqslant f(x_0) \leqslant f(x_0 + 0) \qquad (a < x_0 < b)$$
and that
$$f(a) \leqslant f(a + 0), \qquad f(b - 0) \leqslant f(b).$$

1. MONOTONIC FUNCTIONS

Hence the function $f(x)$ is continuous at the point x_0 if and only if
$$f(x_0-0)=f(x_0)=f(x_0+0).$$
[For $x_0 = a$ ($x_0 = b$), we need to consider only the one-sided limit $f(a+0)$ ($f(b-0)$).]

The numbers
$$f(x_0)-f(x_0-0); \quad f(x_0+0)-f(x_0)$$
are called the left and right *saltus*, respectively, of the function $f(x)$ at the point x_0, and their sum $f(x_0+0)-f(x_0-0)$ is called the saltus of the function $f(x)$ at this point. [For the points a and b, only one-sided saltus are considered.]

LEMMA. *Let an increasing function $f(x)$ be defined on $[a, b]$. Let $x_1, x_2, \ldots, x_n$ be arbitrary points lying in (a, b). Then*
$$[f(a+0)-f(a)] + \sum_{k=1}^{n} [f(x_k+0)-f(x_k-0)] + [f(b)-f(b-0)] \leqslant$$
$$\leqslant f(b)-f(a). \qquad (2)$$

Proof. We may suppose that
$$a < x_1 < x_2 < \ldots < x_n < b.$$
Set $a = x_0$, $b = x_{n+1}$ and choose points $y_0, y_1, \ldots, y_n$ such that
$$x_k < y_k < x_{k+1} \qquad (k = 0, 1, \ldots, n).$$
Then
$$f(x_k+0)-f(x_k-0) \leqslant f(y_k)-f(y_{k-1}) \quad (k=1, 2, \ldots, n)$$
$$f(a+0)-f(a) \leqslant f(y_1)-f(a)$$
$$f(b)-f(b-0) \leqslant f(b)-f(y_n).$$
Adding all these inequalities, we obtain (2).

COROLLARY. *An increasing function $f(x)$ defined on $[a, b]$ can have only a finite number of points of discontinuity at which its saltus is greater than a given positive number σ.*

In fact, if the points $x_1, \ldots, x_n$, lying interior to $[a, b]$, are points of discontinuity with saltus greater than σ, then it follows from (2) that
$$n\sigma \leqslant f(b)-f(a)$$
and hence n cannot be arbitrarily large.

THEOREM 1. *The set of points of discontinuity of an increasing function $f(x)$ defined on $[a, b]$ is at most denumerable. If $x_1, x_2, x_3, \ldots$ are all of the interior points of discontinuity, then*
$$[f(a+0)-f(a)] + \sum_{k=1}^{\infty} [f(x_k+0)-f(x_k-0)] + [f(b)-f(b-0)] \leqslant$$
$$\leqslant f(b)-f(a). \qquad (3)$$

Proof. Designate the set of all points of discontinuity of the function $f(x)$ by H and the set of those points of discontinuity of this function, at which its saltus is greater than $\frac{1}{k}$, by H_k. Obviously,

$$H = H_1 + H_2 + H_3 + \ldots,$$

and the denumerability of H follows from the fact that each H_k is finite.

Inequality (3) follows from (2) upon taking the limit as $n \to \infty$.

Let $f(x)$ be an increasing function defined on $[a, b]$. Define the function $s(x)$ by setting

$$s(a) = 0,$$

$$s(x) = [f(a+0) - f(a)] + \sum_{x_k < x} [f(x_k+0) - f(x_k-0)] + \\ + [f(x) - f(x-0)] \qquad (a < x \leq b).$$

The function $s(x)$ is called the *saltus function* of the function $f(x)$. Clearly, it is also an increasing function.

THEOREM 2. *The difference*

$$\varphi(x) = f(x) - s(x)$$

between an increasing function and its saltus function is an increasing and continuous function.

Proof. Let $a \leq x < y \leq b$. If the inequality (3) is applied to the interval $[x, y]$ instead of to the interval $[a, b]$, we obtain the inequality

$$s(y) - s(x) \leq f(y) - f(x), \qquad (4)$$

as a simple calculation shows. This implies that

$$\varphi(x) \leq \varphi(y)$$

that is, $\varphi(x)$ is increasing.

Further, if we let y approach x in the inequality (4), then in the limit, we obtain

$$s(x+0) - s(x) \leq f(x+0) - f(x). \qquad (5)$$

On the other hand, it follows easily from the definition of the function $s(x)$ that

$$f(x+0) - f(x) \leq s(y) - s(x)$$

for $x < y$. Taking the limit as $y \to x$, we obtain

$$f(x+0) - f(x) \leq s(x+0) - s(x).$$

From this inequality and (5), it is clear that

$$f(x+0) - f(x) = s(x+0) - s(x).$$

Hence,

$$\varphi(x+0) = \varphi(x);$$

by a very similar argument, we can show that $\varphi(x-0) = \varphi(x)$. It follows that $\varphi(x)$ is a continuous function, as we wished to prove.

§ 2. MAPPING OF SETS. DIFFERENTIATION OF MONOTONE FUNCTIONS

Let a function $f(x)$ be defined on a certain abstract set A. Let E be an arbitrary subset of A. We may consider the set of all points y having the form $f(x)$, for $x \in E$, and call this set $f(E)$. Thus the function f induces a mapping of the family of all subsets of A into the family of all subsets of the image set $f(A)$ of A. One may say that $f(E)$, for $E \subset A$, consists exactly of those points y for which the equation $f(x) = y$ has a solution in the set E. The set $f(E)$ is called the *image* of E, and E is called the *inverse image* of the set $f(E)$.

THEOREM 1. *If* $E_1 \subset E_2$, *then* $f(E_1) \subset f(E_2)$. *If* $E = \sum_{n=1}^{\infty} E_n$, *then* $f(E) = \sum_{n=1}^{\infty} f(E_n)$.

This theorem is obvious.

The theory of mappings is particularly simple when the mapping function establishes a one-to-one correspondence between the sets A and $f(A)$. Then there also exists an inverse function $x = g(y)$, defined on the set $f(A)$ and having values lying in the set A. We define $g(y)$ by the condition that $x = g(y)$ if and only if $y = f(x)$. It is easy to see that in this case,

$$f(\prod_{n=1}^{\infty} E_n) = \prod_{n=1}^{\infty} f(E_n).$$

In particular, if two sets E_1 and E_2 are disjoint, then their images $f(E_1)$ and $f(E_2)$ are also disjoint.

As an example of such a well-behaved mapping, we may consider the mapping given by a continuous, strictly increasing function $f(x)$ defined on the closed interval $A = [a, b]$. In this case, $f(A) = [f(a), f(b)]$. The notion of a mapping of sets is very useful in studying differentiation.

DEFINITION. The number λ (finite or infinite) is said to be a *derived number* of the function $f(x)$ at the point x_0 if there exists a sequence $h_1, h_2, h_3, \ldots$ ($h_n \neq 0$), tending to zero, such that

$$\lim_{n \to \infty} \frac{f(x_0 + h_n) - f(x_0)}{h_n} = \lambda.$$

We express in symbols the assertion that λ is a derived number of the function $f(x)$ at the point x_0 as follows:

$$\lambda = Df(x_0).$$

If the (finite or infinite) derivative $f'(x_0)$ exists at the point x_0, then it will be a derived number $Df(x_0)$, and in this case the function $f(x)$ has no other derived numbers at the point x_0.

As an example, let us consider the Dirichlet function $\psi(x)$, which is equal to zero for irrational values of x and is equal to one for rational values of x. Let x_0 be a rational number. Then the ratio

$$\frac{\psi(x_0 + h) - \psi(x_0)}{h}$$

equals zero for rational h and $\frac{-1}{h}$ when h is irrational. From this it follows that the function $\psi(x)$ has three derived numbers at the point x_0: $-\infty$, 0, and $+\infty$. It is easy to verify that the function $\psi(x)$ has no other derived numbers at the point x_0. The same is true when x_0 is irrational.

208   VIII. FUNCTIONS OF FINITE VARIATION. STIELTJES INTEGRAL

THEOREM 2. *If the function $f(x)$ is defined on $[a, b]$, then derived numbers exist at every point $x \in [a, b]$.*

Proof. Let $x_0 \in [a, b]$ and $\{h_n\}$ ($h_n \neq 0$) be a sequence, tending to zero, such that $x_0 + h_n \in [a, b]$. Set

$$\sigma_n = \frac{f(x_0 + h_n) - f(x_0)}{h_n}.$$

If the sequence $\{\sigma_n\}$ is bounded, then, by the Bolzano-Weierstrass theorem, a subsequence can be extracted from it having some limit λ which will be a derived number of the function $f(x)$ at the point x_0: $\lambda = Df(x_0)$. If the sequence $\{\sigma_n\}$ is unbounded, (suppose, for example, that it is unbounded above), then a subsequence $\{\sigma_{n_k}\}$ tending to $+\infty$ can be extracted from it. In this case, $+\infty = Df(x_0)$.

THEOREM 3. *A function $f(x)$ defined on $[a, b]$ has a derivative $f'(x_0)$ at the point $x_0 \in [a, b]$ if and only if all derived numbers of $f(x)$ at this point are equal.*

The necessity of the condition, already noted above, is quite trivial. To prove its sufficiency, suppose that it holds, and let λ be the common value of all derived numbers of $f(x)$ at the point x_0. The existence of the derivative $f'(x_0)$ will be proved once we show that for every sequence $\{h_n\}$ ($h_n \neq 0$) with limit zero, the relation

$$\lim_{n \to \infty} \frac{f(x_0 + h_n) - f(x_0)}{h_n} = \lambda$$

holds. Assume that this is not the case. Then there exists at least one sequence $\{h_n\}$ ($h_n \to 0$ $h_n \neq 0$) for which the ratios

$$\sigma_n = \frac{f(x_0 + h_n) - f(x_0)}{h_n}$$

do not have the limit λ. But this implies [we suppose that $-\infty < \lambda < +\infty$; if $\lambda = \pm \infty$, the reasoning only becomes simpler] the existence of an $\varepsilon > 0$ so that an infinite set of the numbers σ_n lie outside the interval $(\lambda - \varepsilon, \lambda + \varepsilon)$. This infinite set contains a subsequence $\{\sigma_{n_k}\}$, which tends to a finite or infinite limit μ. The number μ is a derived number of the function $f(x)$ at the point x_0 different from λ. The existence of such a derived number contradicts our hypothesis; and thus the theorem is proved.

LEMMA 1. *If the function $f(x)$ is monotonically increasing on $[a, b]$, then all of its derived numbers are non-negative.*

This lemma is obvious.

LEMMA 2. *Let $f(x)$ be a strictly increasing function defined on $[a, b]$. If at every point of the set $E \subset [a, b]$, there exists at least one derived number $Df(x)$ such that*

$$Df(x) \leqslant p, \qquad\qquad (p \geqslant 0)$$

then

$$m^* f(E) \leqslant p \cdot m^* E.$$

Proof. Take any $\varepsilon > 0$ and choose a bounded open set G such that

$$E \subset G, \quad mG < m^* E + \varepsilon.$$

Further, let p_0 be any number such that $p_0 > p$. If $x_0 \in E$, then there exists a sequence $\{h_n\}$ with limit zero such that

$$\lim_{n \to \infty} \frac{f(x_0 + h_n) - f(x_0)}{h_n} = Df(x_0) \leqslant p.$$

2. Mapping of Sets. Differentiation of Monotonic Functions

For all sufficiently large n the closed interval [1] $[x_0, x_0 + h_n]$, is contained entirely in the set G. In addition, for all n sufficiently large,

$$\frac{f(x_0 + h_n) - f(x_0)}{h_n} < p_0.$$

We shall suppose that both of these situations hold for all n.

We next introduce the closed intervals

$$d_n(x_0) = [x_0, x_0 + h_n], \quad \Delta_n(x_0) = [f(x_0), f(x_0 + h_n)].$$

Since the function $f(x)$ is monotonic increasing, it is clear that

$$f[d_n(x_0)] \subset \Delta_n(x_0).$$

The lengths of these intervals are

$$md_n(x_0) = |h_n|, \quad m\Delta_n(x_0) = |f(x_0 + h_n) - f(x_0)|.$$

Therefore

$$m\Delta_n(x_0) < p_0 md_n(x_0).$$

But $h_n \to 0$; this implies that among the intervals $\Delta_n(x_0)$, there is an arbitrarily small one. Since the image $f(E)$ of the set E consists of the points $f(x_0)$ which lie in the interval $\Delta_n(x_0)$, $f(E)$ is covered by all the intervals $\Delta_n(x)$ ($x \in E$) in the sense of Vitali [2]. (See §8, Ch. III.) By Vitali's theorem (Theorem 1, §8, Ch. III), it is possible to select from the family of these intervals a countable sequence of pairwise disjoint intervals $\{\Delta_{n_i}(x_i)\}$ ($i = 1, 2, 3, \ldots$), such that

$$m[f(E) - \sum_{i=1}^{\infty} \Delta_{n_i}(x_i)] = 0.$$

It is clear that

$$m^* f(E) \leqslant \sum_{i=1}^{\infty} m\Delta_{n_i}(x_i) < p_0 \sum_{i=1}^{\infty} md_{n_i}(x_i).$$

Now note that not only the intervals $\Delta_{n_i}(x_i)$ but also the intervals $d_{n_i}(x_i)$ are pairwise disjoint. [3] Therefore

$$\sum_{i=1}^{\infty} md_{n_i}(x_i) = m[\sum_{i=1}^{\infty} d_{n_i}(x_i)].$$

Since

$$\sum_{i=1}^{\infty} d_{n_i}(x_i) \subset G,$$

---

[1] This is for the case $h_n > 0$. If $h_n < 0$, it is necessary to write $[x_0 + h_n, x_0]$. However, we may agree to denote the set of numbers lying *between* α and β by $[\alpha, \beta]$ even if $\alpha > \beta$.

[2] We here use the fact that $f(x)$ is *strictly* increasing. Otherwise some of the intervals $\Delta_n(x)$ could degenerate to points, and it would then be impossible to apply Vitali's theorem.

[3] In fact, if z were in the intersection $d_{n_i}(x_i) \cdot d_{n_k}(x_k)$, then $f(z)$ would be in the intersection $\Delta_{n_i}(x_i) \cdot \Delta_{n_k}(x_k)$. This requires the hypothesis that f is strictly increasing, and hence one-to-one,

we have
$$m^*f(E) < p_0 mG < p_0[m^*E + \varepsilon].$$
As $p_0 \to p$ and $\varepsilon \to 0$, the last inequality becomes in the limit
$$m^*f(E) \leqslant p \cdot m^*E,$$
which is the inequality we set out to establish.

The following lemma has a similar, although technically more complicated, proof.

LEMMA 3. *Let $f(x)$ be a strictly increasing function defined on $[a, b]$. If at every point of the set $E \subset [a, b]$ there exists at least one derived number $Df(x)$ such that*
$$Df(x) \geqslant q, \qquad (q \geqslant 0)$$
then
$$m^*f(E) \geqslant qm^*E.$$

Proof. The lemma is trivial if $q = 0$. We may suppose then that $q > 0$. Let q_0 be any positive number less than q and let ε be any positive number. Let G be a bounded open set such that [4]
$$G \supset f(E), \quad mG < m^*f(E) + \varepsilon.$$

Let S be the set of points $x \in E$ at which $f(x)$ is continuous. The set $E - S$ is at most denumerable, because a monotonic function has at most a denumerable set of points of discontinuity (Theorem 1, §1).

If $x_0 \in E$, we can find a sequence $\{h_n\}$ for which
$$h_n \to 0, \quad \lim_{n \to \infty} \frac{f(x_0 + h_n) - f(x_0)}{h_n} = Df(x_0) \geqslant q.$$

We may suppose that for all n,
$$\frac{f(x_0 + h_n) - f(x_0)}{h_n} > q_0.$$

Therefore, introducing again the intervals
$$d_n(x_0) = [x_0, x_0 + h_n], \quad \Delta_n(x_0) = [f(x_0), f(x_0 + h_n)],$$
we have
$$m\Delta_n(x_0) > q_0 m d_n(x_0).$$

If $x_0 \in S$, the whole interval $[f(x_0), f(x_0 + h_n)]$ will lie entirely in the set G for sufficiently large n. We may suppose that this is the case for all n.

The set S is covered by the intervals $d_n(x)$ (where $x \in S$) in the sense of Vitali. Therefore there exists a denumerable sequence of pairwise disjoint intervals $\{d_{n_i}(x_i)\}$ such that
$$m\left[S - \sum_{i=1}^{\infty} d_{n_i}(x_i)\right] = 0.$$
But then
$$m^*S \leqslant \sum_{i=1}^{\infty} m d_{n_i}(x_i) < \frac{1}{q_0} \sum_{i=1}^{\infty} m\Delta_{n_i}(x_i).$$

---

[4] Note that the set $f(E)$ is bounded: it is in fact contained in the closed interval $[f(a), f(b)]$.

2. MAPPING OF SETS. DIFFERENTIATION OF MONOTONIC FUNCTIONS

The segments $\Delta_{n_i}(x_i)$, as well as the $d_{n_i}(x_i)$, are pairwise disjoint [here we need the hypothesis that the function $f(x)$ is *strictly* increasing]. This implies that

$$\sum_{i=1}^{\infty} m\Delta_{n_i}(x_i) = m\left[\sum_{i=1}^{\infty} \Delta_{n_i}(x_i)\right] \leqslant mG < m^*f(E) + \varepsilon.$$

Thus
$$m^*S < \frac{1}{q_0}[m^*f(E) + \varepsilon].$$

Taking the limit as $\varepsilon \to 0$ and $q_0 \to q$, we find that

$$m^*f(E) \geqslant q m^*S.$$

Since $m^*E \leqslant m^*S + m^*(E - S) = m^*S$, the lemma follows.

COROLLARY. *The set of points at which at least one derived number of an increasing function $f(x)$ is infinite, is of measure zero.*

First of all, suppose that the function is *strictly* increasing. If we had
$$m^*E(Df(x) = +\infty) > 0,$$
the image of this set would necessarily have infinite outer measure, which is an absurdity because this image lies on the segment $[f(a), f(b)]$. This proves the corollary for a strictly increasing function. If $f(x)$ is not strictly increasing, let

$$g(x) = f(x) + x.$$

Then $g(x)$ is a strictly increasing function. But

$$\frac{g(x+h) - g(x)}{h} = \frac{f(x+h) - f(x)}{h} + 1.$$

Hence the set of points where at least one $Df(x) = +\infty$ coincides with the same set for $g(x)$ and therefore has zero measure.

LEMMA 4. *Let $f(x)$ be an increasing function defined on the closed interval $[a, b]$, and let p and q be two numbers such that $p < q$. If at every point x of the set $E_{p,q} \subset [a, b]$, there exist two derived numbers $D_1 f(x)$ and $D_2 f(x)$ such that*

$$D_1 f(x) < p < q < D_2 f(x),$$

then $mE_{p,q} = 0$.

Suppose first that $f(x)$ is strictly increasing. Then both Lemmas 2 and 3 are applicable, and, accordingly we may write

$$m^*f(E_{p,q}) \leqslant p m^*E_{p,q}, \quad m^*f(E_{p,q}) \geqslant q m^*E_{p,q}.$$

It follows that
$$q m^*E_{p,q} \leqslant p m^*E_{p,q}$$

and therefore $m^*E_{p,q} = 0$.

If $f(x)$ is not strictly increasing, then, as above, we set $g(x) = f(x) + x$ and apply the part of the lemma already proved to $g(x)$, substituting $p + 1$ and $q + 1$ for p and q.

We can now establish the main theorem of this section.

THEOREM 4. *Every increasing[5] function $f(x)$ defined on a closed interval $[a, b]$ has a finite derivative $f'(x)$ at almost all points $x \in [a, b]$.*

Proof. Let E denote the set of those points of $[a, b]$ at which the derivative $f'(x)$ does not exist. If $x_0 \in E$, there are two distinct derived numbers $D_1 f(x_0)$ and $D_2 f(x_0)$;

---

[5] Note that we do not suppose $f(x)$ continuous.

suppose that $D_1 f(x_0) < D_2 f(x_0)$. There obviously exist rational numbers p and q such that
$$D_1 f(x_0) < p < q < D_2 f(x_0).$$
It follows that
$$E = \sum_{(p,\,q)} E_{p,\,q},$$
where $E_{p,\,q}$ is the set of those x in $[a, b]$ at which two derived numbers $D_1 f(x)$ and $D_2 f(x)$ exist satisfying the inequalities
$$D_1 f(x) < p < q < D_2 f(x),$$
and the summation is extended over all pairs (p, q) of rational numbers for which $p < q$. According to Lemma 4, every set $E_{p,\,q}$ has measure zero. Since there are clearly only a denumerable number of sets $E_{p,\,q}$, it follows that E is the sum of a denumerable family of sets of measure 0 and hence itself has measure 0.

We have thus shown that the derivative $f'(x)$ exists almost everywhere on $[a, b]$. Since $f'(x) = +\infty$ is possible only on a set of measure 0 (see Corollary to Lemma 3), the theorem is proved.

Henceforth, in referring to the derivative $f'(x)$ of an increasing function, we shall suppose it defined for *all* x in $[a, b]$. To accomplish this, we agree once and for all to define $f'(x) = 0$ at those points x where $f(x)$ has no derivative.

THEOREM 5. *If $f(x)$ is an increasing function defined on $[a, b]$, then its derivative $f'(x)$ is measurable and*
$$\int_a^b f'(x)\,dx \leqslant f(b) - f(a),$$
so that $f'(x)$ is summable.

Proof. Extend the definition of the function $f(x)$, setting
$$f(x) = f(b) \quad \text{if} \quad b < x \leqslant b + 1.$$
Then at every point x such that $f'(x)$ is the derivative of $f(x)$ [except, perhaps, the point $x = b$, where $f'(b)$ was previously only the left-side derivative], we have
$$f'(x) = \lim_{n \to \infty} n \left[f\left(x + \frac{1}{n}\right) - f(x) \right].$$
This implies that $f'(x)$ is the limit of an almost everywhere convergent sequence of measurable[6] functions. Hence $f'(x)$ is a measurable function. Since $f'(x)$ is nonnegative, we can speak of its Lebesgue integral
$$\int_a^b f'(x)\,dx.$$
By Fatou's Theorem [Chapt. VI, §1]
$$\int_a^b f'(x)\,dx \leqslant \sup \left\{ n \int_a^b \left[f\left(x + \frac{1}{n}\right) - f(x) \right] dx \right\}.$$

---

[6] The functions $f(x)$ and $f(x + \frac{1}{n})$ are increasing and are therefore measurable. In fact, $E(f > c)$ is either the void set or an interval.

2. MAPPING OF SETS. DIFFERENTIATION OF MONOTONIC FUNCTIONS

But

$$\int_a^b f\left(x+\frac{1}{n}\right)dx = \int_{a+\frac{1}{n}}^{b+\frac{1}{n}} f(x)\,dx$$

(there is no need to cite the theorem on the change of variable in a Lebesgue integral because $f(x)$ is monotonic, and the integral can be taken in the Riemann sense). Consequently, we have

$$\int_a^b \left[f\left(x+\frac{1}{n}\right)-f(x)\right]dx = \int_b^{b+\frac{1}{n}} f(x)\,dx - \int_a^{a+\frac{1}{n}} f(x)\,dx =$$

$$= \frac{1}{n}f(b) - \int_a^{a+\frac{1}{n}} f(x)\,dx \leqslant \frac{1}{n}[f(b)-f(a)].$$

From this, the inequality

$$\int_a^b f'(x)\,dx \leqslant f(b)-f(a).$$

follows readily.

We are accustomed to think that the integral of the derivative of a function is equal to the difference of the values of the function itself, computed at the ends of the interval of integration:

$$\int_a^b f'(x)\,dx = f(b)-f(a).$$

From this point of view, the inequality obtained in Theorem 5 seems somewhat less than perfect. However, it is impossible always to obtain equality in the inequality of Theorem 5, even if the function $f(x)$ is continuous.

EXAMPLE. Let P_0 be the Cantor perfect set. Its complementary intervals can be divided into groups, putting into the first group the interval $(\frac{1}{3}, \frac{2}{3})$, into the second the two intervals $(\frac{1}{9}, \frac{2}{9})$, $(\frac{7}{9}, \frac{8}{9})$, into the third the four intervals $(\frac{1}{27}, \frac{2}{27})$, $(\frac{7}{27}, \frac{8}{27})$, $(\frac{19}{27}, \frac{20}{27})$, $(\frac{25}{27}, \frac{26}{27})$, and so on. There will be 2^{n-1} intervals in the n-th group.

We define a function $\Theta(x)$ by the following description:

$$\Theta(x) = \frac{1}{2} \quad \text{if} \quad x \in \left(\frac{1}{3}, \frac{2}{3}\right),$$

$$\Theta(x) = \frac{1}{4} \quad \text{if} \quad x \in \left(\frac{1}{9}, \frac{2}{9}\right), \quad \Theta(x) = \frac{3}{4} \quad \text{if} \quad x \in \left(\frac{7}{9}, \frac{8}{9}\right).$$

In the four intervals of the third group, the function $\Theta(x)$ equals $\frac{1}{8}, \frac{3}{8}, \frac{5}{8}, \frac{7}{8}$ respectively. And in general, on the 2^{n-1} intervals of the n-th group, we set the function Θ equal to

$$\frac{1}{2^n}, \frac{3}{2^n}, \frac{5}{2^n}, \ldots, \frac{2^n-1}{2^n}.$$

respectively.

214   VIII. FUNCTIONS OF FINITE VARIATION. STIELTJES INTEGRAL

The function $\Theta(x)$ is defined on the Cantor open set G_0. It is constant on every component interval of this set and is an increasing function on the set G_0 as a whole. [7] We extend the definition of the function $\Theta(x)$ by defining it at the points of the set P_0. To do this, we set

$$\Theta(0) = 0, \quad \Theta(1) = 1.$$

At those points $x_0 \in P_0$ which lie between 0 and 1, we set

$$\Theta(x_0) = \sup \{\Theta(x)\} \qquad (x \in G_0, \ x < x_0).$$

It is easy to see that this definition does not alter the monotonicity of the function $\Theta(x)$, which is now defined everywhere on the closed interval $[0, 1]$. Having established the monotonicity of $\Theta(x)$, we can easily prove that it is continuous. This follows from the fact that the set of values taken on by the function $\Theta(x)$ on the set G_0 is everywhere dense in $[0, 1]$. [In fact, if an increasing function $f(x)$ has a point of discontinuity x_0, then at least one of the intervals $(f(x_0 - 0), f(x_0))$ and $(f(x_0), f(x_0 + 0))$ contains no points $f(x)$.] Thus, $\Theta(x)$ is a continuous increasing function. Besides, almost everywhere on $[0, 1]$, we have

$$\Theta'(x) = 0.$$

(This relation obviously holds at every point of the set G_0.) Therefore

$$\int_0^1 \Theta'(x)\,dx = 0 < 1 = \Theta(1) - \Theta(0).$$

Further on, we shall describe the conditions under which there is equality in (3).

In conclusion, we prove a theorem which is useful in many situations.

THEOREM 6. *For every set E of measure zero on the closed interval $[a, b]$, there exists a continuous increasing function $\sigma(x)$ such that*

$$\sigma'(x) = +\infty$$

at all points $x \in E$.

Proof. For every natural number n, let G_n be a bounded open set such that

$$G_n \supset E, \quad mG_n < \frac{1}{2^n}.$$

Let

$$\psi_n(x) = m\{G_n[a, x]\}.$$

The function $\psi_n(x)$ is increasing, non-negative, continuous, and satisfies the inequality

$$\psi_n(x) < \frac{1}{2^n}.$$

Therefore, the function

$$\sigma(x) = \sum_{n=1}^{\infty} \psi_n(x)$$

---

[7] A simple inductive proof can be given for this fact. It is suggested that the reader work out a detailed proof.

is also increasing, non-negative, and continuous. If $x_0 \in E$, and $|h|$ is sufficiently small, the whole interval $[x_0, x_0 + h]$ lies in the set G_n [for fixed n]. For such h (assuming $h > 0$ for simplicity), we have

$$\psi_n(x_0+h) = m\{G_n \cdot [a, x_0] + G_n \cdot (x_0, x_0+h]\} = \psi_n(x_0) + h.$$

Thus
$$\frac{\psi_n(x_0+h) - \psi_n(x_0)}{h} = 1.$$

It follows that for every natural number N, if $|h|$ is sufficiently small, we have

$$\frac{\sigma(x_0+h) - \sigma(x_0)}{h} \geqslant \sum_{n=1}^{N} \frac{\psi_n(x_0+h) - \psi_n(x_0)}{h} = N,$$

so that
$$\sigma'(x_0) = +\infty.$$

This proves the theorem.

§ 3. FUNCTIONS OF FINITE VARIATION

In this section, we discuss the theory of an important class of functions — functions of finite variation, which are intimately connected with monotonic functions.

Let a function $f(x)$ be defined and finite on the interval $[a, b]$. Subdivide $[a, b]$ into parts by means of the points

$$x_0 = a < x_1 < \ldots < x_n = b$$

and form the sum

$$V = \sum_{k=0}^{n-1} |f(x_{k+1}) - f(x_k)|.$$

DEFINITION 1. *The least upper bound of the set of all possible sums V is called the total variation of the function $f(x)$ on $[a, b]$ and is designated by $\overset{b}{\underset{a}{V}}(f)$*. If

$$\overset{b}{\underset{a}{V}}(f) < +\infty,$$

then $f(x)$ is said to be *a function of finite variation on $[a, b]$*. We also say that $f(x)$ has finite variation on $[a, b]$.

THEOREM 1. *A monotonic function on $[a, b]$ has finite variation on $[a, b]$.*

Proof. It is sufficient to prove the theorem for an increasing function. If $f(x)$ is increasing on $[a, b]$, then all the differences $f(x_{k+1}) - f(x_k)$ are non-negative and

$$V = \sum_{k=0}^{n-1} \{f(x_{k+1}) - f(x_k)\} = f(b) - f(a),$$

This proves the theorem.

Further examples of functions of finite variation are furnished by functions satisfying a Lipschitz condition.

216   VIII. Functions of Finite Variation. Stieltjes Integral

DEFINITION 2. *A finite function $f(x)$ defined on $[a, b]$ is said to satisfy a Lipschitz condition if there exists a constant K such that for any two points x and y in $[a, b]$,*
$$|f(x) - f(y)| \leqslant K|x - y|.$$

If the function $f(x)$ has a derivative $f'(x)$ at every point of $[a, b]$ and if $f'(x)$ is bounded, then, as is clear from the mean value theorem,
$$f(x) - f(y) = f'(z)(x - y) \qquad (x < z < y)$$
and the function $f(x)$ satisfies a Lipschitz condition.

If $f(x)$ satisfies a Lipschitz condition, then
$$|f(x_{k+1}) - f(x_k)| \leqslant K(x_{k+1} - x_k),$$
whence
$$V \leqslant K(b - a)$$
and $f(x)$ is a function of finite variation.

An example of a continuous function with *infinite* total variation is the function
$$f(x) = x \cos \frac{\pi}{2x} \qquad (0 < x \leqslant 1,\ f(0) = 0).$$

If we take the points
$$0 < \frac{1}{2n} < \frac{1}{2n-1} < \ldots < \frac{1}{3} < \frac{1}{2} < 1,$$
for points of division of $[0, 1]$, then it is easy to verify that
$$V = 1 + \frac{1}{2} + \ldots + \frac{1}{n}.$$
Therefore
$$\overset{1}{\underset{0}{V}}(f) = +\infty.$$

THEOREM 2. *Every function of finite variation in $[a, b]$ is bounded in $[a, b]$.*
In fact, for $a \leqslant x \leqslant b$,
$$V = |f(x) - f(a)| + |f(b) - f(x)| \leqslant \overset{b}{\underset{a}{V}}(f).$$
Therefore
$$|f(x)| \leqslant |f(a)| + \overset{b}{\underset{a}{V}}(f).$$

THEOREM 3. *The sum, difference and product of two functions of finite variation are functions of finite variation.*

Proof. Let $f(x)$ and $g(x)$ be functions of finite variation on the closed interval $[a, b]$, and let $s(x)$ be their sum. Then
$$|s(x_{k+1}) - s(x_k)| \leqslant |f(x_{k+1}) - f(x_k)| + |g(x_{k+1}) - g(x_k)|,$$
and it follows that
$$\overset{b}{\underset{a}{V}}(s) \leqslant \overset{b}{\underset{a}{V}}(f) + \overset{b}{\underset{a}{V}}(g).$$

3. FUNCTIONS OF FINITE VARIATION

Therefore $s(x)$ is a function of finite variation. The proof that $f(x) - g(x)$ is of finite variation is very similar to the proof just given, and need not be written out.

Next, suppose that $f(x)$ and $g(x)$ have finite variation and that
$$p(x) = f(x)g(x).$$

Let
$$A = \sup\{|f(x)|\}, \quad B = \sup\{|g(x)|\}$$

both suprema being taken over the closed interval $[a, b]$. Then

$$|p(x_{k+1}) - p(x_k)| \leq |f(x_{k+1})g(x_{k+1}) - f(x_k)g(x_{k+1})| +$$
$$+ |f(x_k)g(x_{k+1}) - f(x_k)g(x_k)| \leq B|f(x_{k+1}) - f(x_k)| + A|g(x_{k+1}) - g(x_k)|.$$

This implies that
$$\overset{b}{\underset{a}{V}}(p) \leq B\overset{b}{\underset{a}{V}}(f) + A\overset{b}{\underset{a}{V}}(g),$$

and accordingly $p(x)$ has finite variation in $[a, b]$.

THEOREM 4. *If $f(x)$ and $g(x)$ are functions of finite variation and if $g(x) \geq \sigma > 0$, then the quotient $\dfrac{f(x)}{g(x)}$ is a function of finite variation.*

The proof is left to the reader.

THEOREM 5. *Let a finite function $f(x)$ be defined on $[a, b]$ and let $a < c < b$. Then*

$$\overset{b}{\underset{a}{V}}(f) = \overset{c}{\underset{a}{V}}(f) + \overset{b}{\underset{c}{V}}(f). \tag{1}$$

Proof. Subdivide each of the intervals $[a, c]$ and $[c, b]$ into subintervals by means of the points
$$y_0 = a < y_1 < \ldots < y_m = c; \quad z_0 = c < z_1 < \ldots < z_n = b$$
and form the sums
$$V_1 = \sum_{k=0}^{m-1} |f(y_{k+1}) - f(y_k)|, \quad V_2 = \sum_{k=0}^{n-1} |f(z_{k+1}) - f(z_k)|.$$

The points $\{y_k\}$ and $\{z_k\}$ subdivide the whole interval $[a, b]$. If V is the sum corresponding to this method of subdivision,
$$V = V_1 + V_2.$$

It follows at once that
$$V_1 + V_2 \leq \overset{b}{\underset{a}{V}}(f)$$

and that
$$\overset{c}{\underset{a}{V}}(f) + \overset{b}{\underset{c}{V}}(f) \leq \overset{b}{\underset{a}{V}}(f). \tag{2}$$

Now subdivide the interval $[a, b]$ by means of the points
$$x_0 = a < x_1 < \ldots < x_n = b,$$
being careful to include the point c as a point of division. Writing $c = x_m$, we can

express the sum V which corresponds to our method of subdivision in the form

$$V = \sum_{k=0}^{m-1} |f(x_{k+1}) - f(x_k)| + \sum_{k=m}^{n-1} |f(x_{k+1}) - f(x_k)|.$$

More briefly,

$$V = V_1 + V_2,$$

where V_1 and V_2 are sums corresponding to the intervals $[a, c]$ and $[c, b]$.
Consequently

$$V \leq \overset{c}{\underset{a}{V}}(f) + \overset{b}{\underset{c}{V}}(f). \tag{3}$$

Inequality (3) has been established only for sums V corresponding to methods of subdivision for which the point c is a point of subdivision. Since the addition of new subdivision points does not decrease the sums V, (3) is true for all sums V. From this it is clear that

$$\overset{b}{\underset{a}{V}}(f) \leq \overset{c}{\underset{a}{V}}(f) + \overset{b}{\underset{c}{V}}(f). \tag{4}$$

Combining (2) and (4), we obtain (1).

COROLLARY 1. *Let $a < c < b$. If the function $f(x)$ has finite variation on $[a, b]$, it has finite variation on each of the intervals $[a, c]$ and $[c, b]$, and conversely.*

COROLLARY 2. *If it is possible to subdivide the segment $[a, b]$ into a finite number of parts on each of which the function $f(x)$ is monotonic, then $f(x)$ has finite variation of $[a, b]$.*

THEOREM 6. *A function $f(x)$ defined and finite on $[a, b]$ is a function of finite variation if and only if it is representable as the difference of two increasing functions.*

Proof. The sufficiency of the condition follows from Theorems 1 and 3. To prove its necessity, we set

$$\pi(x) = \overset{x}{\underset{a}{V}}(f) \qquad (a < x \leq b)$$

$$\pi(a) = 0.$$

By virtue of Theorem 5, the function $\pi(x)$ is an increasing function. Setting

$$\nu(x) = \pi(x) - f(x), \tag{5}$$

we obtain another increasing function $\nu(x)$. In fact, if $a \leq x < y \leq b$, then, by Theorem 5,

$$\nu(y) = \pi(y) - f(y) = \pi(x) + \overset{y}{\underset{x}{V}}(f) - f(y)$$

and hence

$$\nu(y) - \nu(x) = \overset{y}{\underset{x}{V}}(f) - [f(y) - f(x)].$$

However, from the very definition of total variation, it is clear that

$$f(y) - f(x) \leq \overset{y}{\underset{x}{V}}(f),$$

so that

$$\nu(y) - \nu(x) \geq 0.$$

3. FUNCTIONS OF FINITE VARIATION

Therefore $v(x)$ is an increasing function. It remains to write the equality (5) in the form
$$f(x) = \pi(x) - v(x),$$
in order to obtain the desired representation of $f(x)$.

COROLLARY 1. *If a function $f(x)$ has finite variation on $[a, b]$, then at almost every point of $[a, b]$ the derivative $f'(x)$ exists, and is finite. Furthermore, $f'(x)$ is a summable function on $[a, b]$.*

COROLLARY 2. *The set of points of discontinuity of a function of finite variation is at most denumerable. At every point x_0 of discontinuity, both limits*

$$f(x_0 + 0) = \lim_{x \to x_0} f(x) \qquad (x > x_0)$$
$$f(x_0 - 0) = \lim_{x \to x_0} f(x) \qquad (x < x_0)$$

exist.

We now take up the problem of writing a function of finite variation as the sum of a continuous function and a saltus function. Let the sequence

$$x_1, x_2, x_3, \ldots \qquad (a < x_n < b) \qquad (6)$$

consist of all points which are points of discontinuity of at least one of the functions $\pi(x)$ and $v(x)$. Consider the following saltus functions:

$$s_\pi(x) = [\pi(a+0) - \pi(a)] + \sum_{x_k < x} [\pi(x_k+0) - \pi(x_k-0)] + [\pi(x) - \pi(x-0)]$$
$$(a < x \leq b)$$
$$s_v(x) = [v(a+0) - v(a)] + \sum_{x_k < x} [v(x_k+0) - v(x_k-0)] + [v(x) - v(x-0)];$$

we further agree that

$$s_\pi(a) = s_v(a) = 0.$$

(If a point x_k is a point of continuity of one of the functions $\pi(x)$ and $v(x)$, the corresponding saltus vanishes automatically. Moreover, we can show that a point of discontinuity of the function $v(x)$ cannot be a point of continuity of the function $\pi(x)$; but this is only of slight interest.)

Let

$$s(x) = s_\pi(x) - s_v(x).$$

This function $s(x)$ can be written in the following form:

$$s(x) = [f(a+0) - f(a)] + \sum_{x_k < x} [f(x_k+0) - f(x_k-0)] + [f(x) - f(x-0)]$$
$$(a < x < b)$$
$$s(a) = 0.$$

It is clear that $s(x)$ is a function of finite variation. It is called the saltus function of the function $f(x)$. It is self-evident that the definition of the function $s(x)$ is not changed if we remove from the sequence (6) those points at which the function $f(x)$ is continuous,[8] so that we may suppose that the x_k appearing in (6) are all points of discontinuity of $f(x)$.

---

[8] It is possible to show directly that there are no such points in (6). This will be shown in Theorem 1, §5.

It has already been proved (Theorem 2, §1) that the functions
$$\pi(x) - s_\pi(x), \quad v(x) - s_v(x)$$
are continuous and increasing. From this it follows that the difference
$$\varphi(x) = f(x) - s(x)$$
is a continuous function of finite variation.

The result just obtained can be stated in the following form.

THEOREM 7. *Every function of finite variation can be written as the sum of its saltus function and a continuous function of finite variation.*

§ 4. HELLY'S PRINCIPLE OF CHOICE

In this section, we take up a theorem due to E. Helly which has many important applications. We first prove two lemmas.

LEMMA 1. *Let an infinite family of functions $H = \{f(x)\}$ be defined on $[a, b]$. If all the functions of the family are bounded by one and the same number*
$$|f(x)| \leq K, \tag{1}$$
then, for any denumerable subset E of $[a, b]$, it is possible to find a sequence $\{f_n(x)\}$ in the family H which converges at every point of the set E.

Proof. Let $E = \{x_k\}$. Consider the set
$$\{f(x_1)\}$$
of values taken on by the functions of the family H at the point x_1. By (1), this set is bounded and, by the Bolzano-Weierstrass Theorem, we can select a convergent sequence from it:
$$f_1^{(1)}(x_1), \ f_2^{(1)}(x_1), \ f_3^{(1)}(x_1), \ldots; \quad \lim_{n \to \infty} f_n^{(1)}(x_1) = A_1. \tag{2}$$

Now consider the sequence
$$f_1^{(1)}(x_2), \ f_2^{(1)}(x_2), \ f_3^{(1)}(x_2), \ldots$$
of values taken on by the functions of the set $\{f_n^{(1)}(x)\}$ at the point x_2. This sequence is also bounded, and we can apply the Bolzano-Weierstrass Theorem to it. This gives a convergent subsequence
$$f_1^{(2)}(x_2), \ f_2^{(2)}(x_2), \ f_3^{(2)}(x_2), \ldots, \quad \lim_{n \to \infty} f_n^{(2)}(x_2) = A_2, \tag{3}$$
selected from $\{f_n^{(1)}(x_2)\}$. It is essential to note that the relative order of two functions $f_n^{(2)}$ and $f_m^{(2)}$ in the sequence (3) is the same as in the sequence (2). Continuing this process indefinitely, we construct a denumerable set of convergent sequences:

$$f_1^{(1)}(x_1), \ f_2^{(1)}(x_1), \ f_3^{(1)}(x_1), \ \ldots, \quad \lim_{n \to \infty} f_n^{(1)}(x_1) = A_1.$$

$$f_1^{(2)}(x_2), \ f_2^{(2)}(x_2), \ f_3^{(2)}(x_2), \ \ldots, \quad \lim_{n \to \infty} f_n^{(2)}(x_2) = A_2.$$

$$\cdots \cdots \cdots \cdots \cdots \cdots \cdots \cdots \cdots \cdots \cdots$$

$$f_1^{(k)}(x_k), \ f_2^{(k)}(x_k), \ f_3^{(k)}(x_3), \ \ldots, \quad \lim_{n \to \infty} f_n^{(k)}(x_k) = A_k.$$

$$\cdots \cdots \cdots \cdots \cdots \cdots \cdots \cdots \cdots \cdots \cdots$$

4. Helly's Principle of Choice

where each sequence of numbers is a subsequence of the preceding one, and in which the order of elements has not been altered. We now form the sequence of diagonal elements of the infinite matrix just constructed, *i.e.*, the sequence

$$\{f_n^{(n)}(x)\} \qquad (n=1, 2, 3, \ldots).$$

This sequence converges at every point of the set E. In fact, for every fixed k, the sequence

$$\{f_n^{(n)}(x_k)\} \qquad (n \geqslant k)$$

is a subsequence of $\{f_n^{(k)}(x_k)\}$ and converges to A_k.

LEMMA 2. *Let $F = \{f(x)\}$ be an infinite family of increasing functions, defined on the segment $[a, b]$. If all functions of the family are bounded by one and the same number,*

$$|f(x)| \leqslant K, \quad f \in F, \quad a \leqslant x \leqslant b,$$

then there is a sequence of functions $\{f_n(x)\}$ in F which converges to an increasing function $\varphi(x)$ at every point of $[a, b]$.

Proof. Apply Lemma 1 to $\{f(x)\}$, taking for the set E the set consisting of all rational points of $[a, b]$, together with the point a if it is irrational. We thus find a sequence of functions of the family F,

$$\lim_{n \to \infty} f^{(n)}(x_k) \tag{4}$$

such that

$$F_0 = \{f^{(n)}(x)\}$$

exists and is finite at every point $x_k \in E$.

We now define a function $\psi(x)$ by the following procedure. First, we define

$$\psi(x_k) = \lim_{n \to \infty} f^{(n)}(x_k) \qquad (x_k \in E)$$

for all $x_k \in E$. This defines $\psi(x)$ only on E, of course. It is easy to see that $\psi(x)$ is an increasing function on E, that is, if $x_i, x_k \in E$ and $x_k < x_i$, then

$$\psi(x_k) \leqslant \psi(x_i).$$

For $x \in [a, b] - E$, we define $\psi(x)$ by the relation

$$\psi(x) = \sup_{x_k < x} \{\psi(x_k)\} \qquad (x_k \in E).$$

It is obvious that $\psi(x)$ is an increasing function on the closed interval $[a, b]$ and that the set of points Q where $\psi(x)$ is discontinuous is at most denumerable.

We show next that

$$\lim_{n \to \infty} f^{(n)}(x_0) = \psi(x_0) \tag{5}$$

at every point x_0 where $\psi(x)$ is continuous. Let ε be any positive number, and let x_k and x_i be points of E such that

$$x_k < x_0 < x_i, \quad \psi(x_i) - \psi(x_k) < \frac{\varepsilon}{2}.$$

Fixing the points x_k and x_i, select a natural number n_0 such that for $n > n_0$,

$$|f^{(n)}(x_k) - \psi(x_k)| < \frac{\varepsilon}{2}, \quad |f^{(n)}(x_i) - \psi(x_i)| < \frac{\varepsilon}{2}.$$

It is easy to see that
$$\psi(x_0) - \varepsilon < f^{(n)}(x_k) \leqslant f^{(n)}(x_i) < \psi(x_0) + \varepsilon,$$
for $n > n_0$. Since
$$f^{(n)}(x_k) \leqslant f^{(n)}(x_0) \leqslant f^{(n)}(x_i),$$
we have
$$\psi(x_0) - \varepsilon < f^{(n)}(x_0) < \psi(x_0) + \varepsilon,$$
for $n > n_0$. This proves (5). Thus, the equality
$$\lim_{n \to \infty} f^{(n)}(x) = \psi(x) \tag{6}$$
can fail only on the finite or denumerable set Q, where $\psi(x)$ is discontinuous.

We now apply Lemma 1 to the sequence F_0, taking for the set E the set of those points of Q where (6) is not fulfilled. This yields a subsequence
$$\{f_n(x)\}$$
of F_0, which converges at all points of $[a, b]$ (because at points where the sequence $\{f^{(n)}(x)\}$ converges, all subsequences also converge). Setting
$$\varphi(x) = \lim_{n \to \infty} f_n(x),$$
we obtain a function which is obviously an increasing function.

THEOREM (HELLY'S FIRST THEOREM). *Let an infinite family of functions $F = \{f(x)\}$ be defined on the segment $[a, b]$. If all functions of the family and the total variation of all functions of the family are bounded by a single number*
$$|f(x)| \leqslant K, \quad \overset{b}{\underset{a}{V}}(f) \leqslant K,$$
then there exists a sequence $\{f_n(x)\}$ in the family F which converges at every point of $[a, b]$ to some function $\varphi(x)$ of finite variation.

Proof. For every function $f(x)$ of the family F, set
$$\pi(x) = \overset{x}{\underset{a}{V}}(f), \quad \nu(x) = \pi(x) - f(x).$$
Both $\pi(x)$ and $\nu(x)$ are increasing functions. Furthermore,
$$|\pi(x)| \leqslant K, \quad |\nu(x)| \leqslant 2K.$$
Applying Lemma 2 to the family $\{\pi(x)\}$, we find that there is a convergent sequence $\{\pi_k(x)\}$,
$$\lim_{k \to \infty} \pi_k(x) = \alpha(x)$$
in this family. To every function $\pi_k(x)$, there corresponds a function $\nu_k(x)$, extending it to the function $f_k(x)$ of the family F. Applying Lemma 2 to the family $\{\nu_k(x)\}$, we find a convergent subsequence $\{\nu_{k_i}(x)\}$,
$$\lim_{i \to \infty} \nu_{k_i}(x) = \beta(x)$$
of $\{\nu_k(x)\}$. Then the sequence of functions
$$f_{k_i}(x) = \pi_{k_i}(x) - \nu_{k_i}(x),$$

belonging to F, converges to the function
$$\varphi(x) = \alpha(x) - \beta(x).$$
This proves Helly's theorem.

§ 5. CONTINUOUS FUNCTIONS OF FINITE VARIATION

THEOREM 1. *Let a function $f(x)$ of finite variation be defined on the closed interval $[a, b]$. If $f(x)$ is continuous at the point x_0, then the function*
$$\pi(x) = \overset{x}{\underset{a}{V}}(f)$$
is also continuous at x_0.

Proof. Suppose that $x_0 < b$. We shall show that $\pi(x)$ is continuous on the right at the point x_0. For this purpose, taking an arbitrary $\varepsilon > 0$, we subdivide the segment $[x_0, b]$ by means of the points
$$x_0 < x_1 < \ldots < x_n = b$$
so that
$$V = \sum_{k=0}^{n-1} |f(x_{k+1}) - f(x_k)| > \overset{b}{\underset{x_0}{V}}(f) - \varepsilon. \tag{1}$$
Since the sum V only increases when new points are added, we may suppose that
$$|f(x_1) - f(x_0)| < \varepsilon.$$
It follows from (1) that
$$\overset{b}{\underset{x_0}{V}}(f) < \varepsilon + \sum_{k=0}^{n-1} |f(x_{k+1}) - f(x_k)| < 2\varepsilon + \sum_{k=1}^{n-1} |f(x_{k+1}) - f(x_k)| \leqslant 2\varepsilon + \overset{b}{\underset{x_1}{V}}(f).$$
Hence
$$\overset{x_1}{\underset{x_0}{V}}(f) < 2\varepsilon,$$
and consequently
$$\pi(x_1) - \pi(x_0) < 2\varepsilon.$$
This implies that
$$\pi(x_0 + 0) - \pi(x_0) < 2\varepsilon.$$
Since ε is arbitrary, we have
$$\pi(x_0 + 0) = \pi(x_0).$$
It can be shown in like manner that $\pi(x_0 - 0) = \pi(x_0)$, i.e., that $\pi(x)$ is continuous on the *left* (if $x_0 > a$) at the point x_0.

COROLLARY. *A continuous function of finite variation can be written as the difference of two continuous increasing functions.*

In fact, if $f(x)$ is a continuous function of finite variation defined on $[a, b]$, then both of its increasing components
$$\pi(x) = \overset{x}{\underset{a}{V}}(f) \text{ and } \nu(x) = \pi(x) - f(x)$$
are continuous.

VIII. FUNCTIONS OF FINITE VARIATION. STIELTJES INTEGRAL

Let a continuous function $f(x)$ be defined on $[a, b]$. Subdivide $[a, b]$ by means of the points

$$x_0 = a < x_1 < x_2 < \ldots < x_n = b \qquad [\max(x_{k+1} - x_k) = \lambda]$$

and form the sums

$$V = \sum_{k=0}^{n-1} |f(x_{k+1}) - f(x_k)|, \qquad \Omega = \sum_{k=0}^{n-1} \omega_k,$$

where ω_k designates the oscillation of the function $f(x)$ in the interval $[x_k, x_{k+1}]$.

THEOREM 2. *As $\lambda \to 0$, each of the sums V and Ω tends to the total variation $\overset{b}{\underset{a}{V}}(f)$ of the function $f(x)$.* [9] *(We do not suppose that the variation $\overset{b}{\underset{a}{V}}(f)$ is finite).*

Proof. As already noted, the sum V does not decrease when new points of subdivision are added. On the other hand, if this new point falls into the interval between x_k and x_{k+1}, the increase in the sum V due to this point, is not greater than twice the oscillation ω_k of the function $f(x)$ on the segment $[x_k, x_{k+1}]$. Take any number $A < \overset{b}{\underset{a}{V}}(f)$ and find a sum V^* such that

$$V^* > A.$$

This sum corresponds to some method of subdivision, say

$$x_0^* = a < x_1^* < \ldots < x_m^* = b.$$

Now choose $\delta > 0$ so small that

$$|f(x'') - f(x')| < \frac{V^* - A}{4m},$$

provided that

$$|x'' - x'| < \delta.$$

We shall show that $V > A$ for every method of subdivision of $[a, b]$ for which $\lambda < \delta$. Let (I) denote any such subdivision, and let (II) be the subdivision obtained from (I) by adding to (I) all of the points $x_0^*, x_1^*, \ldots, x_m^*$. Let V_0 be the sum corresponding to the method of subdivision (II). It is then clear that

$$V_0 \geqslant V^*. \tag{3}$$

On the other hand, the subdivision (II) is obtained from the subdivision (I) by m repetitions of the process of adding a single point. Each addition of a point of subdivision increases the sum V by an amount less than $\dfrac{V^* - A}{2m}$, and consequently

$$V_0 - V < \frac{V^* - A}{2}.$$

---

[9] It is essential here that we are dealing with a *continuous* function. Let, for instance, $f(x)$ be defined on $[-1, +1]$ as follows: $f(0) = 1, f(x) = 0$ for $x \neq 0$. Then

$$\overset{+1}{\underset{-1}{V}}(f) = 2,$$

but for an arbitrary method of subdividing $[-1, +1]$ for which $x = 0$ is not a point of division,

$$V = 0, \qquad \Omega = 1.$$

5. Continuous Functions of Finite Variation

It follows from this observation and from (3) that

$$V > V_0 - \frac{V^* - A}{2} \geqslant \frac{A + V^*}{2} > A.$$

Therefore the inequality (2) holds for all $\lambda < \delta$. Since

$$V \leqslant \overset{b}{\underset{a}{V}}(f)$$

for all V, it follows that

$$\lim_{\lambda \to 0} V = \overset{b}{\underset{a}{V}}(f).$$

It is now easy to carry out the proof for the sums Ω. On the one hand, it is clear that

$$\Omega \geqslant V. \tag{4}$$

But if we find a sum Ω corresponding to any method of subdivision and then add as new points of subdivision those points at which the function $f(x)$ takes on the values

$$m_k = \min\{f(x)\}, \quad M_k = \max\{f(x)\} \quad (x_k \leqslant x \leqslant x_{k+1}),$$

then the sum V' corresponding to this method of subdivision will obviously be not less than Ω. Consequently

$$\Omega \leqslant \overset{b}{\underset{a}{V}}(f). \tag{5}$$

From (4) and (5), we have

$$\lim_{\lambda \to 0} \Omega = \overset{b}{\underset{a}{V}}(f).$$

The preceding theorem just proved is the basis of a very interesting approach to the study of continuous functions of finite variation due to S. Banach.

Let $f(x)$ be defined and continuous on the segment $[a, b]$ and let

$$m = \min\{f(x)\}, \quad M = \max\{f(x)\}.$$

Introduce the function $N(y)$, defined on the closed interval $[m, M]$, in the following way: $N(y)$ is the number of roots of the equation

$$f(x) = y.$$

If the set of these roots is infinite, then

$$N(y) = +\infty.$$

We will call the function $N(y)$ the *Banach indicatrix*.

THEOREM 3 (S. BANACH). *The Banach indicatrix is measurable and*

$$\int_m^M N(y)\, dy = \overset{b}{\underset{a}{V}}(f).$$

Proof. Subdivide $[a, b]$ into 2^n equal parts and set

$$d_1 = \left[a,\ a + \frac{b-a}{2^n}\right]$$

$$d_k = \left(a + (k-1)\frac{b-a}{2^n},\ a + k\frac{b-a}{2^n}\right] \quad (k = 2, 3, \ldots, 2^n).$$

Further, let the function $L_k(y)$ $(k = 1, 2, \ldots, 2^n)$ be equal to 1 if the equation
$$f(x) = y \tag{6}$$
has at least one root in the interval d_k, and let $L_k(y)$ be equal to 0 if this equation has no root in d_k. If m_k and M_k are, respectively, the greatest lower and least upper bounds of the function $f(x)$ in the interval d_k, then $L_k(y)$ equals 1 in the interval (m_k, M_k) and equals 0 outside the segment $[m_k, M_k]$ so that this function can have no more than two points of discontinuity and is obviously measurable. Note further that
$$\int_m^M L_k(y)\, dy = M_k - m_k = \omega_k,$$
where ω_k is the oscillation of $f(x)$ on the closed interval d_k. Finally, introduce the function
$$N_n(y) = L_1(y) + L_2(y) + \cdots + L_{2^n}(y),$$
equal to the number of those intervals d_k which contain at least one root of equation (6). The function $N_n(y)$ is clearly measurable. Here,
$$\int_m^M N_n(y)\, dy = \sum_{k=1}^{2^n} \omega_k,$$
so that by Theorem 2,
$$\lim_{n \to \infty} \int_m^M N_n(y)\, dy = \overset{b}{\underset{a}{V}}(f).$$

It is easy to see that
$$N_1(y) \leqslant N_2(y) \leqslant N_3(y) \leqslant \cdots.$$
and, hence, the finite or infinite limit
$$N^*(y) = \lim_{n \to \infty} N_n(y),$$
exists and is a measurable function. According to B. Levi's Theorem (Chapt. VI, §1),
$$\int_m^M N^*(y)\, dy = \lim_{n \to \infty} \int_m^M N_n(y)\, dy = \overset{b}{\underset{a}{V}}(f).$$
We now need only to show that
$$N^*(y) = N(y), \tag{7}$$
in order to prove the theorem. First of all, it is perfectly clear that
$$N_n(y) \leqslant N(y),$$
so that
$$N^*(y) \leqslant N(y). \tag{8}$$

Now let q be a natural number not greater than $N(y)$. Then we can find q distinct roots
$$x_1 < x_2 < \cdots < x_q$$
of equation (6). If n is so large that
$$\frac{b-a}{2^n} < \min(x_{k+1} - x_k),$$

all q roots x_k will fall into distinct intervals d_k so that
$$N_n(y) \geqslant q,$$
and therefore
$$N^*(y) \geqslant q. \tag{9}$$

If $N(y) = +\infty$, we can take q arbitrarily large, so that $N^*(y) = +\infty$ also. If $N(y)$ is finite, we can take $q = N(y)$, and then (9) can be written in the form
$$N^*(y) \geqslant N(y).$$
(7) follows from this and (8).

COROLLARY 1. *A continuous function $f(x)$ has finite variation if and only if its Banach indicatrix $N(y)$ is summable.*

COROLLARY 2. *If $f(x)$ is a continuous function of finite variation, then the set of values taken on by it an infinite number of times has (on the axis of ordinates) measure zero.*

In fact, in this case the Banach indicatrix, being summable, is finite almost everywhere.

§ 6. THE STIELTJES INTEGRAL

We now take up a very important generalization of the Riemann integral which is called the Stieltjes integral. Let $f(x)$ and $g(x)$ be finite functions defined on the closed interval $[a, b]$. Subdivide $[a, b]$ into parts by means of the points
$$x_0 = a < x_1 < \ldots < x_n = b,$$
choose a point ξ_k in $[x_k, x_{k+1}]$ for $k = 0, \ldots, n-1$, and form the sum
$$\sigma = \sum_{k=0}^{n-1} f(\xi_k) [g(x_{k+1}) - g(x_k)].$$

If the sum σ tends to a finite limit I as
$$\lambda = \max(x_{k+1} - x_k) \to 0,$$
independently of both the method of subdivision and the choice of the points ξ_k, this limit is called the *Stieltjes integral of the function $f(x)$ with respect to the function $g(x)$* and is designated by
$$\int_a^b f(x) \, dg(x), \quad \text{или} \quad (S) \int_a^b f(x) \, dg(x).$$

The exact meaning of the definition is this: the number I is the Stieltjes integral of the function $f(x)$ with respect to the function $g(x)$ if, for every $\varepsilon > 0$ there exists a $\delta > 0$ such that for an arbitrary method of subdivision for which $\lambda < \delta$, the inequality
$$|\sigma - I| < \varepsilon$$
holds, for all choices of the points ξ_k.

It is clear that the Riemann integral is a special case of the Stieltjes integral, obtained by setting $g(x) = x$.

We list the following obvious properties of the Stieltjes integral.

1. $\int_a^b [f_1(x) + f_2(x)]\, dg(x) = \int_a^b f_1(x)\, dg(x) + \int_a^b f_2(x)\, dg(x).$

2. $\int_a^b f(x)\, d[g_1(x) + g_2(x)] = \int_a^b f(x)\, dg_1(x) + \int_a^b f(x)\, dg_2(x).$

3. If k and l are constants, then
$$\int_a^b kf(x)\, dlg(x) = kl \int_a^b f(x)\, dg(x).$$

(In all three cases, the existence of the right-hand member implies the existence of the left-hand member).

4. If $a < c < b$, and all three integrals involved in the equality
$$\int_a^b f(x)\, dg(x) = \int_a^c f(x)\, dg(x) + \int_c^b f(x)\, dg(x) \quad *$$
exist, then the equality (*) holds.

In order to prove this property of the integral, it is necessary only to see that the point c is included in the points of subdivision of $[a, b]$ when we form the sum σ for the integral $\int_a^b f\, dg$.

It is not difficult to prove that the existence of the integral $\int_a^b f\, dg$ implies the existence of both the integrals $\int_a^c f\, dg$ and $\int_c^b f\, dg$, but we will not stop to discuss this. It is more interesting to note that the converse statement is not true.

EXAMPLE. Let the functions $f(x)$ and $g(x)$ be defined on the segment $[-1, +1]$, where

$$f(x) = \begin{cases} 0 & \text{if } -1 \leqslant x \leqslant 0 \\ 1 & \text{if } 0 < x \leqslant 1 \end{cases}, \quad g(x) = \begin{cases} 0 & \text{if } -1 \leqslant x < 0, \\ 1 & \text{if } 0 \leqslant x \leqslant 1. \end{cases}$$

It is easy to see that the integrals
$$\int_{-1}^0 f(x)\, dg(x), \quad \int_0^1 f(x)\, dg(x)$$

exist (because all of the sums σ equal zero). But at the same time, the integral
$$\int_{-1}^{+1} f(x)\, dg(x)$$

does not exist. In fact, subdivide the segment $[-1, +1]$ into parts so that the point 0 is not a point of subdivision, and form the sum
$$\sigma = \sum_{k=0}^{n-1} f(\xi_k) [g(x_{k+1}) - g(x_k)].$$

It is easy to see that if $x_i < 0 < x_{i+1}$, only the i-th term remains in the sum σ, because if the points x_k and x_{k+1} lie on the same side of the point 0, then $g(x_k) = g(x_{k+1})$. This implies that

$$\sigma = f(\xi_i)[g(x_{i+1}) - g(x_i)] = f(\xi_i).$$

Depending upon whether $\xi_i \leq 0$ or $\xi_i > 0$, we have

$$\sigma = 0 \quad \text{or} \quad \sigma = 1,$$

so that σ has no limit.

5. The existence of one of the integrals $\int_a^b f(x) \, dg(x)$ and $\int_a^b g(x) \, df(x)$ implies the existence of the other. In this case, the equality

$$\int_a^b f(x) \, dg(x) + \int_a^b g(x) \, df(x) = [f(x) g(x)]_a^b \tag{1}$$

holds, where

$$[f(x) g(x)]_a^b = f(b) g(b) - f(a) g(a). \tag{2}$$

The relation (1) is called the *formula for integration by parts*.

Let us prove this formula. Suppose that the integral $\int_a^b g(x) \, df(x)$ exists. Subdivide $[a, b]$ into parts and form the sum

$$\sigma = \sum_{k=0}^{n-1} f(\xi_k) [g(x_{k+1}) - g(x_k)].$$

The sum σ can also be written as

$$\sigma = \sum_{k=0}^{n-1} f(\xi_k) g(x_{k+1}) - \sum_{k=0}^{n-1} f(\xi_k) g(x_k),$$

and this expression is clearly equal to

$$\sigma = -\sum_{k=1}^{n-1} g(x_k) [f(\xi_k) - f(\xi_{k-1})] + f(\xi_{n-1}) g(x_n) - f(\xi_0) g(x_0).$$

Adding and subtracting expression (2) on the right side of the last equality, we find

$$\sigma = [f(x) g(x)]_a^b -$$

$$- \left\{ g(a) [f(\xi_0) - f(a)] + \sum_{k=1}^{n-1} g(x_k) [f(\xi_k) - f(\xi_{k-1})] + g(b) [f(b) - f(\xi_{n-1})] \right\}.$$

The expression in the curly brackets is nothing but the sum formed for the integral $\int_a^b g \, df$, where the points of subdivision of $[a, b]$ are

$$a \leq \xi_0 \leq \xi_1 \leq \xi_2 \leq \ldots \leq \xi_{n-1} \leq b,$$

and the points $a, x_1, x_2, \ldots, x_{n-1}, b$ are points of the closed intervals $[a, \xi_0]$, $[\xi_0, \xi_1]$, $\ldots$, $[\xi_{n-1}, b]$. As $\max(x_{k+1} - x_k)$ approaches zero,

$$\max(\xi_{k+1} - \xi_k),$$

also approaches zero, so that the sum in the curly brackets tends to the integral $\int_a^b g \, df$. This completes the proof.

It is natural to ask the question under what conditions the Stieltjes integral exists.

VIII. Functions of Finite Variation. Stieltjes Integral

We will restrict ourselves to only one theorem in this direction.

THEOREM 1. *The integral*
$$\int_a^b f(x)\, dg(x)$$
exists if the function $f(x)$ is continuous on $[a, b]$ and $g(x)$ is of finite variation on $[a, b]$.

Proof. We may obviously suppose that the function $g(x)$ is increasing, because every function of finite variation is the difference of two increasing functions. Subdivide $[a, b]$ by means of the points
$$x_0 = a < x_1 < \ldots < x_n = b$$
and designate by m_k and M_k, the least and greatest values, respectively, of the function $f(x)$ on $[x_k, x_{k+1}]$. Let
$$s = \sum_{k=0}^{n-1} m_k [g(x_{k+1}) - g(x_k)], \quad S = \sum_{k=0}^{n-1} M_k [g(x_{k+1}) - g(x_k)].$$
It is clear that
$$s \leq \sigma \leq S.$$
for all choices of the points ξ_k in the closed intervals $[x_k, x_{k+1}]$. It is also easy to verify that the sum s does not decrease and S does not increase when new points of subdivision are added. From this it follows that none of the sums s surpasses any of the sums S. In fact, having two methods, I and II, of subdividing the segment $[a, b]$, with corresponding s_1, S_1 and s_2, S_2, respectively, we can form a method III by combining the points of subdivision of both of the methods I and II. If the sums s_3 and S_3 correspond to the method III, we have
$$s_1 \leq s_3 \leq S_3 \leq S_2,$$
so that, in fact, $s_1 \leq S_2$.

With this in mind, we denote the least upper bound of the set $\{s\}$ of all lower sums by the symbol I:
$$I = \sup\{s\}.$$
For every method of subdivision, it is clear that
$$s \leq I \leq S,$$
and consequently, by (3),
$$|\sigma - I| \leq S - s.$$
If we choose an arbitrary $\varepsilon > 0$ and find a $\delta > 0$ such that $|f(x'') - f(x')| < \varepsilon$ whenever $|x'' - x'| < \delta$, then we have
$$M_k - m_k < \varepsilon \qquad (k = 0, 1, \ldots, n-1)$$
for $\lambda < \delta$, and accordingly
$$S - s < \varepsilon [g(b) - g(a)].$$
Clearly
$$|\sigma - I| < \varepsilon [g(b) - g(a)].$$
for $\lambda < \delta$. In other words,
$$\lim_{\lambda \to 0} \sigma = I,$$

so that I is the integral $\int_a^b f(x)\,dg(x)$.

It follows from the theorem just proved that every function of finite variation is integrable with respect to every continuous function.

The problem of computing Stieltjes integrals will be studied in detail in §6, Chapt. IX. We restrict ourselves here to two elementary cases.

THEOREM 2. *If the function $f(x)$ is continuous on $[a, b]$ and if the function $g(x)$ has a Riemann integrable derivative $g'(x)$ at every point of $[a, b]$, then*

$$(S)\int_a^b f(x)\,dg(x) = (R)\int_a^b f(x)g'(x)\,dx. \qquad (4)$$

Proof. It follows from the conditions of the theorem that $g(x)$ satisfies the Lipschitz condition and therefore is of finite variation. Hence the integral on the left side of (4) exists. On the other hand, the function $g'(x)$, and the product $f(x)g'(x)$ with it, are continuous almost everywhere, so that the right member of (4) also exists. It remains to show that the two sides of (4) are equal. For this purpose, subdivide $[a, b]$ by means of the points

$$x_0 = a < x_1 < \ldots < x_n = b,$$

and, to every difference $g(x_{k+1}) - g(x_k)$, apply the mean value theorem:

$$g(x_{k+1}) - g(x_k) = g'(\bar{x}_k)(x_{k+1} - x_k) \qquad (x_k < \bar{x}_k < x_{k+1}).$$

In forming the sum σ for the integral $\int_a^b f\,dg$, we may take the point $\bar{x}_k$ which appears in the mean value theorem for the point ξ_k. The sum σ then becomes

$$\sigma = \sum_{k=0}^{n-1} f(\bar{x}_k)\,g'(\bar{x}_k)(x_{k+1} - x_k).$$

This is a Riemann sum for the function $f(x)\,g'(x)$. Refining the subdivision and taking the limit, we obtain (4).

THEOREM 3. *Let $f(x)$ be a function continuous on $[a, b]$ and let $g(x)$ be constant in each of the intervals $(a, c_1), (c_1, c_2), \ldots, (c_m, b)$, where* [10]

$$a < c_1 < c_2 < \ldots < c_m < b.$$

Then

$$\int_a^b f(x)\,dg(x) = f(a)\,[g(a+0) - g(a)] +$$

$$+ \sum_{k=1}^{m} f(c_k)\,[g(c_k+0) - g(c_k-0)] + f(b)\,[g(b) - g(b-0)]. \qquad (5)$$

Proof. It is easy to see that

$$\overset{b}{\underset{a}{V}}(g) = |g(a+0) - g(a)| + \sum_{k=1}^{m} \{|g(c_k) - g(c_k - 0)| +$$

$$+ |g(c_k + 0) - g(c_k)|\} + |g(b) - g(b-0)|,$$

---

[10] In other words, $g(x)$ is a *step function*.

so that the function $g(x)$ has finite variation on $[a, b]$. Therefore $g(x)$ has finite variation on every subinterval of $[a, b]$. Therefore

$$\int_a^b f(x)\,dg(x) = \sum_{k=0}^m \int_{c_k}^{c_{k+1}} f(x)\,dg(x), \qquad (6)$$

where we write $c_0 = a$, $c_{m+1} = b$. It remains to compute the integral $\int_{c_k}^{c_{k+1}} f(x)\,dg(x)$. Subdividing $[c_k, c_{k+1}]$ and forming the sum σ for the interval $[c_k, c_{k+1}]$ we obviously have

$$\sigma = f(\xi_0)\,[g(c_k + 0) - g(c_k)] + f(\xi_{n-1})\,[g(c_{k+1}) - g(c_{k+1} - 0)],$$

because all other terms vanish. In the limit, therefore,

$$\int_{c_k}^{c_{k+1}} f(x)\,dg(x) = f(c_k)\,[g(c_k + 0) - g(c_k)] + f(c_{k+1})\,[g(c_{k+1}) - g(c_{k+1} - 0)].$$

The equality (5) follows from this and (6).

§ 7. PASSAGE TO THE LIMIT UNDER THE STIELTJES INTEGRAL SIGN.

Theorem 1. *If the function $f(x)$ is continuous on $[a, b]$ and $g(x)$ has finite variation on $[a, b]$, then*

$$\left| \int_a^b f(x)\,dg(x) \right| \leq M(f) \cdot \overset{b}{\underset{a}{V}}(g), \qquad (1)$$

where $M(f) = \max |f(x)|$.

Proof. For an arbitrary method of subdividing $[a, b]$ and an arbitrary choice of the points ξ_k,

$$|\sigma| = \left| \sum_{k=0}^{n-1} f(\xi_k)\,[g(x_{k+1}) - g(x_k)] \right| \leq M(f) \sum_{k=0}^{n-1} |g(x_{k+1}) - g(x_k)| \leq M(f) \cdot \overset{b}{\underset{a}{V}}(g)$$

(1) follows at once.

Theorem 2. *Let $g(x)$ be a function of finite variation defined on the closed interval $[a, b]$, and let $f_n(x)$ be a sequence of continuous functions on $[a, b]$, which converges uniformly to the (necessarily continuous) function $f(x)$. Then*

$$\lim_{n \to \infty} \int_a^b f_n(x)\,dg(x) = \int_a^b f(x)\,dg(x).$$

Proof. Let

$$M(f_n - f) = \max |f_n(x) - f(x)|.$$

Then, by (1),

$$\left| \int_a^b f_n(x)\,dg(x) - \int_a^b f(x)\,dg(x) \right| \leq M(f_n - f) \cdot \overset{b}{\underset{a}{V}}(g)$$

7. Passage to the Limit under the Integral Sign

and it remains merely to note that

$$M(f_n - f) \to 0$$

by hypothesis.

Theorem 3 (Helly's second theorem). *Let $f(x)$ be a continuous function defined on the interval $[a, b]$, and let $\{g_n(x)\}$ be a sequence of functions which converges to a finite function $g(x)$ at every point of $[a, b]$. If*

$$\overset{b}{\underset{a}{V}}(g_n) < K$$

for all n, then

$$\lim_{n \to \infty} \int_a^b f(x)\, dg_n(x) = \int_a^b f(x)\, dg(x). \qquad (2)$$

Proof. We first show that

$$\overset{b}{\underset{a}{V}}(g) \leq K, \qquad (3)$$

so that the limit function also has finite variation. In fact, if we subdivide the interval $[a, b]$ in any way whatsoever, we have

$$\sum_{k=0}^{m-1} |g_n(x_{k+1}) - g_n(x_k)| < K \qquad (n = 1, 2, 3, \ldots).$$

Taking the limit as $n \to \infty$, we find that

$$\sum_{k=0}^{m-1} |g(x_{k+1}) - g(x_k)| \leq K.$$

Since the subdivision used was arbitrary, (3) follows from the last inequality. Now select an arbitrary $\varepsilon > 0$ and subdivide $[a, b]$ by means of the points $\{x_k\}$ ($k = 0, 1, \ldots m$) into subintervals $[x_k, x_{k+1}]$ so small that the oscillation of the function $f(x)$ is less than $\frac{\varepsilon}{3K}$ on every interval $[x_k, x_{k+1}]$. Then

$$\int_a^b f(x)\, dg(x) = \sum_{k=0}^{m-1} \int_{x_k}^{x_{k+1}} f(x)\, dg(x) =$$

$$= \sum_{k=0}^{m-1} \int_{x_k}^{x_{k+1}} [f(x) - f(x_k)]\, dg(x) + \sum_{k=0}^{m-1} f(x_k) \int_{x_k}^{x_{k+1}} dg(x).$$

Now

$$\int_{x_k}^{x_{k+1}} dg(x) = g(x_{k+1}) - g(x_k).$$

Furthermore, the inequality

$$|f(x) - f(x_k)| < \frac{\varepsilon}{3K}$$

holds for all $x \in [x_k, x_{k+1}]$, and therefore

$$\left| \int_{x_k}^{x_{k+1}} [f(x) - f(x_k)] \, dg(x) \right| \leq \frac{\varepsilon}{3K} \bigvee_{x_k}^{x_{k+1}} (g),$$

accordingly,

$$\left| \sum_{k=0}^{m-1} \int_{x_k}^{x_{k+1}} [f(x) - f(x_k)] \, dg(x) \right| \leq \frac{\varepsilon}{3K} \bigvee_a^b (g) \leq \frac{\varepsilon}{3}.$$

We find, therefore, that

$$\int_a^b f(x) \, dg(x) = \sum_{k=0}^{m-1} f(x_k) [g(x_{k+1}) - g(x_k)] + \theta \frac{\varepsilon}{3} \qquad (|\theta| \leq 1).$$

In the same way, we can show that

$$\int_a^b f(x) \, dg_n(x) = \sum_{k=0}^{m-1} f(x_k) [g_n(x_{k+1}) - g_n(x_k)] + \theta_n \frac{\varepsilon}{3} \qquad (|\theta_n| \leq 1).$$

Since $\lim_{n \to \infty} g_n(x) = g(x)$ for all $x \in [a, b]$, it is clear that there exists a natural number n_0 such that the inequality

$$\left| \sum_{k=0}^{m-1} f(x_k) [g_n(x_{k+1}) - g_n(x_k)] - \sum_{k=0}^{m-1} f(x_k) [g(x_{k+1}) - g(x_k)] \right| < \frac{\varepsilon}{3}$$

holds for all $n > n_0$. For all $n > n_0$, therefore, we have

$$\left| \int_a^b f(x) \, dg_n(x) - \int_a^b f(x) \, dg(x) \right| < \varepsilon.$$

This proves the theorem.

With the aid of the preceding theorem we can reduce the problem of evaluating the integral $\int_a^b f(x) \, dg(x)$ ($f(x)$ continuous and $g(x)$ of finite variation) to the case in which $g(x)$ is continuous. Let $g(x)$ be an arbitrary function of finite variation. Consider the saltus function of the function $g(x)$:

$$s(x) = [g(a+0) - g(a)] + \sum_{x_k < x} [g(x_k+0) - g(x_k-0)] + [g(x) - g(x-0)].$$

As proved in Theorem 7, §3, we may write

$$g(x) = s(x) + \gamma(x),$$

where $\gamma(x)$ is a continuous function of bounded variation. Hence,

$$\int_a^b f(x) \, dg(x) = \int_a^b f(x) \, ds(x) + \int_a^b f(x) \, d\gamma(x).$$

The integral $\int_a^b f(x)\,ds(x)$ can be easily evaluated. To do this, first note that the series

$$\sum_{k=1}^{\infty}\{|g(x_k)-g(x_k-0)|+|g(x_k+0)-g(x_k)|\}$$

is convergent.[11] Next, define the functions $s_n(x)$ by setting $s_n(a) = 0$ and

$$s_n(x) = [g(a+0)-g(a)] + \sum_{x_k<x}[g(x_k+0)-g(x_k-0)]+[g(x)-g(x-0)]$$

for $a < x \leqslant b$. (Note that only those points x_k of discontinuity of the function $g(x)$ are considered for which $k \leqslant n$). It is obvious that

$$\lim_{n \to \infty} s_n(x) = s(x)$$

for all $x \in [a, b]$. On the other hand,

$$\overset{b}{\underset{a}{V}}(s_n) = |g(a+0)-g(a)| + \sum_{k=1}^{n}\{|g(x_k)-g(x_k-0)|+$$
$$+|g(x_k+0)-g(x_k)|\}+|g(b)-g(b-0)|,$$

so that the variations of all the functions $s_n(x)$ are all less than a certain fixed value. By Helly's second theorem, therefore,

$$\int_a^b f(x)\,ds(x) = \lim_{n \to \infty}\int_a^b f(x)\,ds_n(x).$$

As the function $s_n(x)$ is constant in the intervals between the points $a, x_1, \ldots x_n, b$, we infer from Theorem 3, §6, that

$$\int_a^b f(x)\,ds_n(x) = f(a)[g(a+0)-g(a)] + \sum_{k=1}^{n} f(x_k)[g(x_k+0)-$$
$$-g(x_k-0)] + f(b)[g(b)-g(b-0)].$$

(It is clear that the saltus of the functions $s_n(x)$ at the points $a, x_1, \ldots x_n, b$ coincide with the saltus of the function $g(x)$.) Hence

$$\int_a^b f(x)\,ds(x) = f(a)[g(a+0)-g(a)] + \sum_{k=1}^{\infty} f(x_k)[g(x_k+0)-$$
$$-g(x_k-0)] + f(b)[g(b)-g(b-0)],$$

---

[11] Write $g(x) = \pi(x) - \nu(x)$, where $\pi(x)$ and $\nu(x)$ are increasing functions. Then each of the non-negative series

$$\sum_{k=1}^{\infty}[\pi(x_k+0)-\pi(x_k-0)], \quad \sum_{k=1}^{\infty}[\nu(x_k+0)-\nu(x_k-0)]$$

obviously converges, and it remains merely to observe that

$$|g(x_k)-g(x_k-0)|+|g(x_k+0)-g(x_k)| \leqslant [\pi(x_k+0)-\pi(x_k-0)]+[\nu(x_k+0)-\nu(x_k-0)].$$

236 VIII. FUNCTIONS OF FINITE VARIATION. STIELTJES INTEGRAL

and to find the integral $\int_a^b f(x)\,dg(x)$ it remains to evaluate $\int_a^b f(x)\,d\gamma(x)$, where $\gamma(x)$ is the continuous component of the function $g(x)$.

We call the reader's attention to the fact that the value $g(x_k)$ itself of the function $g(x)$ at a point x_k of discontinuity such that $a < x_k < b$ has no effect on the value of the integral $\int_a^b f\,dg$. This is quite natural, because we can omit the point x_k as a point of subdivision in forming the sums σ.

§ 8. LINEAR FUNCTIONALS

Let $g(x)$ be a function of finite variation defined on the closed interval $[a, b]$. By means of this function, we can assign the number

$$\Phi(f) = \int_a^b f(x)\,dg(x) \tag{1}$$

to every continuous function $f(x)$ defined on $[a, b]$.

The following properties of this function Φ are evident.
1) $\Phi(f_1 + f_2) = \Phi(f_1) + \Phi(f_2)$
2) $|\Phi(f)| \leq K M(f)$, where $M(f) = \max |f(x)|$ and $K = \overset{b}{\underset{a}{V}}(g)$.

Suppose that an abstract function $\Phi(f)$ is defined for every continuous function on $[a, b]$, for which properties 1) and 2) hold. Such a function is called a bounded linear functional on the space of continuous functions on $[a, b]$. We shall denote this space of continuous functions by the symbol C. Riemann-Stieltjes integrals of the form (1) are thus bounded linear functionals on the space C. It is a famous theorem, first proved by F. Riesz, that these are the *only* bounded linear functionals on the space C.

Before proving this theorem, we observe that if $\Phi(f)$ is a bounded linear functional on C, then

$$\Phi(kf) = k\Phi(f)$$

for all numbers k and $f \in C$. This is proved by the argument employed for the corresponding assertion for bounded linear functionals on the space L_2 (§4, Chapter VII).

THEOREM (F. RIESZ). *Let $\Phi(f)$ be a bounded linear functional defined on the space C of functions $f(x)$, continuous on the segment $[a, b]$. Then there exists a function of finite variation $g(x)$ on $[a, b]$ such that*

$$\Phi(f) = \int_a^b f(x)\,dg(x) \tag{1}$$

for all $f(x) \in C$.

Proof. It is sufficient to consider the case

$$a = 0, \quad b = 1,$$

since the general case can be reduced to this by use of a linear transformation of the argument x. In §5, Chapt. IV, we noted that

$$\sum_{k=0}^{n} C_n^k x^k (1-x)^{n-k} = 1.$$

8. Linear Functionals

For $x \in [0, 1]$, each term of this sum is non-negative. Therefore, if
$$\varepsilon_k = \pm 1 \qquad (k = 0, 1, \ldots, n),$$
then
$$\left| \sum_{k=0}^{n} \varepsilon_k C_n^k x^k (1-x)^{n-k} \right| \leq 1. \qquad (2)$$

We next consider the linear functional $\Phi(f)$ defined for all functions $f(x)$ of the space $C[0, 1]$. By the definition of a linear functional, there exists a K such that
$$|\Phi(f)| \leq K \cdot M(f).$$
From this and (2), we have
$$\left| \sum_{k=0}^{n} \varepsilon_k \Phi[C_n^k x^k (1-x)^{n-k}] \right| \leq K.$$
Upon choosing the numbers ε_k so that all terms of the last sum are non-negative, we find that
$$\sum_{k=0}^{n} |\Phi[C_n^k x^k (1-x)^{n-k}]| \leq K. \qquad (3)$$

Define a step function $g_n(x)$, by setting
$$g_n(0) = 0$$
$$g_n(x) = \Phi[C_n^0 x^0 (1-x)^{n-0}] \qquad \left(0 < x < \frac{1}{n}\right)$$
$$g_n(x) = \Phi[C_n^0 x^0 (1-x)^{n-0}] + \Phi[C_n^1 x^1 (1-x)^{n-1}] \qquad \left(\frac{1}{n} \leq x < \frac{2}{n}\right)$$
$$\cdots \cdots \cdots \cdots \cdots \cdots \cdots \cdots \cdots \cdots \cdots$$
$$g_n(x) = \sum_{k=0}^{n-1} \Phi[C_n^k x^k (1-x)^{n-k}] \qquad \left(\frac{n-1}{n} \leq x < 1\right)$$
$$g_n(1) = \sum_{k=0}^{n} \Phi[C_n^k x^k (1-x)^{n-k}].$$

In view of (3), the functions $g_n(x)$ and their total variations are bounded by a single number. Therefore, on the basis of Helly's first theorem, we infer the existence of a subsequence $\{g_{n_i}(x)\}$ of $\{g_n(x)\}$ which converges at every point of $[0, 1]$ to a function of finite variation $g(x)$.

If $f(x)$ is any continuous function defined on $[0, 1]$, then by Theorem 3, §6,
$$\int_0^1 f(x) \, dg_n(x) = \sum_{k=0}^{n} f\left(\frac{k}{n}\right) \Phi[C_n^k x^k (1-x)^{n-k}],$$
and therefore
$$\int_0^1 f(x) \, dg_n(x) = \Phi[B_n(x)],$$
where
$$B_n(x) = \sum_{k=0}^{n} f\left(\frac{k}{n}\right) C_n^k x^k (1-x)^{n-k}$$

is the Bernstein polynomial of degree n for the function $f(x)$. By Bernstein's theorem of §5, Chapt. IV, we have

$$M(B_n - f) \to 0,$$

and by the definition of a bounded linear functional,

$$|\Phi(B_n) - \Phi(f)| = |\Phi(B_n - f)| \leqslant KM(B_n - f).$$

Accordingly,

$$\Phi(B_n) \to \Phi(f)$$

as $n \to \infty$.

It follows that

$$\lim_{n \to \infty} \int_0^1 f(x) \, dg_n(x) = \Phi(f).$$

By Helly's second theorem, we have

$$\lim_{n \to \infty} \int_0^1 f(x) \, dg_n(x) = \int_0^1 f(x) \, dg(x).$$

Therefore

$$\Phi(f) = \int_0^1 f(x) \, dg(x),$$

as was to be proved.

§ 9. EDITOR'S APPENDIX TO CHAPTER VIII.

For many applications, especially in the theory of probability, it is useful to consider functions of finite variation on the infinite line $(-\infty, \infty)$.

DEFINITION 1. Let $f(x)$ be a function defined for all x, $-\infty < x < \infty$. If $\overset{b}{\underset{a}{V}}(f) < \infty$ for all $a < b$, and if $\sup_{a < b} \overset{b}{\underset{a}{V}}(f)$ is finite, then $f(x)$ is said to be of finite variation on $(-\infty, \infty)$, and the number

$$\sup_{a < b} \overset{b}{\underset{a}{V}}(f) = \overset{\infty}{\underset{-\infty}{V}}(f)$$

is called the total variation of f.

THEOREM 1.
$$\overset{\infty}{\underset{-\infty}{V}}(f) = \lim_{a \to \infty} \overset{a}{\underset{-a}{V}}(f).$$

This theorem is obvious.

DEFINITION 2.
$$\overset{a}{\underset{-\infty}{V}}(f) = \lim_{b \to -\infty} \overset{a}{\underset{b}{V}}(f);$$

$$\overset{\infty}{\underset{a}{V}}(f) = \lim_{b \to \infty} \overset{b}{\underset{a}{V}}(f).$$

THEOREM 2. *For $a < b$,*

$$(1) \qquad \overset{b}{\underset{-\infty}{V}}(f) = \overset{a}{\underset{-\infty}{V}}(f) + \overset{a}{\underset{b}{V}}(f);$$

9. Editor's Appendix

$$(2) \quad \overset{\infty}{\underset{-\infty}{V}}(f) = \overset{a}{\underset{-\infty}{V}}(f) + \overset{\infty}{\underset{a}{V}}(f);$$

$$(3) \quad \overset{\infty}{\underset{a}{V}}(f) = \overset{b}{\underset{a}{V}}(f) + \overset{\infty}{\underset{b}{V}}(f).$$

This theorem is also obvious.

Theorem 3. *Let $f(x)$ have finite variation on $(-\infty,\infty)$. Then* $\lim_{x \to -\infty} \overset{x}{\underset{-\infty}{V}}(f) = 0$ (1)

and $\lim_{x \to \infty} \overset{\infty}{\underset{x}{V}}(f) = 0.$ (2)

Proof. Let us prove, for example, (2). Since $\overset{\infty}{\underset{-\infty}{V}}(f) = \sup \overset{a}{\underset{-\infty}{V}}(f)$, it is clear that for every $\varepsilon > 0$, there exists a number a such that $\overset{\infty}{\underset{-\infty}{V}}(f) - \varepsilon < \overset{a}{\underset{-\infty}{V}}(f)$. Also, $\overset{\infty}{\underset{-\infty}{V}}(f) = \overset{a}{\underset{-\infty}{V}}(f) + \overset{\infty}{\underset{a}{V}}(f)$. It follows that $\overset{\infty}{\underset{a}{V}}(f) < \varepsilon$. This obviously proves (2). The relation (1) is proved similarly.

Theorem 4. *If* $\overset{\infty}{\underset{-\infty}{V}}(f) < \infty$, *then $f(x)$ is bounded.*

Proof. Let a be any number. Then, for every number $x > a$, $|f(x) - f(a)| \leqslant \overset{x}{\underset{a}{V}}(f) \leqslant \overset{\infty}{\underset{-\infty}{V}}(f)$. Similarly, if $x < a$, $|f(a) - f(x)| \leqslant \overset{\infty}{\underset{-\infty}{V}}(f)$. Hence $|f(x)| \leqslant |f(a)| + \overset{\infty}{\underset{-\infty}{V}}(f)$, for all $x \neq a$.

Theorems 3 and 4 of §3 are easily seen to be true of functions having finite variation on $(-\infty, \infty)$.

Theorem 5. *Let $f(x)$ have finite variation on $(-\infty, \infty)$. Then $f(x)$ can be written as the difference of two bounded monotone increasing functions:*

$$f(x) = \pi(x) - \nu(x)$$

and $\pi(x)$ can be chosen so that $\lim_{x \to -\infty} \pi(x) = 0.$

Proof. Let $\pi(x) = \overset{x}{\underset{-\infty}{V}}(f)$, and let $\nu(x) = \pi(x) - f(x)$. Clearly $\pi(x)$ is an increasing function, and by Theorem 3, we have $\lim_{x \to -\infty} \pi(x) = 0$. To show that $\nu(x)$ is monotone, simply repeat the argument of Theorem 6, §3.

Remark. The converse of Theorem 5 is an obvious consequence of Theorem 3, §3.

Corollary. *Let $f(x)$ have finite variation on $(-\infty, \infty)$. Then $\lim_{x \to -\infty} f(x)$ and $\lim_{x \to \infty} f(x)$ exist.*

Corollaries 1 and 2 of Theorem 6, §3, hold for functions of finite variation on $(-\infty, \infty)$.

Theorem 7, §3, is also true for functions $f(x)$ having finite variation on $(-\infty,\infty)$. The construction is, in fact, a bit simpler in the present case. For all x, let

$$s_\pi(x) = \sum_{x_k < x} [\pi(x_k + 0) - \pi(x_k - 0)] + [\pi(x) - \pi(x - 0)],$$

and let

$$s_\nu(x) = \sum_{x_k < x} [\nu(x_k + 0) - \nu(x_k - 0)] + [\nu(x) - \nu(x - 0)].$$

The points $x_1, x_2, x_3, \ldots$ are the points at which at least one of $\pi(x)$ or $\nu(x)$ is discontinuous. Then, putting $s(x) = s_\pi(x) - s_\nu(x)$ and $\varphi(x) = f(x) - s(x)$, we have $f(x) = \varphi(x) + s(x)$, and as before $\varphi(x)$ is a continuous function of finite variation on $(-\infty, \infty)$.

One further reduction is possible at this point. Namely, let us write $a = \lim\limits_{x \to -\infty} \nu(x)$, and $\rho(x) = \nu(x) - a$. Then we have

$$f(x) = \pi(x) - \rho(x) - a,$$

where both $\pi(x)$ and $\rho(x)$ are monotone increasing functions with limit zero at $-\infty$.

The results of §§ 4, 5, and 6 have obvious extensions for functions of finite variation on $(-\infty, \infty)$. We mention explicitly only the definition of the Stieltjes integral $\int_{-\infty}^{\infty} f(x)\, dg(x)$ for *bounded* continuous functions $f(x)$ and functions of finite variation $g(x)$, on $(-\infty, \infty)$. We make the definition

$$\int_{-\infty}^{\infty} f(x)\, dg(x) = \lim_{\substack{a \to -\infty \\ b \to \infty}} \int_a^b f(x)\, dg(x).$$

The theorems of §6 have obvious analogues for this integral, except for the formula for integration by parts, in which care must be exercised in taking the limits $a \to -\infty$, $b \to \infty$.

Helly's second theorem is not true as it stands for functions on $(-\infty, \infty)$. For example, let

$$\gamma(x) = \begin{cases} 0 & x \leq 0, \\ x & 0 \leq x \leq 1, \\ 1 & x > 1, \end{cases}$$

and let $g_n(x) = \gamma(x - n)$ $(n = 1, 2, 3, \ldots)$. Then $g(x) = \lim\limits_{n \to \infty} g_n(x) = 0$ for all x, and yet

$$\lim_{n \to \infty} \int_{-\infty}^{\infty} 1\, dg_n(x) = 1 \neq 0 = \int_{-\infty}^{\infty} 1\, dg(x).$$

To obtain the analogue of Helly's second theorem, we must introduce a new function class. Let C_∞ denote the class of all continuous functions $f(x)$ defined $-\infty < x < \infty$ and having the property that $\lim\limits_{x \to \infty} f(x) = \lim\limits_{x \to -\infty} f(x) = 0$.

THEOREM 6. *Let $\{g_n(x)\}$ be a sequence of functions on $-\infty < x < \infty$ such that $\overset{\infty}{\underset{-\infty}{V}}(g_n) < K$ for all n, and such that $\lim\limits_{n \to \infty} g_n(x)$ exists for all x, $-\infty < x < \infty$. For all $f(x) \in C_\infty$, we then have*

$$\lim_{n \to \infty} \int_{-\infty}^{\infty} f(x)\, dg_n(x) = \int_{-\infty}^{\infty} f(x)\, dg(x).$$

Proof. This is a simple corollary of Helly's second theorem (Theorem 3, §7). It is first clear that $\overset{\infty}{\underset{-\infty}{V}}(g) \leq K$. Now let ε be an arbitrary positive number. Then there exists a number A such that $|f(x)| < \frac{\varepsilon}{8K}$ for $|x| \geq A$. We then have

$$\left| \int_{-\infty}^{-A} f(x)\, d\varphi(x) \right| < \frac{\varepsilon}{8K} K = \frac{\varepsilon}{8} \text{ and } \left| \int_A^{\infty} f(x)\, d\varphi(x) \right| < \frac{\varepsilon}{8}$$

if φ is any function with variation $< K$.

It follows that

$$\left|\int_{-\infty}^{\infty} f(x)\, dg_n(x) - \int_{-\infty}^{\infty} f(x)\, dg(x)\right| \leqslant$$

$$\left|\int_{-A}^{A} f(x)\, dg_n(x) - \int_{-A}^{A} f(x)\, dg(x)\right| +$$

$$\left|\int_{-\infty}^{-A} f(x)\, dg_n(x)\right| + \left|\int_{-\infty}^{-A} f(x)\, dg(x)\right| + \left|\int_{A}^{\infty} f(x)\, dg_n(x)\right| +$$

$$\left|\int_{A}^{\infty} f(x)\, dg(x)\right| < \left|\int_{-A}^{A} f(x)\, dg_n(x) - \int_{-A}^{A} f(x)\, dg(x)\right| + \frac{\varepsilon}{2}.$$

It remains only to apply Helly's second theorem.

We mention in conclusion the analogue of Riesz's theorem of §8, for functions of finite variation on $(-\infty, \infty)$.

THEOREM 7. *Let Φ be a bounded linear functional on C_∞, i.e., a function $\Phi(f)$ defined for all $f \in C_\infty$ such that*

$$\Phi(f_1 + f_2) = \Phi(f_1) + \Phi(f_2)$$

and

$$|\Phi(f)| \leqslant A \max_x |f(x)|.$$

Then there exists a function $g(x)$ of finite variation on $(-\infty, \infty)$ such that

$$\Phi(f) = \int_{-\infty}^{\infty} f(x)\, dg(x)$$

for all $f(x) \in C_\infty$.

We omit the proof.

Exercises for Chapter VIII

1. A necessary and sufficient condition that the function $f(x)$ be of finite variation is that there exist an increasing function $\varphi(x)$ such that

$$f(x'') - f(x') \leqslant \varphi(x'') - \varphi(x')$$

for $x' < x''$.

2. If at every point of the set E, the derivative $f'(x)$ of a finite function $f(x)$ exists and $|f'(x)| \leqslant K$, then

$$m^* f(E) \leqslant K \cdot m^* E.$$

3. A function $f(x)$ is said to satisfy a Lipschitz condition of order $\alpha > 0$ if $|f(x'') - f(x')| \leqslant K|x'' - x'|^\alpha$. Show that for $\alpha > 1$, $f(x) \equiv$ const. Construct an example of a function of finite variation which satisfies no Lipschitz condition. Construct a function satisfying a Lipschitz condition of given order $\alpha < 1$ and having infinite total variation.

4. The integral $\int_a^b f(x)\, dg(x)$ exists if $f(x)$ satisfies a Lipschitz condition of order α, and $g(x)$ satisfies a Lipschitz condition of order β, where $\alpha + \beta > 1$. (V. Kondurar').

5. If (x) is continuous and $g(x)$ is of finite variation, then $\int_a^x f(x)\, dg(x)$ is a function of finite variation, continuous at all points of continuity of $g(x)$.

6. Let $\mu_0, \mu_1, \mu_2, \ldots$ be a given sequence of numbers. Set $\Delta^0 \mu_n = \mu_n$, $\Delta^{k+1}\mu_n = \Delta^k \mu_n - \Delta^k \mu_{n+1}$. A necessary and sufficient condition that there exist an increasing function $g(x)$ for which

$$\int_0^1 x^n \, dg(x) = \mu_n \qquad (n = 0, 1, 2, \ldots) \qquad (1)$$

is that

$$\Delta^k \mu_n \geqslant 0$$

for all k and n (F. Hausdorff).

7. (Notation as in Exercise 6.) A necessary and sufficient condition that a function of bounded variation $g(x)$ exist and satisfy condition (1) is that

$$\sum_{k=0}^n C_n^k |\Delta^{n-k} \mu_k| \leqslant K$$

(F. Hausdorff).

8. Show that Riesz's theorem of §8 is a corollary of Hausdorff's theorem stated in Exercise 7.

9. The set $F = \{f(x)\}$ consists of *equicontinuous* functions if to every $\varepsilon > 0$ there corresponds a $\delta > 0$ such that $|f(x'') - f(x')| < \varepsilon$ for $|x'' - x'| < \delta$ and for all functions of F. If all the functions of such an infinite set F are bounded in absolute value by a single number, then a uniformly convergent sequence can be found in F (C. Arzelà—J. Ascoli).

10. Prove the equality $\overset{b}{\underset{a}{V}}(f) = \overset{c}{\underset{a}{V}}(f) + \overset{b}{\underset{c}{V}}(f)$ for continuous functions $f(x)$, using Banach's theorem of §5.

CHAPTER IX

ABSOLUTELY CONTINUOUS FUNCTIONS
THE INDEFINITE LEBESGUE INTEGRAL

§ 1. ABSOLUTELY CONTINUOUS FUNCTIONS

We now take up a special class of functions of finite variation, the class of absolutely continuous functions. These functions are important for a number of applications, and are in addition interesting on their own account.

DEFINITION. Let $f(x)$ be a finite function defined on the closed interval $[a, b]$. Suppose that for every $\varepsilon > 0$, there exists a $\delta > 0$ such that

$$\sum_{k=1}^{n} (b_k - a_k) < \delta \tag{1}$$

for all numbers $a_1, b_1, \ldots, a_n, b_n$ such that $a_1 < b_1 \leqslant a_2 < b_2 \leqslant \ldots \leqslant a_n < b_n$ and

$$\left| \sum_{k=1}^{n} \{f(b_k) - f(a_k)\} \right| < \varepsilon. \tag{2}$$

Then the function $f(x)$ is said to be *absolutely continuous*.

It is evident that every absolutely continuous function is continuous, since the case $n = 1$ is not excluded in the above definition. We shall show in the sequel that there are, however, continuous functions which are not absolutely continuous.

Without altering the sense of the definition, we can replace condition (2) by the stronger condition

$$\sum_{k=1}^{n} |f(b_k) - f(a_k)| < \varepsilon. \tag{3}$$

In fact, let the number $\delta > 0$ be such that the inequality

$$\left| \sum_{k=1}^{n} \{f(b_k) - f(a_k)\} \right| < \frac{\varepsilon}{2}$$

follows from (2).

Then, taking an arbitrary system of pairwise disjoint open intervals $\{(a_k, b_k)\}$ $(k = 1, 2, \ldots, n)$, for which (2) holds, we divide this system into parts A and B; we put into A those intervals (a_k, b_k) for which $f(b_k) - f(a_k) \geqslant 0$, and into B all remaining intervals of the system. By virtue of the obvious relations

$$\sum_{A} |f(b_k) - f(a_k)| = \left| \sum_{A} \{f(b_k) - f(a_k)\} \right| < \frac{\varepsilon}{2},$$

$$\sum_{B} |f(b_k) - f(a_k)| = \left| \sum_{B} \{f(b_k) - f(a_k)\} \right| < \frac{\varepsilon}{2},$$

it is clear that (3) holds.

244 IX. ABSOLUTELY CONTINUOUS FUNCTIONS. THE INDEFINITE LEBESGUE INTEGRAL

Since all terms of the sum (3) are non-negative, and their number is arbitrary, it is clear that to each $\varepsilon > 0$ there corresponds a $\delta > 0$ such that for an arbitrary finite or denumerable system of pairwise disjoint open intervals $\{(a_k, b_k)\}$, for which

$$\sum_k (b_k - a_k) < \delta,$$

the inequality

$$\sum_k |f(b_k) - f(a_k)| < \varepsilon$$

holds.

It is possible to replace the increments $|f(b_k) - f(a_k)|$ in the definition of absolute continuity by the oscillations of $f(x)$ in the intervals $[a_k, b_k]$. Let us prove this assertion. Let m_k and M_k be the least and greatest values, respectively, of the function $f(x)$ in the interval $[a_k, b_k]$.

Then there exist points α_k and β_k in $[a_k, b_k]$ such that

$$f(\alpha_k) = m_k, \quad f(\beta_k) = M_k.$$

Since the sum of the lengths of the intervals (α_k, β_k) is less than or equal to the sum of the lengths of the intervals (a_k, b_k), it is obvious that

$$\sum_k [f(\beta_k) - f(\alpha_k)] < \varepsilon.$$

Hence, if the function $f(x)$ is absolutely continuous, to every $\varepsilon > 0$ there corresponds a $\delta > 0$ such that for an arbitrary finite or denumerable system of pairwise disjoint intervals $\{(a_k, b_k)\}$ for which

$$\sum_k (b_k - a_k) < \delta,$$

the inequality

$$\sum_k \omega_k < \varepsilon$$

holds. (ω_k designates as usual the oscillation of $f(x)$ in $[a_k, b_k]$.)

A function $f(x)$ satisfying the Lipschitz condition

$$|f(x'') - f(x')| \leq K |x'' - x'|$$

is a simple example of an absolutely continuous function.

THEOREM 1. *If the functions $f(x)$ and $g(x)$ are absolutely continuous, then their sum, difference, and product are also absolutely continuous. If $g(x)$ vanishes nowhere, then the quotient $\dfrac{f(x)}{g(x)}$ is absolutely continuous.*

Proof. The absolute continuity of the sum and difference follows at once from the fact that

$$|\{f(b_k) \pm g(b_k)\} - \{f(a_k) \pm g(a_k)\}| \leq |f(b_k) - f(a_k)| + |g(b_k) - g(a_k)|.$$

Furthermore, if A and B are upper bounds for $|f(x)|$ and $|g(x)|$, then

$$|f(b_k) g(b_k) - f(a_k) g(a_k)| \leq |g(b_k)| \cdot |f(b_k) - f(a_k)| +$$
$$+ |f(a_k)| \cdot |g(b_k) - g(a_k)| \leq B |f(b_k) - f(a_k)| + A |g(b_k) - g(a_k)|,$$

from which the absolute continuity of $f(x)g(x)$ follows. Finally, if $g(x)$ vanishes nowhere, then $|g(x)| \geq \sigma > 0$, from which it follows that

$$\left|\frac{1}{g(b_k)} - \frac{1}{g(a_k)}\right| \leq \frac{|g(b_k) - g(a_k)|}{\sigma^2}.$$

The function $\frac{1}{g(x)}$ is therefore absolutely continuous and the function $\frac{f(x)}{g(x)} = f(x) \cdot \frac{1}{g(x)}$ is absolutely continuous, being the product of two absolutely continuous functions.

If $f(x)$ is absolutely continuous on $[a, b]$ and $F(y)$ is absolutely continuous on $[\min f, \max f]$, then the composite function $F(f(x))$ may or may not be absolutely continuous. We shall return to this subject later, and for the time being content ourselves with pointing out two simple conditions under which $F(f(x))$ is absolutely continuous.

THEOREM 2. *Let $f(x)$ be an absolutely continuous function defined on the closed interval $[a, b]$. Let the values of $f(x)$ lie in the closed interval $[A, B]$. If $F(y)$ is a function defined on the segment $[A, B]$ which satisfies the Lipschitz condition, then the composite function $F[f(x)]$ is absolutely continuous.*

Proof. If $|F(y'') - F(y')| \leq K|y'' - y'|$, then for an arbitrary system of pairwise disjoint open intervals (a_k, b_k), the inequality

$$\sum_{k=1}^{n} |F[f(b_k)] - F[f(a_k)]| \leq K \sum_{k=1}^{n} |f(b_k) - f(a_k)|$$

holds. The right-hand member of this equality becomes arbitrarily small together with

$$\sum_{k=1}^{n} (b_k - a_k).$$

THEOREM 3. *Let $f(x)$ be an absolutely continuous function defined on $[a, b]$ and suppose that $f(x)$ is strictly increasing. If $F(y)$ is absolutely continuous on $[f(a), f(b)]$, then the function $F[f(x)]$ is absolutely continuous on $[a, b]$.*

Proof. Let ε be an arbitrary number > 0, and let $\delta > 0$ have the property that for an arbitrary system of pairwise disjoint open intervals (A_k, B_k), for which

$$\sum_{k=1}^{n} (B_k - A_k) < \delta,$$

the inequality

$$\sum_{k=1}^{n} |F(B_k) - F(A_k)| < \varepsilon$$

holds. Then, for this δ, there exists a number $\eta > 0$ such that the inequality

$$\sum_{k=1}^{m} (b_k - a_k) < \eta$$

implies the inequality

$$\sum_{k=1}^{m} [f(b_k) - f(a_k)] < \delta,$$

provided that the intervals (a_k, b_k) are pairwise disjoint. Next, select an arbitrary system of intervals (a_k, b_k), which are pairwise disjoint and for which the sum of the lengths is less than η. The intervals $(f(a_k), f(b_k))$ are also pairwise disjoint (this fact is essential) and the sum of their lengths is less than δ. Hence

$$\sum_{k=1}^{m} |F[f(b_k)] - F[f(a_k)]| < \varepsilon.$$

This completes the proof.

§ 2. DIFFERENTIAL PROPERTIES OF ABSOLUTELY CONTINUOUS FUNCTIONS

THEOREM 1. *Every absolutely continuous function has finite variation.*[1]

Proof. Let $f(x)$ be an absolutely continuous function defined on the closed interval $[a, b]$. Choose a $\delta > 0$ such that for every system of pairwise disjoint open intervals $\{(a_k, b_k)\}$ for which $\sum_{k=1}^{n} (b_k - a_k) < \delta$, the inequality

$$\sum_{k=1}^{n} |f(b_k) - f(a_k)| < 1.$$

obtains. Subdivide $[a, b]$ by means of points

$$c_0 = a < c_1 < c_2 < \ldots < c_N = b$$

into parts such that

$$c_{k+1} - c_k < \delta. \qquad (k = 0, 1, \ldots, N-1).$$

Then, for every subdivision of the segment $[c_k, c_{k+1}]$, the sum of the absolute increments for $f(x)$ on these parts is less than 1. It follows that

$$\overset{c_{k+1}}{\underset{c_k}{V}}(f) \leq 1, \quad \text{and therefore} \quad \overset{b}{\underset{a}{V}}(f) \leq N,$$

COROLLARY. *Let $f(x)$ be a function which is absolutely continuous on $[a, b]$. Then the derivative $f'(x)$ exists and is finite at almost every point of $[a, b]$. Furthermore, the function $f'(x)$ is summable on $[a, b]$.*

THEOREM 2. *If the derivative $f'(x)$ of an absolutely continuous function $f(x)$ is zero almost everywhere, then the function $f(x)$ is constant.*

Proof. Let E be the set of those points of (a, b) for which $f'(x) = 0$. Let $\varepsilon > 0$. If $x \in E$, then for all sufficiently small $h > 0$,

$$\frac{|f(x+h) - f(x)|}{h} < \varepsilon. \qquad (*)$$

It is clear that the closed intervals $[x, x+h]$ for which $h > 0$ and condition $(*)$

---

[1] This theorem implies the existence of continuous functions which are not absolutely continuous (for example, $x \cos \dfrac{\pi}{2x}$ is such a function; see §3, Chapt. VIII).

2. Differential Properties of Absolutely Continuous Functions

is satisfied cover the set E in Vitali's sense. Hence we can select from them a finite number of pairwise disjoint closed intervals

$$d_1 = [x_1, x_1 + h_1], \quad d_2 = [x_2, x_2 + h_2], \ldots, \quad d_n = [x_n, x_n + h_n],$$

lying in the open interval (a, b) and such that the outer measure of the part of the set E not covered by them is less than an arbitrary preassigned number $\delta > 0$. We may suppose without loss of generality that $x_k < x_{k+1}$.

If

$$[a, x_1), \quad (x_1 + h_1, x_2), \ldots, (x_{n-1} + h_{n-1}, x_n), (x_n + h_n, b] \tag{1}$$

are the intervals which remain after removing from $[a, b]$ all intervals d_k ($k = 1, 2, \ldots n$), then the sum of the lengths of these intervals will necessarily be less than δ. This follows from the fact that

$$b - a = mE \leqslant \sum_{k=1}^{n} m\, d_k + m^* \left[E - \sum_{k=1}^{n} d_k\right] < \sum_{k=1}^{n} md_k + \delta.$$

This in turn implies

$$\sum_{k=1}^{n} md_k > b - a - \delta.$$

Recall now that the function $f(x)$ is absolutely continuous. Hence, the number δ can be chosen so small that the sum of the increments of $f(x)$ on the intervals (1) is less than ε :

$$\left|\{f(x_1) - f(a)\} + \sum_{k=1}^{n-1} \{f(x_{k+1}) - f(x_k + h_k)\} + \{f(b) - f(x_n + h_n)\}\right| < \varepsilon. \tag{2}$$

On the other hand, the definition of the segments d_k shows that

$$|f(x_k + h_k) - f(x_k)| < \varepsilon h_k,$$

from which we infer that

$$\left|\sum_{k=1}^{n} \{f(x_k + h_k) - f(x_k)\}\right| < \varepsilon (b - a) \tag{3}$$

(since $\sum h_k = \sum m d_k \leqslant b - a$). Upon combining (2) and (3), we have

$$|f(b) - f(a)| < \varepsilon (1 + b - a)$$

since ε is arbitrary, it follows that

$$f(b) = f(a).$$

This reasoning can be carried out for every interval $[a, x]$ such that $a < x \leqslant b$. For arbitrary $x \in [a, b]$, therefore, we have

$$f(x) = f(a),$$

and $f(x)$ is a constant.[2]

---

[2] It follows from the theorem just proved that the continuous function $\Theta(x)$ constructed in §2, Chapter VIII is not absolutely continuous.

248   IX. Absolutely Continuous Functions. The Indefinite Lebesgue Integral

COROLLARY. *If the derivatives $f'(x)$ and $g'(x)$ of two absolutely continuous functions $f(x)$ and $g(x)$ are equivalent, then the difference of these functions is constant.*

In fact, if we remove from $[a, b]$ the set of points (of zero measure) at which at least one of the functions $f(x)$ or $g(x)$ does not have a finite derivative or for which their derivatives are not equal, then for every remaining point, we have

$$[f(x) - g(x)]' = 0.$$

§ 3. CONTINUOUS MAPPINGS

In §2, Chapter VIII, we had occasion to consider images of point sets under certain mappings. We consider here a number of more refined properties of various image sets under *continuous* mappings. To avoid repetition, we agree once for all that $f(x)$ is a continuous function defined on the closed interval $[a, b]$.

THEOREM 1. *The image $f(F)$ of a closed set F is a closed set.*

Proof. Let y_0 be a limit point of the set $f(F)$,

$$y_0 = \lim_{n \to \infty} y_n \qquad [y_n \in f(F)].$$

For every point y_n, let x_n be a point in F such that

$$f(x_n) = y_n.$$

Since the sequence $\{x_n\}$ is bounded, there exists a convergent subsequence $\{x_{n_k}\}$:

$$\lim x_{n_k} = x_0.$$

Since the set F is closed, it follows that

$$x_0 \in F$$

and so

$$f(x_0) \in f(F).$$

On the other hand, since $f(x)$ is a continuous function,

$$\lim y_{n_k} = \lim f(x_{n_k}) = f(x_0),$$

so that

$$y_0 = f(x_0)$$

and $y_0 \in f(F)$. The set $f(F)$ therefore contains all of its limit points and is closed.

Combining this theorem with Theorem 1, §2, Chapter VIII we have the following fact.

COROLLARY. *If E is a set of type F_σ, then its map $f(E)$ is a set of type F_σ.*

Now let us consider the question of whether the property of measurability is invariant under a continuous mapping. To answer this question, the following definition, due to N. N. Luzin, will be helpful.

DEFINITION. *If the map $f(e)$ of every set e of measure zero is also a set of measure zero, then the function $f(x)$ is said to have the property (N).*

THEOREM 2. *In order that the map $f(E)$ of every measurable set E be a measurable set, it is necessary and sufficient that the function $f(x)$ possess the property (N).*

Proof. Let $f(x)$ possess the property (N) and let E be a measurable set lying in $[a, b]$. Then

$$E = A + e,$$

3. CONTINUOUS MAPPINGS

where A is a set of type F_σ and e is a set of measure 0.[3] Therefore,
$$f(E) = f(A) + f(e)$$
and consequently the set $f(E)$ is measurable, being the sum of an F_σ and a set of measure zero.

Suppose now that the function $f(x)$ does not possess the property (N). Then we can find a set e_0 of measure zero lying in the closed interval $[a, b]$ such that the outer measure of its image under f is positive:
$$m^*f(e_0) > 0.$$
Under these conditions, the set $f(e_0)$ contains a non-measurable subset B.[4] For every $y \in B$, consider an element $x \in e_0$ such that $f(x) = y$. The set of all such points x comprises a set A contained in e_0 such that $f(A) = B$. It is obvious that A is measurable. In fact, A is a subset of e_0 and hence has outer measure zero. At the same time, $f(A) = B$ is non-measurable, so that the function $f(x)$ carries at least one measurable set into a non-measurable set.

THEOREM 3. *All absolutely continuous functions possess property* (N).

Proof. Let the function $f(x)$ be absolutely continuous and let E be a set of measure zero. We shall show that
$$mf(E) = 0.$$
We first suppose that the points a and b do not belong to E, so that
$$E \subset (a, b).$$
Taking an arbitrary $\varepsilon > 0$, let $\delta > 0$ have the property that for an arbitrary finite or denumerable system of pairwise disjoint intervals $\{(a_k, b_k)\}$, the sum of the lengths of which is less than δ, the inequality
$$\sum_k (M_k - m_k) < \varepsilon$$
holds, where as usual
$$m_k = \min\{f(x)\}, \quad M_k = \max\{f(x)\} \qquad (x \in [a_k, b_k]).$$
(See remarks following the definition in §1.) Since $mE = 0$, there exists a bounded open set G such that
$$E \subset G, \quad mG < \delta.$$
We may suppose that $G \subset (a, b)$, since E by hypothesis is contained in the open interval (a, b). Now, G is the sum of its component intervals (a_k, b_k), the sum of the lengths of which is less than δ. This implies that
$$f(E) \subset f(G) = \sum_k f[(a_k, b_k)] \subset \sum_k f([a_k, b_k]),$$

---

[3] To show this, let F_n be a closed subset of E such that $mF_n > mE - \frac{1}{n}$, for $n = 1, 2, 3, \ldots$. Then set
$$A = \sum_{n=1}^{\infty} F_n.$$

[4] If $f(e_0)$ is non-measurable, set $B = f(e_0)$; otherwise apply the statement proved at the end of §6, Chapt. III.

250   IX. Absolutely Continuous Functions. The Indefinite Lebesgue Integral

Therefore
$$m^*f(E) \leq \sum_k m^*f([a_k, b_k]).$$

It is also clear that
$$f([a_k, b_k]) = [m_k, M_k]$$
and consequently
$$m^*f(E) \leq \sum_k (M_k - m_k) < \varepsilon.$$

It follows from this that $mf(E) = 0$, since ε is arbitrary.

To establish the general case, it suffices to note that omitting the points a and b from the set E implies the removal from the set $f(E)$ of not more than two points, $f(a)$ and $f(b)$; this obviously has no effect on the measure of the set $f(E)$.

COROLLARY. *An absolutely continuous function maps measurable sets into measurable sets.*

We have now proved that every absolutely continuous function has finite variation and possesses the property (N). It turns out that these two properties characterize the class of absolutely continuous functions.

THEOREM 4 (S. BANACH AND M. A. ZARECKI). *If $f(x)$ is a continuous function of finite variation possessing the property (N), then it is absolutely continuous.*

Proof. Suppose that $f(x)$ is *not* absolutely continuous. Then there exists a positive number ε_0 such that for every $\delta > 0$, there exists a family of pairwise disjoint open intervals $\{(a_k, b_k)\}_{k=1}^n$ for which
$$\sum_{k=1}^n (b_k - a_k) < \delta$$
and having the additional property that
$$\sum_{k=1}^n (M_k - m_k) \geq \varepsilon_0.$$

Let
$$\sum_{i=1}^\infty \delta_i$$
be a convergent series of positive terms, and for every δ_i, let $(a_k^{(i)}, b_k^{(i)})$ $(k = 1, 2, \ldots, n_i)$ be a collection of pairwise disjoint open intervals for which
$$\sum_{k=1}^{n_i} (b_k^{(i)} - a_k^{(i)}) < \delta_i,$$
and
$$\sum_{k=1}^{n_i} (M_k^{(i)} - m_k^{(i)}) \geq \varepsilon_0.$$

[As usual, $M_k^{(i)}$ and $m_k^{(i)}$ are the maximum and minimum values, respectively, of the function $f(x)$ in the interval $(a_k^{(i)}, b_k^{(i)})$.]

Set
$$E_i = \sum_{k=1}^{n_i} (a_k^{(i)}, b_k^{(i)}), \qquad A = \prod_{n=1}^\infty \sum_{i=n}^\infty E_i.$$

3. Continuous Mappings

It is easy to see that $mA = 0$; it then follows from our hypotheses that

$$mf(A) = 0. \tag{1}$$

We next define functions $L_k^{(i)}(y)$ $[k = 1, 2, \ldots, n_i; \ i = 1, 2, 3, \ldots]$ by the following rule. $L_k^{(i)}(y) = 1$ if there is at least one x in the open interval $(a_k^{(i)}, b_k^{(i)})$ for which

$$f(x) = y. \tag{2}$$

Otherwise, $L_k^{(i)}(y) = 0$. Plainly $L_k^{(i)}(y) = 1$ for all y in the open interval $(m_k^{(i)}, M_k^{(i)})$ and $L_k^{(i)}(y) = 0$ for y not in the closed interval $[m_k^{(i)}, M_k^{(i)}]$. Therefore

$$\int_m^M L_k^{(i)}(y) \, dy = M_k^{(i)} - m_k^{(i)}. \tag{3}$$

Let

$$N_i(y) = \sum_{k=1}^{n_i} L_k^{(i)}(y).$$

It is clear that $N_i(y)$ is the number of those intervals $(a_k^{(i)}, b_k^{(i)})$ containing at least one x satisfying equation (2). Hence

$$N_i(y) \leq N(y), \tag{4}$$

where $N(y)$ is the Banach indicatrix of the function $f(x)$. Because of (3),

$$\int_m^M N_i(y) \, dy \geq \varepsilon_0. \tag{5}$$

To complete the proof, we shall show that for almost all y in $[m, M]$, the equality

$$\lim_{i \to \infty} N_i(y) = 0 \tag{6}$$

holds. Since the Banach indicatrix $N(y)$ is summable, it will follow from (4) and (6) and Lebesgue's convergence theorem that

$$\lim_{i \to \infty} \int_m^M N_i(y) \, dy = 0,$$

this contradicts inequality (5). Let B be the set of y for which (6) does not hold, and let C be the set of y for which $N(y) = \infty$. Since $N(y)$ is a summable function, $mC = 0$, and to prove the theorem it is sufficient to verify that

$$B - C \subset f(A). \tag{7}$$

Let $y_0 \in B - C$. Since the functions $N_i(y)$ assume only non-negative integral values, there is a sequence $\{i_r\}$ of natural numbers such that

$$N_{i_r}(y_0) \geq 1 \qquad (r = 1, 2, 3, \ldots).$$

For every r, accordingly, there exists a point x_{i_r} such that

$$f(x_{i_r}) = y_0, \qquad x_{i_r} \in E_{i_r}.$$

Since $N(y_0) < +\infty$, there are only a finite number of distinct points among the points x_{i_r}. Hence, one of them, say x_0, occurs an infinite number of times in the sequence

$\{x_{i_r}\}$. The point x_0 belongs to the infinite number of the sets E_i, and clearly
$$f(x_0) = y_0.$$
It is then clear that $x_0 \in A$ and that $f(x_0) = y_0 \in f(A)$. This verifies the inclusion (7) and completes the proof.

THEOREM 5 (G. M. FICHTENHOLZ). *Let $F(y)$ and $f(x)$ be two absolutely continuous functions such that the values of $f(x)$ all lie in the closed interval on which $F(y)$ is defined. The composite function $F[f(x)]$ is absolutely continuous if and only if it has finite variation.*[6]

Proof. The necessity of the stated condition is obvious. To prove its sufficiency, we need only note that if f and F both possess property (N), then the composite function $F[f(x)]$ also possesses property (N). Then apply Theorem 4.

§4. THE INDEFINITE LEBESGUE INTEGRAL

Let $f(t)$ be a summable function defined on the closed interval $[a, b]$. The function
$$\Phi(x) = C + \int_a^x f(t)\,dt$$
is called an indefinite Lebesgue integral of the function $f(t)$, for every choice of the constant C. The term 'indefinite' refers to the variable upper limit of the integral.

THEOREM 1. *The indefinite integral $\Phi(x)$ is an absolutely continuous function.*

Proof. For every $\varepsilon > 0$, there exists, in view of Theorem 8, §2, Chapt. VI, a $\delta > 0$ such that for every measurable set e of measure $me < \delta$, the inequality
$$\left| \int_e f(t)\,dt \right| < \varepsilon$$
holds. In particular, if the sum of the lengths of a finite system (a_k, b_k) of pairwise disjoint open intervals is less than δ, then
$$\left| \sum_{k=1}^n \int_{a_k}^{b_k} f(t)\,dt \right| < \varepsilon.$$
Since
$$\int_{a_k}^{b_k} f(t)\,dt = \Phi(b_k) - \Phi(a_k),$$
we infer
$$\left| \sum_{k=1}^n \{\Phi(b_k) - \Phi(a_k)\} \right| < \varepsilon.$$
That is, $\Phi(x)$ is absolutely continuous.

The preceding theorem, together with the corollary to Theorem 1, §2, implies that $\Phi(x)$ has a finite derivative almost everywhere, which is itself a summable function of x. This derivative can be identified completely, as the following theorem shows.

---

[6] This theorem was discovered by G. M. Fichtenholz in 1922. In 1925, a new proof was given by M. A. Zarecki, who at the same time obtained Theorem 4 above. Theorem 4 was also proved by S. Banach.

4. The Indefinite Lebesgue Integral

Theorem 2. *Let $f(x)$ be a summable function on $[a, b]$. The derivative $\Phi'(x)$ of the indefinite integral*

$$\Phi(x) = \int_a^x f(t)\, dt$$

is equal to the function $f(x)$ almost everywhere on $[a, b]$.

Proof. Let p and q be two real numbers such that $p < q$. Let $E_{p,q}$ be the set of those points of $[a, b]$ where the function $\Phi(x)$ is differentiable and where its derivative $\Phi'(x)$ satisfies the inequalities

$$\Phi'(x) > q > p > f(x).$$

It is clear that the set $E_{p,q}$ is measurable. Our first problem is to prove that

$$mE_{p,q} = 0. \tag{1}$$

For this purpose, let ε be an arbitrary positive number, and let $\delta > 0$ have the properties that $\delta > \varepsilon$ and that

$$\left| \int_e f(t)\, dt \right| < \varepsilon,$$

whenever $me < \delta$. Let G^7 be an open set such that $G \subset [a, b]$ and

$$G \supset E_{p,q}, \quad mG < mE_{p,q} + \delta.$$

If $x \in E_{p,q}$, then

$$\frac{\Phi(x+h) - \Phi(x)}{h} > q \tag{2}$$

for all sufficiently small $h > 0$. It is clear that the set $E_{p,q}$ is covered by the closed intervals $[x, x + h]$ [for positive h satisfying condition (2)] in the sense of Vitali. We may suppose that all of the intervals $[x, x + h]$ are contained in G. By Vitali's Theorem, there exists a sequence

$$[x_1, x_1 + h_1], \quad [x_2, x_2 + h_2], \ldots,$$

of these intervals which are pairwise disjoint and for which

$$m\left\{ E - \sum_{k=1}^{\infty} [x_k, x_k + h_k] \right\} = 0.$$

Because of (2), we have

$$\frac{1}{h_k} \int_{x_k}^{x_k + h_k} f(t)\, dt > q.$$

Set $S = \sum_{k=1}^{\infty} [x_k, x_k + h_k]$. Then the last inequality implies that

$$\int_S f(t)\, dt > q \cdot mS,$$

or, equivalently,

$$\int_S f(t)\, dt > q\, [mE_{p,q} + \theta\varepsilon] \qquad (0 \leq \theta \leq 1). \tag{3}$$

---

[7] We may suppose that the points a and b are not in $E_{p,q}$.

On the other hand, we have $S \subset G$, and therefore
$$S - E_{p,q} \subset G - E_{p,q}.$$
Hence $m[S - E_{p,q}] < \delta$, and
$$\int_{S - E_{p,q}} f(t)\,dt < \varepsilon.$$
Consequently,[8]
$$\int_S f(t)\,dt < \int_{E_{p,q}} f(t)\,dt + \varepsilon. \tag{4}$$

On the set $E_{p,q}$, we have $f(t) < p$ and hence
$$\int_{E_{p,q}} f(t)\,dt \leqslant p \cdot mE_{p,q}. \tag{5}$$
Relations (3), (4) and (5) imply that
$$q[mE_{p,q} + \theta\varepsilon] < pmE_{p,q} + \varepsilon,$$
from which it follows that
$$qmE_{p,q} \leqslant pmE_{p,q},$$
since ε is arbitrary. This is possible only if
$$mE_{p,q} = 0.$$
The equality (1) is thus established.

Now let E be the set of those points of $[a, b]$ at which the function $\Phi(x)$ is differentiable and
$$\Phi'(x) > f(x).$$
Then
$$E = \sum_{(p,\,q)} E_{p,\,q},$$
where the summation runs through all pairs (p, q) of rational numbers for which $p < q$. By virtue of (1), we have
$$mE = 0.$$
In other words, if A is the set of points at which the derivative $\Phi'(x)$ exists, then
$$\Phi'(x) \leqslant f(x) \tag{6}$$
almost everywhere on A.

Finally, set
$$g(x) = -f(x), \quad \Gamma(x) = \int_a^x g(t)\,dt.$$
It is easy to see that $\Gamma(x) = -\Phi(x)$, so that $\Gamma'(x)$ exists at all points of A.

---

[8] Note that $m(E_{p,q} - S) = 0$. Therefore
$$\int_{E_{p,q}} f\,dt = \int_{SE_{p,q}} f\,dt.$$

4. The Indefinite Lebesgue Integral

Applying the assertion just proved to the function $\Gamma(x)$, we find that
$$\Gamma'(x) \leqslant g(x),$$
almost everywhere on A, or, equivalently, that
$$\Phi'(x) \geqslant f(x) \qquad (7)$$
for almost all $x \in A$. From (6) and (7), it follows that
$$\Phi'(x) = f(x)$$
for almost all $x \in A$, and hence for almost all $x \in [a, b]$. This completes the proof.

Theorem 3. *Every absolutely continuous function is an indefinite integral of its own derivative.*

Proof. Let $F(x)$ be an absolutely continuous function. Its derivative $F'(x)$ exists almost everywhere and is summable.

Write
$$\Phi(x) = F(a) + \int_a^x F'(t)\,dt.$$

The function $\Phi(x)$ is also absolutely continuous and, as proved in the preceding theorem
$$\Phi'(x) = F'(x)$$
almost everywhere. In view of the corollary of Theorem 2, §2, we infer that the difference $F(x) - \Phi(x)$ is constant; since this difference is 0 for $x = a$, the functions $F(x)$ and $\Phi(x)$ must be identical.

Theorem 2 can be considerably sharpened. We first state a definition.

Definition. If
$$\lim_{h \to 0} \frac{1}{h} \int_x^{x+h} |f(t) - f(x)|\,dt = 0$$
at the point x, the point x is said to be a *Lebesgue point* of the function $f(t)$.

Theorem 4. *Let x be a Lebesgue point of the function $f(t)$. The indefinite integral $\Phi(x) = \int_a^x f(t)\,dt$ is differentiable at the point x, and $\Phi'(x) = f(x)$.*

Proof. It is easy to show that
$$\frac{\Phi(x+h) - \Phi(x)}{h} - f(x) = \frac{1}{h} \int_x^{x+h} \{f(t) - f(x)\}\,dt;$$
therefore
$$\left| \frac{\Phi(x+h) - \Phi(x)}{h} - f(x) \right| \leqslant \frac{1}{h} \int_x^{x+h} |f(t) - f(x)|\,dt,$$
and this proves the theorem. We note that the converse statement is not true, in general.

Theorem 5. *If the function $f(x)$ is summable on $[a, b]$, then almost every point of $[a, b]$ is a Lebesgue point of $f(x)$.*

Proof. Let r be a rational number. The function $|f(t) - r|$ is summable on $[a, b]$ and hence for almost all points $x \in [a, b]$, we have
$$\lim_{h \to 0} \frac{1}{h} \int_x^{x+h} |f(t) - r|\,dt = |f(x) - r|. \qquad (8)$$

256 IX. Absolutely Continuous Functions. The Indefinite Lebesgue Integral

Let $E(r)$ be the set of those points of $[a, b]$ at which (8) does not hold. It is clear that $mE(r) = 0$. We enumerate all rational numbers as a sequence $r_1, r_2, r_3, \ldots$, and put

$$E = \sum_{n=1}^{\infty} E(r_n) + E(|f| = +\infty).$$

Then $mE = 0$, and it is sufficient to prove that all points of the set $[a, b] - E$ are Lebesgue points of the function $f(t)$. Let $x_0 \in [a, b] - E$, and let ε be an arbitrary positive number. Let r_n be a rational number such that

$$|f(x_0) - r_n| < \frac{\varepsilon}{3}.$$

Then it is clear that

$$\Big| |f(t) - r_n| - |f(t) - f(x_0)| \Big| < \frac{\varepsilon}{3}$$

and we have

$$\left| \frac{1}{h} \int_{x_0}^{x_0+h} |f(t) - r_n|\, dt - \frac{1}{h} \int_{x_0}^{x_0+h} |f(t) - f(x_0)|\, dt \right| \leq \frac{\varepsilon}{3}.$$

Since $x_0 \overline{\in} E$, we have

$$\left| \frac{1}{h} \int_{x_0}^{x_0+h} |f(t) - r_n|\, dt - |f(x_0) - r_n| \right| < \frac{\varepsilon}{3},$$

for $|h| < \delta(\varepsilon)$, i.e.,

$$\frac{1}{h} \int_{x_0}^{x_0+h} |f(t) - r_n|\, dt < \frac{2}{3}\varepsilon,$$

and hence for $h < \delta(\varepsilon)$,

$$\frac{1}{h} \int_{x_0}^{x_0+h} |f(t) - f(x_0)|\, dt < \varepsilon.$$

THEOREM 6. *Every point of continuity of a summable function $f(t)$ is a Lebesgue point of $f(t)$.*

Proof. Let $f(t)$ be continuous at the point x. Then to every $\varepsilon > 0$, there corresponds a $\delta > 0$ such that

$$|f(t) - f(x)| < \varepsilon$$

for $|t - x| < \delta$. For $|h| < \delta$, we have

$$\frac{1}{h} \int_{x}^{x+h} |f(t) - f(x)|\, dt < \varepsilon,$$

and the theorem follows.

Theorems 1 and 3 imply that for a function $\Phi(x)$ to be the indefinite integral of a summable function, it is necessary and sufficient that it be absolutely continuous. We may also try to characterize functions which are indefinite integrals of functions in L_p for $p > 1$.

4. The Indefinite Lebesgue Integral

Theorem 7 (F. Riesz). *A function $F(x)$ ($a \leqslant x \leqslant b$) can be represented in the form*

$$F(x) = C + \int_a^x f(t)\, dt, \tag{9}$$

where $f(t) \in L_p$ ($p > 1$) if and only if, for every subdivision of $[a, b]$ by points

$$a = x_0 < x_1 < x_2 < \ldots < x_n = b$$

the inequality

$$\sum_{k=0}^{n-1} \frac{|F(x_{k+1}) - F(x_k)|^p}{(x_{k+1} - x_k)^{p-1}} \leqslant K \tag{10}$$

holds, where K is independent of the manner of subdividing $[a, b]$. [9]

Proof. The necessity of condition (10) is almost obvious. In fact, by Hölder's inequality [Chapter 7, §6, Formula (1)],

$$|F(x_{k+1}) - F(x_k)| = \left| \int_{x_k}^{x_{k+1}} f(t)\, dt \right| \leqslant \sqrt[q]{x_{k+1} - x_k} \cdot \sqrt[p]{\int_{x_k}^{x_{k+1}} |f(t)|^p\, dt},$$

where $q = \dfrac{p}{p-1}$. Hence

$$\frac{|F(x_{k+1}) - F(x_k)|^p}{(x_{k+1} - x_k)^{p-1}} \leqslant \int_{x_k}^{x_{k+1}} |f(t)|^p\, dt,$$

and (10) holds, where the number K is the integral $\int_a^b |f(t)|^p dt$.

The sufficiency of condition (10) is more difficult to prove. First of all, we note that condition (10) is only strengthened if some of its left-hand terms are omitted. Hence for an arbitrary finite system of pairwise disjoint open intervals (a_k, b_k) ($k = 1, 2, \ldots, n$) contained in $[a, b]$, we have

$$\sum_{k=1}^{n} \frac{|F(b_k) - F(a_k)|^p}{(b_k - a_k)^{p-1}} \leqslant K.$$

By Hölder's inequality for sums [Chapt. VII, §6, Formula (8)], we have

$$\sum_{k=1}^n |F(b_k) - F(a_k)| = \sum_{k=1}^n \frac{|F(b_k) - F(a_k)|}{(b_k - a_k)^{\frac{p-1}{p}}} (b_k - a_k)^{\frac{1}{q}} \leqslant$$

$$\leqslant \sqrt[p]{\sum_{k=1}^n \frac{|F(b_k) - F(a_k)|^p}{(b_k - a_k)^{p-1}}} \cdot \sqrt[q]{\sum_{k=1}^n (b_k - a_k)}.$$

---

[9] If $p = 1$, (10) is the condition that $F(x)$ have finite variation. Hence, this condition remains necessary but ceases to be sufficient for $F(x)$ to be representable in the form (9) for $f(t) \in L$.

258   IX. ABSOLUTELY CONTINUOUS FUNCTIONS. THE INDEFINITE LEBESGUE INTEGRAL

Therefore

$$\sum_{k=1}^{n} |F(b_k) - F(a_k)| \leqslant \sqrt[p]{K} \cdot \sqrt[q]{\sum_{k=1}^{n} (b_k - a_k)},$$

and accordingly, $F(x)$ is absolutely continuous. Therefore $F(x)$ is representable in the form (9) with $f(t) \in L$. It remains to show that $f(t) \in L_p$. For this purpose, we subdivide $[a, b]$ into n equal parts by the points $x_k^{(n)} = a + \frac{k}{n}(b-a)$ $(k = 0, 1, \ldots, n)$, and introduce the function $f_n(t)$, setting

$$f_n(t) = \frac{F(x_{k+1}^{(n)}) - F(x_k^{(n)})}{x_{k+1}^{(n)} - x_k^{(n)}}, \quad \text{for} \quad x_k^{(n)} < t < x_{k+1}^{(n)}.$$

At the points of subdivision, we set $f_n(x_k^{(n)}) = 0$.

It is easy to see that

$$\lim_{n \to \infty} f_n(t) = f(t)$$

almost everywhere. [The equality just stated may fail to be true at points of the form $x_k^{(n)}$ and at points where $F'(x) \neq f(x)$.] In fact, if x is not a point of subdivision and if $F'(x)$ exists and is finite, then x lies in some open interval $(x_{k_n}^{(n)}, x_{k_n+1}^{(n)})$ for all natural numbers n. Since $x_{k_n+1}^{(n)} - x_{k_n}^{(n)} = \frac{b-a}{n} \to 0$ as $n \to \infty$, it follows that each of the expressions

$$\frac{F(x_{k_n+1}^{(n)}) - F(x)}{x_{k_n+1}^{(n)} - x}, \quad \frac{F(x) - F(x_{k_n}^{(n)})}{x - x_{k_n}^{(n)}} \qquad (*)$$

converges to $F'(x)$ as $n \to \infty$. However, $f_n(x) = \frac{F(x_{k_n+1}^{(n)}) - F(x_{k_n}^{(n)})}{x_{k_n+1}^{(n)} - x_{k_n}^{(n)}}$, and accordingly the number $f_n(x)$ lies between the two numbers $(*)$. Accordingly $\lim_{n \to \infty} f_n(x) = F'(x)$. Fatou's theorem now implies that

$$\int_a^b |f(t)|^p \, dt \leqslant \sup \left\{ \int_a^b |f_n(t)|^p \, dt \right\}.$$

We obtain an upper bound for the right-hand side of the preceding expression, as follows:

$$\int_a^b |f_n(t)|^p \, dt = \sum_{k=0}^{n-1} \int_{x_k^{(n)}}^{x_{k+1}^{(n)}} |f_n(t)|^p \, dt = \sum_{k=0}^{n-1} \frac{|F(x_{k+1}^{(n)}) - F(x_k^{(n)})|^p}{(x_{k+1}^{(n)} - x_k^{(n)})^{p-1}} \leqslant K.$$

Therefore

$$\int_a^b |f(t)|^p \, dt < +\infty.$$

This completes the proof.

In conclusion, we compute the total variation of an indefinite integral.

4. The Indefinite Lebesgue Integral

THEOREM 8. *Let $f(t)$ be a summable function defined on $[a, b]$. If*

$$F(x) = \int_a^x f(t)\, dt, \quad \text{then} \quad \overset{b}{\underset{a}{V}}(F) = \int_a^b |f(t)|\, dt,$$

i.e., the total variation of an absolutely continuous function is the integral of the absolute value of its derivative.

Proof. If $x_0 = a < x_1 < x_2 < \ldots < x_n = b$ is any subdivision of $[a, b]$, then

$$\sum_{k=0}^{n-1} |F(x_{k+1}) - F(x_k)| = \sum_{k=0}^{n-1} \left| \int_{x_k}^{x_{k+1}} f(t)\, dt \right| \leqslant \sum_{k=0}^{n-1} \int_{x_k}^{x_{k+1}} |f(t)|\, dt = \int_b^b |f(t)|\, dt.$$

Accordingly,

$$\overset{b}{\underset{a}{V}}(F) \leqslant \int_a^b |f(t)|\, dt.$$

In order to establish the reverse inequality, we set $(a, b) = E$ and let

$$P = E(f \geqslant 0), \quad N = E(f < 0).$$

Then

$$\int_a^b |f(t)|\, dt = \int_P f(t)\, dt - \int_N f(t)\, dt.$$

Let ε be an arbitrary positive number. Since the integral is absolutely continuous, there exists a $\delta > 0$ such that for every measurable set $e \subset [a, b]$ with measure $me > \delta$, the inequality

$$\int_e |f(t)|\, dt < \varepsilon$$

holds.

Let $F(P)$ and $F(N)$ be closed sets, contained in P and N respectively, such that

$$m[P - F(P)] < \delta, \quad m[N - F(N)] < \delta.$$

Then

$$\int_a^b |f(t)|\, dt < \int_{F(P)} f(t)\, dt - \int_{F(N)} f(t)\, dt + 2\varepsilon.$$

In accordance with the separation theorem for disjoint closed sets (Theorem 2, §4, Chapter II), one can find open sets $\Gamma(P)$ and $\Gamma(N)$ such that

$$\Gamma(P) \supset F(P), \quad \Gamma(N) \supset F(N), \quad \Gamma(P) \cdot \Gamma(N) = 0,$$

where the sets $\Gamma(P)$ and $\Gamma(N)$ are contained in (a, b). Furthermore, there exist bounded open sets $A(P)$ and $A(N)$, containing $F(P)$ and $F(N)$, respectively, and such that $m[A(P) - F(P)] < \delta$, $m[A(N) - F(N)] < \delta$. Now set

$$G(P) = A(P) \cdot \Gamma(P), \quad G(N) = A(N) \cdot \Gamma(N).$$

$G(P)$ and $G(N)$ are disjoint open sets contained in (a, b), containing $F(P)$ and $F(N)$ respectively and such that $m[G(P) - F(P)] < \delta$, $m[G(N) - F(N)] < \delta$. Hence

$$\int_a^b |f(t)|\,dt < \int_{G(P)} f(t)\,dt - \int_{G(N)} f(t)\,dt + 4\varepsilon.$$

The set $G(P)$ is the sum of its component intervals. Taking a sufficiently large finite number of these intervals, we obtain a set $B(P)$ whose measure differs from that of $G(P)$ by less than δ. Then we have

$$\int_{G(P)} f(t)\,dt - \int_{B(P)} f(t)\,dt < \varepsilon.$$

Let $B(P) = \sum_{k=1}^{n} (\lambda_k, \mu_k)$. Then

$$\int_{B(P)} f(t)\,dt = \sum_{k=1}^{n} \int_{\lambda_k}^{\mu_k} f(t)\,dt = \sum_{k=1}^{n} [F(\mu_k) - F(\lambda_k)].$$

Hence

$$\int_{G(P)} f(t)\,dt < \sum_{k=1}^{n} [F(\mu_k) - F(\lambda_k)] + \varepsilon.$$

In an analogous manner we can find a finite number of component intervals of the set $G(N)$, say $(\sigma_1, \tau_1), (\sigma_2, \tau_2), \ldots, (\sigma_m, \tau_m)$, such that

$$\int_{G(N)} f(t)\,dt > \sum_{i=1}^{m} [F(\tau_i) - F(\sigma_i)] - \varepsilon.$$

Combining all these statements, we see that

$$\int_a^b |f(t)|\,dt < \sum_{k=1}^{n} [F(\mu_k) - F(\lambda_k)] - \sum_{i=1}^{m} [F(\tau_i) - F(\sigma_i)] + 6\varepsilon,$$

and necessarily

$$\int_a^b |f(t)|\,dt < \sum_{k=1}^{n} |F(\mu_k) - F(\lambda_k)| + \sum_{i=1}^{m} |F(\tau_i) - F(\sigma_i)| + 6\varepsilon.$$

Since the intervals (λ_k, μ_k) and (σ_i, τ_i) are pairwise disjoint, it follows that

$$\sum_{k=1}^{n} |F(\mu_k) - F(\lambda_k)| + \sum_{i=1}^{m} |F(\tau_i) - F(\sigma_i)| \leq \overset{b}{\underset{a}{V}}(F).$$

Thus

$$\int_a^b |f(t)|\,dt < \overset{b}{\underset{a}{V}}(F) + 6\varepsilon.$$

Since ε is arbitrary, the theorem follows.

§ 5. POINTS OF DENSITY. APPROXIMATE CONTINUITY

Let E be a given measurable set. Taking an arbitrary point x_0 and a number $h > 0$, we set

$$E(x_0, h) = E \cdot [x_0 - h, x_0 + h].$$

5. POINTS OF DENSITY. APPROXIMATE CONTINUITY

This set also is measurable. Let us consider the number

$$\frac{mE(x_0, h)}{2h}. \tag{1}$$

It is natural to consider this the "mean density" of the set E on the closed interval

$$[x_0 - h, x_0 + h].$$

DEFINITION 1. *The limit of* (1) *as* $h \to 0$ *is called the* density *of the set* E *at the point* x_0 *and is denoted by*

$$D_{x_0} E.$$

If $D_{x_0} E = 1$, *then* x_0 *is a* point of density *of the set* E, *and if* $D_{x_0} E = 0$, x_0 *is a* point of rarefaction *of* E.

In stating this definition, we do not assume that $x_0 \in E$. Furthermore, a measurable set should not be expected to have a defined density at every point of the line.

However, the following theorem is valid.

THEOREM 1. *Almost all points of a measurable set* E *are points of density of* E.

Proof. Let the set E be measurable. Take an arbitrary closed interval $[\alpha, \beta]$ containing the set E, and let $a = \alpha - 1$, $b = \beta + 1$. Then, for $x \in E$ and $h \leq 1$, it is certain that the closed interval $[x - h, x + h]$ is contained in $[a, b]$. If the contrary is not specified, we shall suppose that $h \leq 1$.

Consider the characteristic function $\varphi(x)$ of the set E,

$$\varphi(x) = \begin{cases} 1 & \text{if } x \in E \\ 0 & \text{if } x \bar{\in} E \end{cases}$$

taken only on $[a, b]$. This function is measurable and bounded. Let

$$\Phi(x) = \int_a^x \varphi(t) \, dt.$$

Then $\Phi'(x) = \varphi(x)$ almost everywhere on $[a, b]$, by Theorem 2, §4, and in particular

$$\Phi'(x) = 1 \tag{2}$$

almost everywhere on E.

We shall show that points for which (2) holds are points of density of the set E. In fact at every such point,

$$\lim_{h \to 0} \frac{\Phi(x+h) - \Phi(x)}{h} = \lim_{h \to 0} \frac{\Phi(x) - \Phi(x-h)}{h} = 1,$$

and hence

$$\lim_{h \to 0} \frac{\Phi(x+h) - \Phi(x-h)}{2h} = 1.$$

But

$$\Phi(x+h) - \Phi(x-h) = \int_{x-h}^{x+h} \varphi(t) \, dt = mE(x, h),$$

so that

$$D_x E = \lim_{h \to 0} \frac{mE(x, h)}{2h} = 1,$$

as was to be proved.

IX. ABSOLUTELY CONTINUOUS FUNCTIONS. THE INDEFINITE LEBESGUE INTEGRAL

There is an important generalization of the concept of a continuous function which is closely connected with the concept of points of density.

DEFINITION 2. *Let $f(x)$ be a function $f(x)$ defined on the closed interval $[a, b]$, and let $x_0 \in [a, b]$. If there exists a measurable subset E of $[a, b]$ having the point x_0 as a point of density[11], such that $f(x)$ is continuous at the point x_0 with respect to E, then $f(x)$ is said to be approximately continuous at x_0.*

It is clear that every point of continuity of the function is a point of approximate continuity of the function. Of course, a measurable function may have no points of continuity at all. Such a function, for example, is the function equal to 0 at irrational points and 1 at rational points.

On the other hand, the following theorem is true.

THEOREM 2. *If $f(x)$ is a measurable function defined on the closed interval $[a, b]$ and finite almost everywhere, then it is approximately continuous at almost all points of $[a, b]$.*

Proof. Let ε be an arbitrary positive number. Using Luzin's Theorem (Theorem 4, §5, Chapter IV), we find a continuous function $\varphi(x)$ such that

$$mE(f \neq \varphi) < \varepsilon.$$

Let A be the set of all points of density of the set $E(f = \varphi)$ which belong to $E(f = \varphi)$. By the preceding theorem,

$$mA = mE(f = \varphi) > b - a - \varepsilon.$$

If $x_0 \in A$, then $f(x)$ obviously is approximately continuous at this point, since we can take the set $E(f = \varphi)$ for the set E in definition 2. Hence, the set H of all the points of approximate continuity of $f(x)$ has interior measure

$$m_* H \geqslant mA > b - a - \varepsilon,$$

and, since ε is arbitrary,

$$m_* H \geqslant b - a.$$

Furthermore, $H \subset [a, b]$, so that

$$b - a \leqslant m_* H \leqslant m^* H \leqslant b - a.$$

H is therefore measurable, and $mH = b - a$, as was to be proved.

REMARK. The concept of density given above can be generalized. Namely, we may define the density of the set E at the point x_0 as the limit of the ratio

$$\frac{mE(x_0, h_1, h_2)}{h_1 + h_2},$$

as $h_1 > 0$ and $h_2 > 0$ tend to zero independently one of the other, where $E(x_0, h_1, h_2)$ is some subset of the set E contained in $[x_0 - h, x_0 + h]$. However, this generalization alters neither the set of points of rarefaction nor the set of points of density of the set E. In fact, let x_0 be a point of rarefaction of the set E in the sense of definition 1. Taking

---

[11] If $x_0 = a$, then instead of requiring the set E to have x_0 as a point of density, we must require the right-hand density of the set to be one at x_0; i.e.,

$$\lim_{h \to 0} \frac{m\{E \cdot [a, a + h]\}}{h} = 1.$$

For the point b, the definition of approximate continuity must be modified in an analogous way by using the left-hand density.

numbers $h_1 > 0$ and $h_2 > 0$, we let h be the greater of the two. Then
$$E(x_0, h_1, h_2) \subset E(x_0, h),$$
and hence
$$\frac{mE(x_0, h_1, h_2)}{h_1 + h_2} \leqslant 2 \cdot \frac{mE(x_0, h)}{2h}.$$

Since the right-hand member of this inequality tends to zero together with h, it follows that
$$\lim_{\substack{h_1 \to 0 \\ h_2 \to 0}} \frac{mE(x_0, h_1, h_2)}{h_1 + h_2} = 0$$

and x_0 is a point of rarefaction of the set E in the sense of the generalized definition. The converse statement is obvious. It is just for this reason that we gave the definition of density set forth in Definition 2. It is clear, for example, that the definition of a point of approximate continuity does not depend on the definition of density used in establishing it.

§ 6. SUPPLEMENT TO THE THEORY OF FUNCTIONS OF FINITE VARIATION AND STIELTJES INTEGRALS

Let $f(x)$ ($a \leqslant x \leqslant b$) be a continuous function of finite variation. Its derivative $f'(x)$ exists almost everywhere and is summable. We set

Then
$$\varphi(x) = f(a) + \int_a^x f'(t)\, dt, \quad r(x) = f(x) - \varphi(x).$$
$$f(x) = \varphi(x) + r(x),$$

where $\varphi(x)$ is an absolutely continuous function [with $\varphi(a) = f(a)$], and $r(x)$ is a continuous function of bounded variation, whose derivative obviously equals zero almost everywhere. It is clear that $r(x)$ vanishes only when $f(x)$ itself is absolutely continuous.

DEFINITION. A non-constant continuous function of finite variation whose derivative equals zero almost everywhere is called a *singular* function.

It is clear that a singular function cannot be absolutely continuous, for otherwise (Theorem 2, §2) it would be constant. An example of a singular function is the function $\Theta(x)$ constructed at the end of §2, Chapt. VIII.

THEOREM 1. *A continuous function $f(x)$ of finite variation can be uniquely represented in the form*
$$f(x) = \varphi(x) + r(x),$$

where $\varphi(x)$ is absolutely continuous, $\varphi(a) = f(a)$, and $r(x)$ is a singular function or zero.

Proof. The possibility of such a representation was established above. Let us demonstrate its uniqueness. If there are two such representations,
$$f(x) = \varphi(x) + r(x) = \varphi_1(x) + r_1(x),$$
we have
$$\varphi(x) - \varphi_1(x) = r_1(x) - r(x).$$

IX. ABSOLUTELY CONTINUOUS FUNCTIONS. THE INDEFINITE LEBESGUE INTEGRAL

Hence, the derivative of the difference $\varphi(x) - \varphi_1(x)$ is 0 almost everywhere, and since this difference is absolutely continuous, it is constant. But $\varphi(a) = \varphi_1(a) = f(a)$. This implies

$$\varphi(x) \equiv \varphi_1(x),$$

and therefore $r(x) \equiv r_1(x)$ as well.

THEOREM 2. *If $f(x)$ is an increasing function, then both of its components $\varphi(x)$ and $r(x)$ are increasing functions.*

Proof. It is obvious that $f'(x) \geq 0$ wherever this derivative exists. This implies that the function

$$\varphi(x) = \varphi(a) + \int_a^x f'(t)\,dt$$

is an increasing function. Furthermore, Theorem 5, §2, Chapt. VIII, implies that

$$\int_x^y f'(t)\,dt \leq f(y) - f(x) \qquad (y > x).$$

Therefore

$$\varphi(y) - \varphi(x) \leq f(y) - f(x),$$

or $r(x) \leq r(y)$.

COROLLARY. *A necessary and sufficient condition that an increasing continuous function $f(x)$ be absolutely continuous is that*

$$\int_a^b f'(x)\,dx = f(b) - f(a). \qquad (1)$$

The necessity of condition (1) is obvious. Conversely, suppose that $f(x)$ is not absolutely continuous and let $\varphi(x)$ and $r(x)$ be the absolutely continuous and singular components of $f(x)$. Then

$$f(b) - f(a) = \varphi(b) - \varphi(a) + r(b) - r(a),$$

or

$$f(b) - f(a) = \int_a^b f'(x)\,dx + r(b) - r(a). \qquad (2)$$

But $r(x)$ is an increasing, non-constant function. Therefore $r(b) > r(a)$, and (1) is not satisfied. This proves that the condition of the theorem is sufficient.

In §3, Chapt. VIII, we saw that every function of finite variation can be written as the sum of its saltus function and a continous function of finite variation. Combining this with Theorem 1, we see that every function of bounded variation can be written

$$f(x) = \varphi(x) + r(x) + s(x),$$

where $\varphi(x)$ is an absolutely continuous function, $r(x)$ is a singular function and $s(x)$ is a saltus function (some terms can be absent).*

In §7, Chapt. VIII we raised the question of evaluating the Stieltjes integral

$$\int_a^b f(x)\,dg(x)$$

in the case where $g(x)$ is continuous. We see that the case $g(x)$ absolutely continuous can be reduced to the computation of an ordinary Lebesgue integral.

---

*A similar resolution can be obtained for functions $f(x)$ of finite variation on $(-\infty, \infty)$.—E. H.

6. SUPPLEMENT TO THE THEORY OF FUNCTIONS OF FINITE VARIATION 265

THEOREM 3. *If $f(x)$ is continuous and $g(x)$ is absolutely continuous on $[a, b]$, then*

$$(S) \int_a^b f(x) \, dg(x) = (L) \int_a^b f(x) g'(x) \, dx.$$

Proof. It is obvious that both integrals exist. To show that they are equal, we evaluate the difference between the sum

$$\sigma = \sum_{k=0}^{n-1} f(\xi_k) [g(x_{k+1}) - g(x_k)]$$

and the integral

$$\int_a^b f(x) g'(x) \, dx.$$

Since

$$g(x_{k+1}) - g(x_k) = \int_{x_k}^{x_{k+1}} g'(x) \, dx,$$

we have

$$\sigma - \int_a^b f(x) g'(x) \, dx = \sum_{k=0}^{n-1} \int_{x_k}^{x_{k+1}} [f(\xi_k) - f(x)] g'(x) \, dx. \tag{3}$$

If the oscillation of the function $f(x)$ on $[x_k, x_{k+1}]$ is written as ω_k, then (3) implies that

$$\left| \sigma - \int_a^b f(x) g'(x) \, dx \right| \leq \sum_{k=0}^{n-1} \omega_k \int_{x_k}^{x_{k+1}} |g'(x)| \, dx \leq \alpha \int_a^b |g'(x)| \, dx,$$

where $\alpha = \max \{\omega_k\}$. If the lengths of the intervals $[x_k, x_{k+1}]$ tend to 0, then $\alpha \to 0$ also; therefore σ approaches the integral $\int_a^b f(x) g'(x) \, dx$. Since $\lim \sigma$ is the integral $\int_a^b f(x) \, dg(x)$, the theorem is proved.

We have shown that computing a Stieltjes integral $\int_a^b f(x) \, dg(x)$ involves only summation of an infinite series and evaluating an ordinary Lebesgue integral, unless the function $g(x)$ has a singular part.*

Certain properties of Lebesgue integrals can be established with the aid of Theorem 3, as the following example shows.

---

*The computation of $\int_a^b f(x) dg(x)$ when $g(x)$ is a singular function can be very complicated indeed, and can lead to very curious results. For example, one can show that

$$\int_0^1 \cos 2\pi x \, d\theta(x) = - \prod_{j=1}^{\infty} \cos\left(\frac{2\pi}{3^j}\right)$$

where $(\theta)(x)$ is the singular function defined in §3, Chapter VIII.—E. H.

266   IX. ABSOLUTELY CONTINUOUS FUNCTIONS. THE INDEFINITE LEBESGUE INTEGRAL

THEOREM 4 (INTEGRATION BY PARTS). *If $f(x)$ and $g(x)$ are absolutely continuous, then*
$$\int_a^b f(x) g'(x) \, dx + \int_a^b g(x) f'(x) \, dx = [f(x) g(x)]_a^b.$$

To prove this, it is sufficient to write the left-hand member in the form
$$\int_a^b f(x) \, dg(x) + \int_a^b g(x) \, df(x)$$
and apply Formula (1), §6, Chapt. VIII.

§ 7. RECONSTRUCTION OF THE PRIMITIVE FUNCTION

§5, Chapt. V, we solved the problem of reconstructing a continuous function $f(x)$ from its derivative $f'(x)$ if the latter exists everywhere and is bounded. Here we ask, does the equality
$$f(x) = f(a) + \int_a^x f'(t) \, dt \tag{1}$$
hold when $f'(x)$ exists everywhere but is not necessarily bounded? It is perfectly clear that this is so if $f(x)$ is absolutely continuous. In this case, it suffices to suppose only that $f'(x)$ exists almost everywhere, which is insufficient in general for (1) to hold even when $f(x)$ is an increasing continuous function whose derivative $f'(x)$ equals zero almost everywhere.[12] However, we will formulate conditions for the validity of equality (1) in terms, not of the function $f(x)$ itself but of its derivative $f'(x)$.

THEOREM 1. *If the derivative $f'(x)$ exists everywhere, is finite, and is summable, then (1) holds.*

The proof will be based on two lemmas.

LEMMA 1. *Let the function $\Phi(x)$ be defined and finite on $[a, b]$. If at every point of $[a, b]$ all derived numbers of $\Phi(x)$ are non-negative, then $\Phi(x)$ is an increasing function.*

Proof. Let ε be an arbitrary positive number, and let
$$\Phi_1(x) = \Phi(x) + \varepsilon x.$$
Assume that
$$\Phi_1(b) < \Phi_1(a). \tag{2}$$
Then, if $c = \dfrac{a+b}{2}$, at least one of the differences
$$\Phi_1(b) - \Phi_1(c), \quad \Phi_1(c) - \Phi_1(a)$$
is negative. Let $[a_1, b_1]$ be the interval $[a, c]$ or $[c, b]$ for which
$$\Phi_1(b_1) < \Phi_1(a_1),$$
and set $c_1 = \dfrac{a_1 + b_1}{2}$. At least one of the differences
$$\Phi_1(b_1) - \Phi_1(c_1), \quad \Phi_1(c_1) - \Phi_1(a_1)$$
is negative. Let $[a_2, b_2]$ be the interval $[a_1, c_1]$ or $[c_1, b_1]$ for which
$$\Phi_1(b_2) < \Phi_1(a_2).$$

---

[12] This is clear from the example given by the function $\Theta(x)$ (§2, Chapt. VIII).

7. RECONSTRUCTION OF THE PRIMITIVE FUNCTION

Continuing this process, we construct a sequence of nested closed intervals $\{[a_n, b_n]\}$ for which
$$\Phi_1(b_n) < \Phi_1(a_n).$$
Let x_0 be a point lying in all of the intervals $[a_n, b_n]$. Then, for each n, one of the differences
$$\Phi_1(b_n) - \Phi_1(x_0), \quad \Phi_1(x_0) - \Phi_1(a_n)$$
is negative. Put $h_n = b_n - x_0$ if $\Phi_1(b_n) < \Phi_1(x_0)$ and $h_n = a_n - x_0$ if $\Phi_1(b_n) \geq \Phi_1(x_0)$. It is clear that
$$\Delta_n = \frac{\Phi_1(x_0 + h_n) - \Phi_1(x_0)}{h_n} < 0.$$

Selecting a subsequence $\{\Delta_{n_k}\}$ having a (finite or infinite) limit, we obtain a derived number
$$D\Phi_1(x_0) \leq 0,$$
This inequality cannot hold, since
$$D\Phi_1(x) \geq \varepsilon.$$
for all points $x \in [a, b]$. Thus (2) is impossible. This implies that
$$\Phi_1(b) \geq \Phi_1(a),$$
or, equivalently,
$$\Phi(b) + \varepsilon b \geq \Phi(a) + \varepsilon a.$$
The number ε is arbitrary, and we infer that
$$\Phi(b) \geq \Phi(a).$$

This completes the proof, since we could have taken an arbitrary subinterval $[x, y]$ in place of $[a, b]$.

LEMMA 2. *Let $\varphi(x)$ be a function which is defined and finite throughout $[a, b]$. Suppose that all derived numbers of $\varphi(x)$ are non-negative at almost every point of $[a, b]$ and that no derived number of $\varphi(x)$ is equal to $-\infty$ at any point of $[a, b]$. Then $\varphi(x)$ is an increasing function.*

Proof. Let E be the set of points of $[a, b]$ where at least one derived number of $\varphi(x)$ is negative. By hypothesis,
$$mE = 0.$$
By Theorem 6, §2, Chapt. VIII, there exists a continuous increasing function $\sigma(x)$ such that
$$\sigma'(x) = +\infty.$$
at all points of the set E. Let
$$\Phi(x) = \varphi(x) + \varepsilon \sigma(x),$$
where ε is an arbitrary positive number. No derived number of $\Phi(x)$ is negative at any point of $[a, b]$. In fact, since $\sigma(x)$ is an increasing function,
$$\frac{\Phi(x+h) - \Phi(x)}{h} \geq \frac{\varphi(x+h) - \varphi(x)}{h},$$

and therefore
$$D\Phi(x) \geqslant 0$$
for $x \bar{\in} E$. If $x \in E$, then $\Phi'(x)$ exists and equals $+\infty$, since for $h_n \to 0$, the ratio
$$\frac{\varphi(x+h_n)-\varphi(x)}{h_n}$$
is bounded below (for otherwise there would be a derived number $D\varphi(x) = -\infty$), and $\sigma'(x) = +\infty$. We have accordingly,
$$D\Phi(x) \geqslant 0$$
everywhere. By the preceding lemma, $\Phi(x)$ is an increasing function, i.e.,
$$\Phi(x) \leqslant \Phi(y)$$
for $x < y$, or, in other terms,
$$\varphi(x) + \varepsilon\sigma(x) \leqslant \varphi(y) + \varepsilon\sigma(y).$$
Taking the limit as $\varepsilon \to 0$, we obtain
$$\varphi(x) \leqslant \varphi(y),$$
as was to be proved.

Proof of Theorem 1. We introduce the function $\varphi_n(x)$, by the definition
$$\varphi_n(x) = \begin{cases} f'(x) & \text{if } f'(x) \leqslant n, \\ n, & \text{if } f'(x) > n. \end{cases}$$
It is easy to see that
$$|\varphi_n(x)| \leqslant |f'(x)|, \tag{3}$$
and that $\varphi_n(x)$ is summable. Write
$$R_n(x) = f(x) - \int_a^x \varphi_n(t)\,dt.$$
We shall show that $R_n(x)$ is an increasing function. To do this, note first that
$$R_n'(x) = f'(x) - \varphi_n(x) \geqslant 0$$
almost everywhere, so that the set of points at which some derived number of the function $R_n(x)$ is negative has measure zero. On the other hand, we have $\varphi_n(x) \leqslant n$, so that
$$\frac{1}{h}\int_x^{x+h} \varphi_n(t)\,dt \leqslant n$$
and
$$\frac{R_n(x+h)-R_n(x)}{h} \geqslant \frac{f(x+h)-f(x)}{h} - n.$$
This makes it clear that no derived number of the function $R_n(x)$ is $-\infty$. Hence, by Lemma 2, $R_n(x)$ increases. This implies that
$$R_n(b) \geqslant R_n(a)$$

7. Reconstruction of the Primitive Function

or, in other terms,

$$f(b) - f(a) \geq \int_a^b \varphi_n(x)\, dx.$$

Since

$$\lim_{n \to \infty} \varphi_n(x) = f'(x),$$

it follows from (3) that

$$\lim_{n \to \infty} \int_a^b \varphi_n(x)\, dx = \int_a^b f'(x)\, dx.$$

Consequently,

$$f(b) - f(a) \geq \int_a^b f'(x)\, dx.$$

The same reasoning applied to the function $-f(x)$ yields the inequality

$$f(b) - f(a) \leq \int_a^b f'(x)\, dx.$$

Therefore

$$f(b) = f(a) + \int_a^b f'(x)\, dx,$$

which completes the proof, since any $x \in (a, b]$ can take the rôle of b.

In conclusion, we mention two examples.

I. Let the function

$$f(x) = x^{\frac{3}{2}} \sin \frac{1}{x} \qquad (x > 0)$$

$$f(0) = 0.$$

be defined on $[0, 1]$. This function has a finite derivative everywhere

$$f'(x) = \frac{3}{2} x^{\frac{1}{2}} \sin \frac{1}{x} - x^{-\frac{1}{2}} \cos \frac{1}{x} \qquad (x > 0)$$

$$f'(0) = 0.$$

This derivative is summable, since

$$|f'(x)| \leq \frac{3}{2} + \frac{1}{\sqrt{x}}.$$

Hence, the function $f(x)$ satisfies all conditions of Theorem 1. However, it is easy to see that $f'(x)$ is not bounded, so that the theorem of §5, Chapt. V is not applicable.

II. Let the function

$$f(x) = x^2 \cos \frac{\pi}{x^2} \qquad (x > 0)$$

$$f(0) = 0$$

be defined on $[0, 1]$. This function also has a finite derivative everywhere, but the derivative is not summable. In fact, if $0 < \alpha < \beta \leq 1$, the derivative $f'(x)$ is bounded on the segment $[\alpha, \beta]$ and we have

$$\int_\alpha^\beta f'(x)\,dx = \beta^2 \cos\frac{\pi}{\beta^2} - \alpha^2 \cos\frac{\pi}{\alpha^2}.$$

In particular, for

$$\alpha_n = \sqrt{\frac{2}{4n+1}}, \qquad \beta_n = \frac{1}{\sqrt{2n}}$$

we have

$$\int_{\alpha_n}^{\beta_n} f'(x)\,dx = \frac{1}{2n}.$$

The intervals $[\alpha_n, \beta_n]$ $(n = 1, 2, \ldots)$ are pairwise disjoint; writing $E = \sum_{n=1}^\infty [\alpha_n, \beta_n]$, we have $\int_E |f'(x)|\,dx \geq \sum_{n=1}^\infty \frac{1}{2n} = +\infty$ and $f'(x)$ is not summable. Hence Lebesgue integration does not furnish a complete solution to the problem of reconstructing a function from its derivative. A complete solution of the problem is given by the process of Perron-Denjoy integration, which generalizes the Lebesgue integral. We cannot discuss this generalized integral here.*

Exercises for Chapter IX

1. A summable function is approximately continuous at each of its Lebesgue points. The converse is not true.

2. Let $f(x)$ be a bounded measurable function. A point x_0 is a Lebesgue point for $f(x)$ if and only if it is a point of approximate continuity.

3. It is possible for a function to be equal to the derivative of its indefinite integral at a point x_0 without being approximately continuous at x_0.

4. If all derived numbers of the function $f(x)$ satisfy the inequality $|Df(x)| \leq K$, then $f(x)$ satisfies the Lipschitz condition $|f(x) - f(y)| \leq k|x - y|$.

5. If the function $F[f(x)]$ is absolutely continuous for every absolutely continuous $f(x)$, then $F(x)$ satisfies a Lipschitz condition. (G. M. Fichtenholz).

6. Let $f(x)$ be defined on $[a, b]$. If to every $\varepsilon > 0$, there corresponds a $\delta > 0$ such that for every finite system of intervals $\{(a_k, b_k)\}$, the sum of whose lengths is less than δ, we have

$$\left| \sum_{k=1}^n \{f(b_k) - f(a_k)\} \right| < \varepsilon,$$

then $f(x)$ satisfies the Lipschitz condition.[13] (G. M. Fichtenholz).

7. Prove the following particular case of the Banach-Zarecki Theorem directly. If a continuous and strictly increasing function possesses the property (N), then it is absolutely continuous.

---

*The interested reader is referred to S. Saks, *Theory of the Integral*, 2nd edition, Monografie Matematyczne, Warsaw, 1937.—E. H.

[13] This result shows that it is impossible to discard the requirement that the intervals (a_k, b_k) be pairwise disjoint in the definition of absolute continuity.

8. Let $f(x)$ be continuous on $[a, b]$ and let E be the set of points at which at least one derived number of $f(x)$ is non-positive. If the map $f(E)$ of the set E contains no closed interval, then $f(x)$ is an increasing function. (A. Zygmund).

9. Using the preceding result, generalize Lemma 2, §7 in the following way. If $f(x)$ is continuous on $[a, b]$, if all derived numbers of $f(x)$ are non-negative at almost all points of $[a, b]$, and if the set of points at which at least one derived number equals $-\infty$, is finite or denumerable, then $f(x)$ is an increasing function.

10. Let $f(x)$ be continuous and let $f'(x)$ exist everywhere and be summable. If the set $E(|f'|=+\infty)$ is finite or denumerable, then $f(x)$ is absolutely continouous. (Apply the preceding exercise.)

11. A function having a finite derivative everywhere possesses property (N).

12. A necessary and sufficient condition that a continuous, strictly increasing function $f(x)$ be absolutely continuous is that the map $f(E)$ of the set E of points at which $f'(x) = +\infty$ have measure zero. (M. A. Zarecki).

13. A necessary and sufficient condition that a function which is the inverse of a continuous and strictly increasing function $f(x)$ be absolutely continuous, is that $mE(f' = 0) = 0$. (M. A. Zarecki)

INDEX

Absolute continuity of the integral, 149
Absolutely continuous functions, 243
  differential properties of, 246
Accumulation point, *see* Point of accumulation
Aggregate, 11
Aleksandrov, 107
Aleph-nought, 19
Algebraic numbers, 21
Almost everywhere, 90
Approximate continuity, 149, 260, 262
Arzelà, 242
Ascoli, 242
Assemblage, 11
At most denumerable, 17

Bad element, 29
Baire functions, 129
  measurability of, 131
Banach, 81, 252
  indicatrix, 225
Banach's theorem, 80, 225
Banach-Zarecki theorem, 250
Bari, 203
Bernstein, 107
  polynomials, 108
Bernstein's theorem, 108
Bessel's
  identity, 177
  inequality, 177
Binary expansion, 23
  unique representation of, 24
Bolzano-Cauchy property, 171
Bolzano-Weierstrass theorem, 35, 36
Borel
  covering theorem, 39
  set, 76, 86
Borel's theorem, 39, 104

Bounded
  function, 102
  sequence, 36
  variation, *see* Functions of finite variation

Cantor, 11, 27, 60
  sets G_o and P_o, 49, 56, 60, 75, 76, 213
Carathéodory, 88
Cardinal number, 27
Cauchy, 116
  sequence, 169
Cauchy-Bunyakovski-Schwarz inequality, 165
CBS inequality, *see* Cauchy-Bunyakovski-Schwarz inequality
Characteristic function, 93, 112
Class
  of bounded measurable functions, M, 172
  of continuous functions, C, 172
  of measurable sets, 75
  of polynomials, P, 172
  of step functions, S, 172
Closed
  interval, 13
  set, 36, 37
  orthogonal system, 177
Closure of a set, 37
Collection of objects, 11
Comparison
  of powers, 27
  of Riemann and Lebesgue integrals, 129
Complement of a set, 42
Complementary intervals, 49

273

INDEX

Complete
  additivity, 57
  space, 171
  system, 181
Component interval, 47
Condensation point, *see* Point of condensation
Congruent sets, 75
Continued fraction expansion, 23
Continuity, 102
  of the norm, 169
Continuous
  functions of finite variation, 223
  mappings, 248
  real-valued functions, 32
Continuum hypothesis, 28
Convergence
  in measure, 95, 96
  in the mean, *see* Mean convergence
Convergent subsequence, 36
Countable additivity, 57, 67, 86, 122
  of the integral, 142
Countable set, 17
Counting elements, 15

Darboux sum, 132
Decreasing function, 204
De la Vallée-Poussin, 88, 164
De la Vallée-Poussin's theorem, 159
Dense in itself, 37
Density of a set, 261
Denumerable, 17
  set, 17
  subset, 18
Derived
  number, 207
  set, 37
Difference of sets, 14
Differentiation of monotone functions, 207
Difficult problem of the theory of measure, 79
Dirichlet function, 116, 133
Disjoint sets, 14
Distance, 44
Dominated convergence, 161

Easy problem of the theory of measure, 79, 80
Egorov, 96
Egorov's theorem, 99, 112
Element, 11
Enumeration of sets, 17
Equality of sets, 12
Equi-absolutely continuous integral, 151, 152
Equicontinuous functions, 242
Equivalent
  functions, 90, 181
  sets, 15
Euclidean space R_n, 185
Everywhere dense set, 171

Faddeyev, 163
Family of sets, 13
Fatou's theorem, 140, 152, 160
Fichtenholz, 96, 203, 252, 270
Fichtenholz's theorem, 158, 252
Finite
  additivity of the integral, 145
  set, 18
  variation, *see* Functions of finite variation
First law of the mean for integrals, 121
Fixed point, 16
For almost all points, *see* Almost everywhere
Fourier, 176
  cofficients, 176
  series, 176
Fraenkel, 11
Fréchet's theorem, 106, 110
Functions
  continuous at a point, 102
  of finite variation, 204, 215, 223, 238
  with summable square, *see* Square-summable functions
Functionals, *see* Linear functionals
Fundamental
  lemma, 130
  sequence, *see* Cauchy sequence

Gavurin, 163
Good element, 29
Gram determinant, 193
Greater power, 28
Greatest lower bound, 44

Half-open interval, 13
Hardy, 23
Hausdorff, 11, 242
Hausdorff's theorem, 80
Helly's theorem (principle of choice), 220, 222, 233
Hilbert, 167
Hilbert space, 167, 187
    completeness of, 171
Hildebrandt, 270
Hobson, 133
Hölder's inequality, 197

Image of a set, 71, 207
Inclusion, 12
Increasing
    function, 204
    indices, 36
Indefinite Lebesgue integral, 243, 252
Index conjugate to p, 197
Infimum, 39
Infinite
    matrix, 26
    set, 11, 15, 18, 20
    subset, 18
Inner
    measure of a bounded set, 63, 64
    product, 184
Integrable
    (L), 120, 144
    (R), 116
Integration by parts, 229, 266
Interior point, 41
Intersection of sets, 13
Interval, 13
Invariance of measurability and measure under isometries, 71
Inverse image of a set, 207
Inverse isometry, 73
Irrational numbers, 23

Isolated point, 34, 37
Isometry, 71

Kaczmarz's theorem, 182
Kantorovič, 115, 163
Kolmogorov, 107, 163, 203
Kondurar', 241

Lebesgue, 163, 202
    measure, 67
    point, 255
Lebesgue integral, 116, 119, 136, 144
    fundamental properties of, 121
Lebesgue's theorem, 95, 96, 112, 127, 149, 153
    on dominated convergence, 161
Length of a vector, 186, 189
Legendre polynomials, 195
Levi's theorem, 141, 161
Limit
    point, 34, 35
    of a sequence, 167
Linear functional, 236
Linearly
    dependent system, 192
    independent system, 192
Lipschitz condition, 216
Lower
    Baire function, 129
    Lebesgue sum, 118
l_2-space, 186
l_p-space, 196
L_2-space, 165
L_p-space, 196
Luzin, 248
Luzin's theorem, 106, 107, 113, 114

Mapping of sets, 207
Mean convergence, 167, 168
    of order p, 199
Measurable
    (B), 76
    in the Lebesgue sense, 67
    (L), 67
    set, 55, 66, 84

Measurable functions, 89, 90
    properties of, 93
Measure
    of a bounded closed set, 60
    of a bounded open set, 55, 56
    of an interval, 55
    of arbitrary sets, 56, 66, 84
    of a set, 66, 84
Metric space, 167
Minkowski's inequality, 166, 198
Monotonic functions, 204
    differentiation of, 207
Mutually exclusive relations, 32

Natanson, 203
Non-denumerable set, 17
Non-measurable set, 76, 77
Non-void set, 13
Norm, 167, 186, 199, 200
Normalized measurable function, 175
Number of elements in a set, 15
Numerical sequence, 168

One-element set, 14
One-to-one correspondence, 15
Open
    interval, 13
    set, 41
Orlicz, 202
Orthogonal system, 175
Orthonormal system, 175
    closure and completeness in, 181
    linear independence of, 192
Outer measure of a bounded set, 63

Pairwise disjoint, 14
Parseval's identity, 177
Passage to the limit under the integral sign, 127, 149, 232
Perfect set, 37
Perron-Denjoy integration, 270
Point
    in R_3-space, 185
    of accumulation, 34
    of condensation, 50, 52
    of density, 261
    of rarefaction, 261
    set, 34
Points in Euclidean space R_2, 184
Power of a set, 15, 27
    symbol for, 27
Power
    a, 17
    c, 22, 27
    f, 29
    of a closed set, 50
    of the continuum, 21, 22
Primitive function, 133
Principle of monotonicity, 80, 81
Problem of measure, 79
Proper subset, 12
Property (N), 248

Rademacher's theorem, 183
Rational numbers, 19
Real numbers, 23
Real-valued functions, 28
Real variable, 11
Reconstruction of primitive function, 133, 266
Reflection in the origin, 71
Regular solution, 81
Relations, 32
Riemann, 116
Riemann integral, 116
Riesz, 106, 114, 202, 236
Riesz-Fischer theorem, 179, 187
Riesz's theorem, 98, 112, 127, 150, 152, 190, 236, 257
Rule of association, 15
Russell, 11
Russell's paradox, 11, 27

Saks, 271
Saltus, 205
    function, 206, 219
Schmidt's theorem, 194
Section of a function by the number N, 136
Separability, 45
Separation, 44
    property, 46
    theorem, 45

Sequence
    of measurable functions, 95
    of natural numbers, 23
Set, 11
    dense in itself, 37
    inclusion, *see* Inclusion
    of irrational numbers, 23
    of natural numbers, 36
    of real-valued functions, 28
    of square-summable functions, *see* L_2-space
    of type F_σ, 75
    of type G_δ, 75
Single-valued mapping, 71
Singular function, 263
Sliding hump method, 156
Smaller power, 28
Smallest closed interval containing a set, 43
Square-summable function, 165
Steklov's theorem, 178
Step functions, 91, 112
Stieltjes integral, 204, 227
Stone, 167, 202
Strictly monotonic function, 204
Structure
    of bounded closed sets, 47
    of bounded open sets, 47
    of measurable functions, 101
Subsequence, 36
Subset, 12
Sum of sets, 12
Summable function, 136, 144, 149
    of arbitrary sign, 143
Supremum, 39

Suslin, 76
System, 13

Ternary expansion, 50
Theory
    of functions, 11
    of sets, 11
Titchmarsh, 202
Total variation, 215, 238
Transcendental numbers, 23
Transitive relation, 32
Translation, 71
Trichotomic property, 32
Trigonometric polynomials, 110

Union of sets, 12
Upper
    Baire function, 129
    Lebesgue sum, 118

Vector in R_3-space, 185
Vitali covering, 81
Vitali's theorem, 81, 83, 152, 157, 209
Void
    intersection, 14
    set, 12

Weak convergence, 174, 200
Weierstrass's theorems, 107, 109, 111, 173, 174
Wright, 23

Zarecki, 252, 271
Zygmund, 270